Low Temperature Metamorphism

LOW TEMPERATURE METAMORPHISM

edited by
M. FREY
Professor of Geology
Mineralogisch-Petrographisches Institut
University of Basel

Blackie

Glasgow and London

Published in the USA by
Chapman and Hall
New York

Blackie & Son Limited,
Bishopbriggs, Glasgow G64 2NZ
7 Leicester Place
London WC2H 7BP

Published in the USA by
Chapman and Hall
in association with Methuen, Inc.
29 West 35th Street, New York, NY 10001

British Library Cataloguing in Publication Data

Low temperature metamorphism.
1. Metamorphism (Geology)
1. Frey, M.
552'.4   QE475.A2
ISBN 0-216-92011-6

Library of Congress Cataloging-in-Publication Data

Main entry under title:
Low temperature metamorphism.
Includes bibliographies and index.
1. Metamorphism (Geology)   2. Diagenesis.
I. Frey, Martin, 1940–
QE475.A2L68   1987   552   86-2234
ISBN   0-412-01341-X   (Chapman and Hall)

Phototypesetting by Thomson Press (India) Ltd., New Delhi
Printed in Great Britain by Bell & Bain (Glasgow) Ltd.

Preface

Books currently available on metamorphic petrology restrict their discussion of sub-greenschist facies conditions to volcanic and volcaniclastic rocks, while clastic sedimentary rocks at this metamorphic grade are largely neglected. Also missing from other books is a discussion of the maturation of organic matter. This book intends to fill these gaps and is addressed to researchers, lecturers and advanced students in the earth sciences.

Chapter one discusses some problems concerning the definition of very low-grade metamorphism and explains the terminology used in the book to define metamorphic grade. Chapter two describes various approaches to the determination of metamorphic grade in very low-grade shales and sandstones: (i) textures as observed in thin-sections, (ii) 'crystallinities', (iii) polytypism and polymorphism in sheet silicates, and (iv) mineral assemblages. Chapter three covers very low-grade metamorphism of volcanic and volcaniclastic rocks, and deals with (i) characteristic metabasaltic–andesitic assemblages, (ii) critical reactions involving facies transitions, (iii) T–X and P–X stabilities for index mineral assemblages, and (iv) a quantitative petrogenetic grid for low-temperature metamorphism of basaltic rocks. Chapter four describes the changes that take place in organic matter during very low-grade metamorphism. The processes of coalification, anthracitization and graphitization with their different stages, rank parameters and geological causes are treated separately. Special emphasis is placed on optical methods which can be applied to finely dispersed organic material in rocks other than coal. Chapter five reviews the principles of fluid inclusion in quartz found in fissures and in very low-grade metamorphic rocks. Three fluid zones are distinguished with increasing metamorphic grade, and the use of fluid inclusions for geothermometry and geobarometry is evaluated. Chapter six deals with the use of radiogenic isotopes in very lowgrade metamorphism. Possibilities and limitations of the various dating systems (K–Ar, ^{40}Ar–^{39}Ar, Rb–Sr, fission track, U–Th–Pb, and Pb–Pb) are discussed, and several examples of the economic importance of radiogenic isotope work in very low-grade metamorphic terranes are given. Chapter seven provides a correlation between different indicators of very low-grade metamorphism, with special emphasis on illite 'crystallinity', coal rank and mineral facies of metabasites.

We are pleased to acknowledge the secretarial help of Mrs S. Tobler and the advice and encouragement of the publishers. Most of the illustrations were re-drawn to a high standard by my son, Mathias. Finally, I cordially thank my wife Anja for her moral support during this project.

M.F.

Contributors

Moonsup Cho　　　　　Department of Geology, Stanford University,
Stanford, CA 94305, USA

Martin Frey　　　　　Mineralogisch-Petrographisches Institut der
Universität Basel, CH-4056 Basel, Switzerland

Johannes C. Hunziker　Laboratorium für Isotopengeologie der
Universität Bern, Erlachstrasse 9a, CH-3012
Bern, Switzerland

Hanan J. Kisch　　　　Department of Geology, University of the
Negev, P.O. Box 2053, Beer-Sheva, Israel

Juhn G. Liou　　　　Department of Geology, Stanford University,
Stanford, CA 94305, USA

Shigenori Maruyama　　Department of Earth Sciences, Toyoma
University, Toyoma, Japan

Josef Mullis　　　　Mineralogisch-Petrographisches Institut der
Universität Basel, CH-4056 Basel, Switzerland

Marlies Teichmüller　　Geologisches Landesamt, Nordrhein-Westfalen,
De Greiffstrasse 195, D-4150 Krefeld, Federal
Republic of Germany

Contents

4 Organic material and very low-grade metamorphism
MARLIES TEICHMÜLLER

1 Scope of subject

MARTIN FREY and HANAN J. KISCH

In this chapter some problems concerning the definition of very low-grade metamorphism are discussed, the terminology used in this book is defined, and two approaches to the study of very low-grade metamorphic rocks in deep boreholes and cross-sections are compared.

1.1 Problems of definition of very low-grade metamorphism

This book deals with transformations of minerals and organic material in the approximate temperature range of 150–200 to 350–400 °C.

There are several reasons that contribute to problems in terminology of very low-grade metamorphism, and these are summarized below.

(i) Temperatures (and pressures) at which transformations set in are strongly dependent on the material under investigation. Important transformations of evaporites, of vitreous material, and of organic material, for example, begin to take place at considerably lower temperatures than such transformations of most silicate rocks.

(ii) In many rocks phase transformations begin shortly after sedimentation and continue to take place with increasing burial. Whether such transformations are called 'diagenetic' or 'metamorphic' is largely arbitrary.

(iii) In the past, incipient metamorphism has been investigated by researchers with different backgrounds (e.g. sedimentologists, coal petrologists, metamorphic petrologists) and hence using a different terminology. In addition, many terms introduced to designate the degree of transformation or metamorphic grade changed their meaning with time, even when used by the same author.

Basically, three different kinds of definition of metamorphism—specifically with respect to the beginning of metamorphism and its delimitation from diagenesis—can be distinguished.

(1) *Definitions following Turner* that recognize as metamorphic 'the mineralogical and structural adjustment of solid rocks to physical and chemical conditions which have been imposed at depths below the surface zones of weathering and cementation, and which differ from the conditions under which the rock in question originated' (Turner, 1948, 1968, p. 3, 1981, p. 3; Turner and Verhoogen, 1960, p. 460).

Coombs (1961, p. 214) agrees with Pettijohn (1957) that no distinction between diagenesis and metamorphism is possible. He draws the onset of metamorphism—the line of demarcation from diagenesis in the restricted sense (i.e. changes before complete lithification)—'at the stage at which mineralogical or textural modification first sets in, in response to a temperature differing appreciably from that under which a sediment has been deposited or a volcanic rock has consolidated'.

Coombs thus incorporates changes after lithification, as included in the 'epigenetic diagenesis' of Packham and Crook (1960), the 'late diagenesis' of US authors, and the 'epigenesis', 'catagenesis', and 'metagenesis' of Russian authors within his definition of metamorphism.

Later, Turner (1981, p. 2) states: 'For convenience of discussion, arbitrary limits have been set to the range of rock transformations that are to be included within the scope of metamorphism. Purely surface changes such as weathering, leaching and cementation by meteoric waters, and the processes of diagenesis operating close to the Earth's surface are excluded. But the boundary is artificially drawn; the student of diagenesis and the petrologist interested in the incipient stages of metamorphism may find themselves studying the same or closely related phenomena'.

(2) *Definitions based on the degree of chemical and textural reconstitution of the rock.* Thus, Turner (1958, pp. 215–216; 1968, p. 264; 1981, p. 299) classed minor changes in the clay matrix and crystallization of cement minerals in sandstones as diagenetic, and processes extensively involving coarse clastic grains so that the rock becomes substantially recrystallized as metamorphic. Schistosity is regarded as a criterion for metamorphism. For the undeformed Taringatura section Turner (1958, pp. 216–217) accepted Coombs' earlier (1954) 'arbitrary line between diagenesis and metamorphism (...where albitization of plagioclase becomes extensive, and the principal zeolite is laumontite)', classing reactions such as crystallization of glass to heulandite and analcime, that were considered to have taken place independently, as diagenetic, and localized but interdependent reactions–such as the laumontitization of heulandite and analcime by lime and alumina set free elsewhere by the albitization of plagioclase—as metamorphic. Thus he regards the prevalence of heulandite rather than laumontite as marking the upward transition to the zone of diagenesis.

However, with respect to the low-temperature assemblages formed in chemically unstable rocks without the aid of deformation, Turner (1968, p. 264) concludes: 'whether the resulting mineral facies (in this case the zeolite facies) are termed diagenetic or metamorphic is of no great significance'.

(3) *Definitions that attempt to delimit diagenesis from metamorphism* by the first appearance of specific minerals or mineral assemblages postulated to be 'metamorphic' in rocks of suitable bulk composition.

Miyashiro (1973, p. 21) expressed the following view on the low-temperature limit of metamorphism: 'We may define diagenesis to include those changes taking place at essentially the same temperature as that of the original deposition, and metamorphism to include those changes taking place at essentially higher temperatures (Coombs, 1961). The boundary should be defined by definite mineralogical changes. Since the stability of a mineral or a mineral assemblage is partly related to the bulk chemical composition of the rocks, the temperature for such a boundary should vary with the composition. Our present knowledge on the mineralogy of low-temperature metamorphic rocks is not ample enough for effective application of this definition'. According to this author '... the low-temperature limit of metamorphism is probably around 150 °C'. This limit of 150 °C refers to a specific reaction of zeolites and other minerals—such as analcime + quartz = albite + water, taken to take place in the middle of the zeolite facies (*ibid.*, p. 91)—which is postulated to represent the beginning of metamorphism.

Winkler (1979, p. 11) used the following definition to designate the beginning of metamorphism: 'Metamorphism has begun and diagenesis has ended when a mineral assemblage is formed which cannot originate in a sedimentary environment'. As an example, he regarded laumontite as the only zeolite mineral of metamorphic origin, while all other zeolites were attributed to the diagenetic realm. Other minerals taken to indicate the beginning of metamorphism are lawsonite, glaucophane, paragonite, or pyrophyllite. It is clear that such a classification is purely arbitrary.

However, no matter how the boundary between diagenesis and very low-grade metamorphism is defined by petrologists, the traditional inclusion by many sedimentologists and clay petrologists of very low-grade processes in sedimentary rocks beyond the completion of lithification, up to the appearance of slaty cleavage, within diagenesis in the broad sense (as 'late diagenesis', 'burial diagenesis') is likely to continue, thus paralleling the range of processes in volcanic rocks now widely considered as metamorphic by petrologists.

1.2 Terminology

There exist many different methods to determine metamorphic grade of incipient metamorphism, some of which will be described in subsequent chapters. Almost every method possesses its own nomenclature to designate the progress of transformation. This is a useful concept because, then, the terminology automatically implies which method has been used. A warning should be expressed at this point: mixing the terminology from different methods is not recommended. As an example, the 'anchizone' should be delimited exclusively on the basis of illite-'crystallinity' data (see below) and not on the basis of coal rank data (for the onset of the anchizone) as proposed by Teichmüller *et al.* (1979, Figure 30).

We have desisted here from introducing new terms, but some will be defined below.

Very low-grade metamorphism is used here for the lowest grade of metamorphism as defined by Coombs and Turner up to the beginning of the greenschist facies or of low-grade metamorphism of Winkler. Very low-grade metamorphism thus widely overlaps with 'late diagenesis' and 'burial metamorphism', and includes 'catagenesis', 'epigenesis', 'metagenesis', and 'epigenetic diagenesis' of various authors.

In an orogenic belt, regional metamorphism is followed by uplift and erosion, yielding a clockwise pressure (P)–temperature (T) path (England and Richardson, 1977; England and Thompson, 1984). Along this path there are three possible ways of passing through the P–T field of very low-grade metamorphism (Figure 1.1). (A) During prograde metamorphism: in most cases, however, transformations at higher temperature will obliterate the imprint of very low-grade metamorphism. (B) During retrograde metamorphism. Such 'late' overprints are important in unravelling a P–T path and some examples derived from fluid inclusion studies will be described in Chapter 5. (C) Maximum P and T are reached within the realm of very low-grade metamorphism. Transformations described in Chapters 2–4 deal mainly with the last case.

Diagenesis is used here in the restricted sense ('early diagenesis') to include all changes taking place in a sediment between sedimentation and the completion of lithification or cementation—i.e. the onset of very low-grade metamorphism—except those caused by weathering.

It is also used 'traditionally' in a broad sense to include changes in sedimentary rocks (rarely in volcanic and volcaniclastic rocks) after the

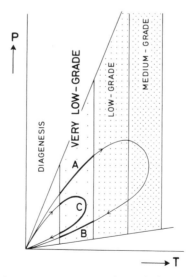

Figure 1.1 Hypothetical pressure–temperature–time paths for two (out of an infinite number of) regionally metamorphosed rock bodies. Transects through very low-grade metamorphism are shown in heavy lines: *A*, during prograde metamorphism; *B*, during retrograde metamorphism; *C*, during culmination of metamorphism. See text for discussion.

completion of lithification, up to the appearance of slaty cleavage, and then overlaps the low-grade part of very low-grade metamorphism as defined; in this broad sense it includes 'late diagenesis', 'epigenetic diagenesis', 'burial diagenesis', 'deep diagenesis', 'catagenesis' (in the restricted sense), etc., which are regarded by the authors (following Coombs) as virtual synonyms of 'burial metamorphism' as applied to sedimentary rocks (cf. Kisch, 1983, p. 296).

Development of the concept of diagenesis in general has been reviewed by Dunoyer de Segonzac (1968) and Larsen and Chilingar (1979), and the nature and definition of the high-grade limit of diagenesis has been discussed by Kisch (1983, pp. 294–297).

Usage of mineral facies: subdivision of very low-grade metamorphism with aid of volcanic and volcaniclastic rocks is based on mineral assemblages defining various subgreenschist facies and is extensively used in Chapter 3. The mineral facies concept will not be applied here to clastic sedimentary rocks because critical mineral assemblages rarely occur and/or because phase relations are still unsatisfactorily known.

Anchimetamorphism or anchizone is recognized only on the basis of illite-'crystallinity' data, and this method is used as the main indicator of metamorphic grade for very low-grade metaclastites as detailed in Chapter 2. Kübler (1967*a*, 1968) introduced the concept of anchimetamorphic zone or anchizone intervening between late diagenesis and the epizone (see below). Kübler's approach makes use of the increase in 'crystallinity' of illite as monitored by the concomitant decrease of the width of the illite (10 Å) X-ray diffraction peak at half peak height. The anchizone is defined by limiting values of the illite 'crystallinity' index (see section 2.5.1).

Epimetamorphism or epizone will be used here only in connection with illite 'crystallinity' data indicating that metamorphic grade has surpassed that of the anchizone. The term epizone was originally introduced by Grubenmann (1904) in his concept of depth-zones as the uppermost zone of regional metamorphism. Epizone has continued to be used, mainly in Alpine studies, as synonymous with greenschist facies but not implying that the essential controlling factor is depth.

Burial metamorphism and incipient regional metamorphism: These constitute descriptive sub-categories of very low-grade metamorphism.

Burial metamorphism (or burial diagenesis) is used after Coombs (1961, p. 214) for 'mineralogical and less spectacular textural reconstitution... in thick sedimentary columns. ... The fabric of silicate-rich rocks is not modified by the development of schistosity. ...The metamorphism appears to follow burial...and is not accompanied by significant penetrative movements. Diagenetic processes in the most restricted use of that term, that is, occurring essentially at the temperature of deposition, are excluded'. The characteristic absence of the establishment of schistosity distinguishes burial metamorphism from regional metamorphism, which is 'characterized by the common development of schistosity'. The term has gained wide acceptance, though some authors have favoured slight modifications to Coombs' classification by stressing different attributes.

Winkler (4th edition, 1976, pp. 4–5) stipulates the absence of a genetic relationship of 'regional burial metamorphism' to orogenesis, delegating the general absence of 'penetrative movement producing a schistose structure' to a secondary characteristic.

Turner (1981) rejects the separation of burial metamorphism from regional metamorphism on the basis of lack of foliation, regarding burial metamorphism as a type or subcategory of regional metamorphism 'whose degree and mineralogical imprint can be correlated on field evidence with stratigraphically (or tectonically) controlled depth gradients'.

Coombs (1961, p. 214) noted that '...many slates and slightly sheared graywackes such as those of the Chl. 1 subzone of Otago mark transitions from the products of burial metamorphism to those of regional metamorphism'. These rocks have since been shown to be metamorphosed in the prehnite–pumpellyite facies (Landis and Coombs, 1967; Bishop, 1972), and to show anchimetamorphic or even epimetamorphic illite 'crystallinities' (Kisch, 1981b; see also Chapter 7). In view of the appearance of slaty cleavage in pelitic rocks, and the relation to orogenesis they cannot be referred to burial metamorphism following any of the definitions of the term. Kisch (1983, p. 309) therefore used incipient metamorphism 'as a general term for the more advanced stages of mineral modification as characterized by the appearance of the attributes of anchi-metamorphism'. Since these rocks commonly show slaty cleavage or schistosity and have a genetic relation to orogenesis, incipient regional metamorphism would be a more suitable general term. The term has very generalized mineral-facies connotations: in most terranes the prehnite–pumpellyite, pumpellyite–actinolite, and glaucophane–lawsonite facies, the anchizone, and anthracitic coal ranks would be included within incipient regional metamorphism.

Coal rank indicates the degree of transformation of organic matter, and different rank scales are detailed in Chapter 4.

Catagenesis, epigenesis, and metagenesis: Catagenesis and metagenesis are used to designate two stages in the evolution of hydrocarbon source rocks as defined in section 4.3.1.

The term catagenesis was originally introduced by Fersman (1922) for 'changes occurring in an already formed sedimentary rock buried by a distinct (though sometimes thin) covering layer, characterized by pressure–temperature conditions that are much different from those of deposition'. According to Bates and Jackson (1980, p. 98) 'the term is more or less equivalent to epigenesis as applied by Russian geologists'.

The term 'metagenesis', originally a synonym of 'epigenesis' (Loewinson-Lessing and Struve, 1937, as quoted by Teodorovich, 1961, p. 9), was re-introduced with two different meanings: (i) in a broader sense by Strakhov

(1957) to cover both epigenesis (or catagenesis) and 'protometamorphism', and (ii) in a more restricted sense as 'initial metamorphism' by Kossovskaya *et al.* (1957; also Kossovskaya and Shutov, 1958, 1961, and 1963) and Vasoievich (1962) for the higher-grade part of this range, i.e. as an intermediate stage between epigenesis (or catagenesis) in a restricted sense and metamorphism (see also Dunoyer de Segonzac, 1968, Table II).

Teodorovich (1961, p. 9) considers the term superfluous—rightly so, as would appear to the writers—and several Russian authors accordingly continue to use 'catagenesis' in the extended sense, some using 'metagenesis' for low-grade metamorphism (e.g. Neruchev *et al.*, 1976, quoted by Karpova *et al.*, 1981). However, metagenesis in the restricted sense has been widely used after Kossovskaya and Shutov, and seems to have become a permanent feature in Russian literature.

Kossovskaya *et al.* (1957; also Kossovskaya and Shutov, 1958) divided the range into four 'stages', original (or early) epigenesis, deep epigenesis, early metagenesis, and late metagenesis, which they correlated with the textural zones earlier described from the Upper Yana region (Kossovskaya and Shutov, 1955). Unfortunately, they changed this correlation in 1961 and subsequent papers (1963 and 1970, Table I); to compound the confusion, the text in the 1970 (p. 16) paper refers to the new correlation, but the section heading to the old correlation—probably due to printing errors (see Kisch, 1974, p. 88, footnote; 1983, p. 302). When referring to the textural zones and the 'stages of regional epigenesis and metagenesis' of Kossovskaya and Shutov it is therefore imperative to indicate to which of their two correlations reference is being made.

Fluid zone is used to indicate the degree of transformation of fluids as deduced from fluid inclusion studies based on the occurrence of characteristic fluid species. The different fluid zones are defined in Chapter 5.

1.3 Sampling in deep boreholes and in cross-sections

Diagenetic transformations have been traditionally studied on samples from deep boreholes, whereas surface samples from cross-sections have furnished much information on very low-grade metamorphism. In recent years, however, important results on hydrothermal metamorphism from boreholes in geothermal areas have been published. In this section some advantages and disadvantages of sampling in deep boreholes and cross-sections will be summarized.

Boreholes allow continuous sampling, provide direct temperature determinations (though these temperatures may be minimum values in case of uplift), and the fluid phase can be extracted. On the other hand different lithologies will often be encountered and this may restrict the evaluation of diagenetic and metamorphic processes. Also note that a single borehole represents 'a one-dimensional outcrop'.

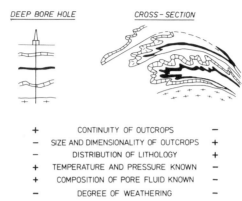

+	CONTINUITY OF OUTCROPS	–
–	SIZE AND DIMENSIONALITY OF OUTCROPS	+
–	DISTRIBUTION OF LITHOLOGY	+
+	TEMPERATURE AND PRESSURE KNOWN	–
+	COMPOSITION OF PORE FLUID KNOWN	–
–	DEGREE OF WEATHERING	–

Figure 1.2 Advantages (+) and disadvantages (–) of sampling in deep boreholes *v.* sampling in cross-sections. See text for details.

In cross-sections the same formation may be followed from the foreland (with possibly diagenetic conditions) towards the inner part of an orogenic belt (with possibly very low-grade metamorphic conditions). However, outcrops are not always continuous and surface samples are exposed to weathering.

Advantages and disadvantages of the two sampling environments are summarized in Figure 1.2.

2 Very low-grade metamorphism of clastic sedimentary rocks

MARTIN FREY

2.1 Introduction

This chapter deals with shales and sandstones with strong emphasis on the former. Greywackes, that often contain volcanogenic constituents, will be treated in Chapter 3. Clastic sedimentary rocks may be classified further on a chemical and mineralogical basis into pelites and marls (carbonate-bearing pelites).

During the last fifteen years a number of valuable texts and reviews on this subject have appeared in the literature (Dunoyer de Segonzac, 1970; Zen and Thompson, 1974; Velde, 1977, 1985a, Kisch, 1983; Kübler, 1984).

This chapter describes various approaches to determining metamorphic grade in very low-grade metaclastites; that is (i) textures as observed in thin sections; (ii) 'crystallinities'; (iii) polytypism and polymorphism in sheet silicates; and (iv) mineral assemblages.

2.2 General characteristics of very low-grade metaclastites

Shales and slates are very fine-grained, which makes them unattractive for optical investigations. However, there exist today several possibilities for overcoming this difficulty and appropriate methods will be mentioned in the following section.

Many very low-grade metaclastites have retained their original sedimentary structures, and these rocks, therefore, appear in the field to be unmetamorphosed. Also, early diagenetic textures may survive under greenschist facies conditions (Fairchild, 1985).

Clastic sediments entering the realm of very low-grade metamorphism may contain diagenetically newly-formed, inherited detrital, and partly to completely transformed detrital minerals. The latter group is particularly well represented by clay minerals evolving into mica and chlorite (Millot, 1964, 1970). The alteration of clastic biotite, muscovite and plagioclase has been discussed by Kisch (1983, p. 340). It is noteworthy that detrital muscovite may persist to medium-grade greenschist facies (e.g. Dietrich, 1983).

The treatment of mineral assemblages from a phase petrological standpoint (as will be done in section 2.7) presupposes the attainment of chemical equilibrium. Opinions in the literature regarding the stability of clay minerals,

a major mineral group of very low-grade metaclastites, are strongly con-
troversial. According to Kossovskaya and Shutov (1965, p. 1163) mineral
associations 'of regional epigenesis and metagenesis should be regarded not as
systems in equilibrium, but rather as metastable formations, more or less
approaching equilibrium'. By mainly thermodynamic reasoning, Lippmann
(1977, 1982) concluded that 'virtually all clay minerals are more or less
metastable or even completely unstable' (1982, p. 484). During a study of layer
silicate minerals in a borehole of the Salton Sea geothermal system, McDowell
and Elders (1983) described the formation of a series of metastable intermediate mineral phases, although these phases exist over a finite temperature range.
The results of several hydrothermal experiments summarized by Eberl (1984,
p. 251) 'suggest that many clays in nature also could be metastable, forming in
response to Ostwald's step rule rather than in response to equilibrium
conditions'. A more optimistic viewpoint is held by Velde (1983, p. 263) who
considers clay mineral assemblages above 100 °C 'to be the equivalent of
metamorphic rocks as far as the phase rule is concerned'. Also, Hutcheon
(1983, p. 8) states that authigenic minerals in the pore space of sandstones
often represent a chemical system in thermodynamic equilibrium. This short
review of the literature is completed by a statement by Zen and Thompson
(1974, p. 197), with which the present author fully agrees: 'The concept of phase
equilibrium is a helpful limiting model, whereby the actual mineral assemblages may be compared and understood'.

2.3   How to study very low-grade metaclastites

In this section some advice for the study of very low-grade metaclastites will be
given. Additional relevant methods will be described in Chapters 4, 5 and 6.

(i)   *Sampling.* Haematite-free siltstones and sandstones are well suited for
textural studies (section 2.4) while most shales and slates can be used to
determine 'crystallinity' indices (section 2.5) and polytypes of sheet silicates
(section 2.6). One disadvantage of sampling very low-grade metaclastites is
that the fine grain size prevents the field determination of mineral contents.
An interesting mineral assemblage will show up only after laboratory
investigations. In this respect sampling is like buying a pig in a poke.

(ii)   *Thin-section analysis.* Insist on obtaining a thin section, even of a fissile
slate which will need some impregnation with an epoxy. The fine grain size will
in many cases not allow identification of the matrix minerals under the
microscope, but the recognition of clastic mica, for example, is important for
the interpretation of 'crystallinity' indices and polytypes, as will be discussed
later.

(iii)   *X-ray powder diffraction.* This is the most rapid method for mineral identification of very low-grade metaclastites. Depending on the lithology, whole rock powders, decarbonated powders or several clay size fractions may be used. The identification of the sheet silicates is usually made by X-ray diffractometry, but the complementary Guinier camera technique is recommended. The X-ray mounts are prepared by sedimentation, and are air-dried, glycolated or heat-treated where necessary.

(iv) *Scanning electron microscopy.* Over the past two decades the scanning electron microscope has been successfully applied to sandstone diagenesis. Most of this work has been undertaken using the secondary electron mode, but in the last few years attention has been drawn to the advantages of imaging with backscattered electrons (e.g. Robinson and Nickel, 1979; Hall and Lloyd, 1981; Krinsley et al., 1983; Huggett, 1984; Pye and Krinsley, 1984; White et al., 1984, 1985). Using backscatter it is possible to identify individual minerals in polished thin sections by virtue of their atomic number contrast and differential hardness (relief). The amount of detail observable is far greater than with optical microscopy, and this new technique will become important for studying textural relationships in shales and slates. In addition, the use of energy- and/or wavelength-dispersive spectrometers in conjunction with atomic-number contrast should provide quantitative information on mineral chemistry in very low-grade metaclastites.

(v) *Electron microprobe.* This instrument has been revolutionary in metamorphic petrology but has seen little application to mineral chemical analysis in very low-grade metaclastites (e.g. Boles and Franks, 1979; McDowell and Elders, 1980, 1983; Nicot, 1981; Velde, 1983; Lee et al., 1984; Hunziker et al., 1986). Since a minimum grain size of 5–10 μm is required, polyphase mixed analyses will often be obtained.

(vi) *Transmission electron microscopy.* The study of ion-thinned specimens with a transmission electron microscope allows characterization of the specific

Table 2.1 Important methods of study of mineralogy of very low-grade metaclastites (not including methods described in Chapters 4, 5, and 6).

Method	Ability
Thin-section	(Texture, mineral assemblage)[1]
X-ray diffraction	Mineral identification; determination of 'crystallinity' indices, polytypes or polymorphs
Scanning electron microscope	Texture, (mineral chemistry)
Electron microprobe	Mineral chemistry
Transmission electron microscope	Texture at Å scale, structural identification, mineral chemistry

[1] Where a method provides limited information only, this is indicated by brackets.

nature of individual layers of sheet silicates at an Ångstrom level. This method yields a combination of detailed textural information with structural identification of individual phyllosilicate particles. In addition, energy-dispersive analysis provides accurate analytical data with a spatial resolution of up to 300 Å, whereby ratios of element concentrations relative to Si are measured. Application of these techniques to shales and slates can be found, for example, in the following studies: Phakey *et al.*, 1972; Lee and Peacor, 1983, 1985; Lee *et al.*, 1984, 1985; Ahn and Peacor, 1985. Unfortunately, the high analytical cost will prevent this method from becoming a routine analytical tool.

The methods described above and their potential are summarized in Table 2.1.

2.4   Textural zones

In the late fifties and early sixties, Kossovskaya and co-workers described four textural zones in sandy and clay rocks of the Russian and Siberian platforms based on optical observations on thin sections. The results were introduced to the Western literature mainly in translated papers (Kossovskaya, 1961; Kossovskaya and Shutov, 1958, 1970) and also presented in tabular form by Kisch (1983, p. 301). A brief characterization of the main microscopic features of these four zones follows, based on the above-mentioned literature and complemented by our own observations. Corresponding thin-section drawings are displayed in Figure 2.1.

(i) *Zone of unaltered clay matrix.* The original sedimentary texture is essentially preserved while some unstable detrital minerals (pyroxene, amphibole, intermediate and calcic plagioclase) gradually disappear. Clearcut grain boundaries between clastic quartz grains and the clay matrix testify to the absence of any reaction between the two.

(ii) *Zone of altered clay matrix.* The gradual transformation of clay minerals into hydromica (= illite) and chlorite produces fine-grained matrix quartz. Clastic quartz grains appear serrulated due to incipient pressure solution processes which produce additional authigenic matrix quartz.

(iii) *Zone of quartzitic structures and hydromica–chlorite matrix.* Increasing pressure solution develops a quartzitic texture in sandstones. The original clay matrix is transformed to hydromica, chlorite and quartz, and phyllosilicates are oriented through stress.

(iv) *Zone of spiny-like structures and muscovite-chlorite matrix.* The typical feature of this zone consists of spiny or prickly structures in siltstones as a result of extensive pressure solution. Phyllosilicates are coarser-grained than in zone (iii), and pelitic rocks are represented by slates and phyllites.

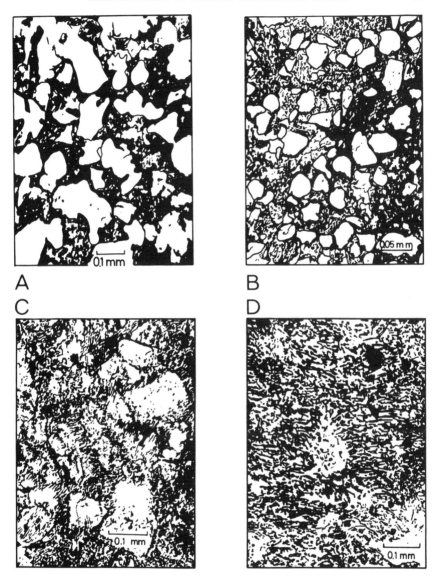

Figure 2.1   Thin-section textures in the transition zone from diagenesis to metamorphism. *A:* 'Zone of unaltered clay matrix' in marlstone. *B:* 'Zone of altered clay matrix' in shale. *C:* 'Zone of quartzitic structures and hydromica-chlorite matrix' in slate. *D:* 'Zone of spiny-like structures and muscovite-chlorite matrix' in slate to phyllite. (Modified from Frey, 1970, Figure 9.)

These textural zones were assigned to 'stages of regional epigenesis and metagenesis' by Kossovskaya and Shutov (1958, 1963, 1965, 1970). As discussed by Kisch (1983, p. 302), different interpretations are found in the literature. This writer follows here the correlation proposed by Kossovskaya and Shutov

Table 2.2 Correlation of textural zones, stages of regional epigenesis or metagenesis and metamorphic facies

Textural zone	Stage	Metamorphic facies
(i)	Initial epigenesis	Zeolite
(ii) ⎫ (iii) ⎭	Deep epigenesis	Zeolite
(iv)	Early metagenesis	Prehnite–pumpellyite

(1963; 1970, Table 1), in agreement with Kisch (1983, p. 301), and this is shown in Table 2.2.

According to Kossovskaya and co-workers these textural zones are mappable on a regional scale (Figure 2.2) and the method has been applied to many regions in the USSR (for references see Kossovskaya and Shutov, 1970, p. 14; see also Simanovich, 1972). In the Western world, however, this method has met little response and the following two examples are the only ones known to this author. Frey (1969a, 1970) recognized textural changes from zone (ii)–(iv) in anchi- and epimetamorphic pelitic and marly platform sediments, but individual textural zones were not mapped. Thum and Nabholz (1972) documented textural changes in anchimetamorphic and early epimetamorphic flysch sandstones. Textural zone (ii) was encountered in the early anchizone. No characteristic examples of textural zone (iii) were noted. Textural zone (iv) was generally found near the anchizone–epizone boundary, but sometimes also in tectonized regions at lower grade.

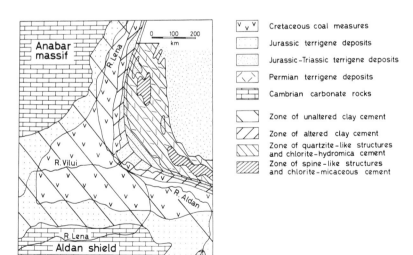

Figure 2.2 Zones of 'epigenesis' and 'metagenesis' within the Verkhoiansk mountain range and the adjoining sections of the Siberian platform. (Modified from Kossovskaya and Shutov, 1958, Figure 9; 1963, Figure 1.)

A short note will be made here of references which refer to textural variations in very low-grade metaclastites not based on the sequence of Kossovskaya and co-workers. Such studies, which use criteria similar to those of Turner (1938), Blake *et al.* (1967) or Dapples (1967, 1979) include Schamel (1973), Cassinis *et al.* (1978) and Helmhold *et al.* (1982).

Conclusion. The great advantage of mapping textural zones in very low-grade metaclastites is the simplicity of analytical procedure, that is, thin-section study with a petrographic microscope. The main disadvantage is the difficulty of quantifying gradual textural changes. An unequivocal assignment of a texture to one of the four classes may be somewhat arbitrary. Also, the textural evolution will vary according to lithology and, therefore, a homogeneous protolith should be chosen for study. It should be remembered that several of the following parameters probably determine the textural changes described above: mineralogy of the protolith, grain size, temperature, stress, pore-fluid composition, amount of pore-fluid flow, and perhaps time. This enumeration suggests that correlations of textural zones with metamorphic facies may vary from one region to another.

2.5   Crystallinity

Crystallinity usually signifies the amount of crystalline material in any substance. Kübler (1967a, p. 106), on the other hand, defined crystallinity as the degree of 'ordering' in a crystalline lattice, but he did not specify what is meant by ordering. The term is commonly applied to illite, but crystallinity indices can also be determined for other sheet silicates, for example, chlorite, kaolinite or pyrophyllite (see below). Recently, Kübler (1984, p. 575) proposed replacing the term 'illite crystallinity' by the term 'Scherrer width' (*largeur de Scherrer*), which is defined by the Scherrer equation (Scherrer, 1918; see e.g. Klug and Alexander, 1974, p. 687; Brindley, 1980, p. 131, see also Árkai and Toth, 1983):

$$B = \Delta\,^\circ 2\theta = \frac{K \cdot \lambda}{N \cdot d \cdot \cos\theta}$$

where

B = angular width at half-maximum intensity measured in radians 2θ
K = a constant approximately equal to unity (values of 0.89 and 0.94 are given by Klug and Alexander, 1974, p. 689)
λ = radiation wavelength
d = spacing of the lattice planes
N = number of planes
$N \cdot d$ = size of coherently diffracting domains (domain size)
θ = Bragg angle of a specific reflection.

Two reasons were put forward by Kübler for this change in terminology: (i) illite does not exist, from a strict crystallographic standpoint, and (ii) crystallinity is not believed to be an appropriate term for the measured property B. The present author agrees that the original definition of illite crystallinity has some weaknesses. However, because the term is so well established in the literature, we favour retaining it, but in quotation marks (following a suggestion by Kisch, 1983, p. 344).

2.5.1   Illite 'crystallinity' indices

The following enumeration is in chronological order.

(i)   *The Weaver index.* Weaver (1960) was the first to determine a relationship between the shape of the illite 10 Å diffraction peak and metamorphic grade of shales. To do so, he defined a so-called sharpness ratio, that is the intensity ratio at 10.0 Å and 10.5 Å (Weaver index $= H(10.0 \text{ Å})/h(10.5)$ Å, see Figure 2.3a). The numerical value of this sharpness ratio increases with enhanced 'crystallinity', i.e. values less than 2.3 are taken to indicate the diagenetic zone, values from 2.3 to 12.1 the anchizone, and values greater than 12.1 the epizone (Weaver, 1960; Kübler, 1968).

A non-linear correlation curve for illite 'crystallinity' values determined by the Weaver index and the Kübler index (see below) was published by Kübler (1968, Figure 3). The latter pointed out that the method proposed by Weaver

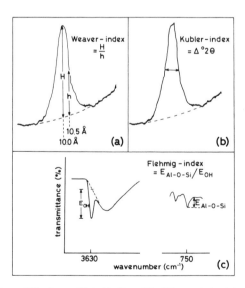

Figure 2.3   Definition of illite 'crystallinity' indices. The Weaver index (*a*) and the Kübler index (*b*) are determined from the shape of the first illite basal reflection on X-ray diffractograms. The Flehmig index (*c*) is derived from the extinction ratio of two absorption bands on infrared spectra.

works well in unmetamorphosed sediments, but is less well suited to the anchizone, due to the increasing error in determining the sharpness ratio. This limitation of the Weaver technique has led to rather restricted use in very low-grade metamorphic terrains (e.g. Weaver, 1960; Kübler, 1964; Gill et al., 1977; Brazier et al., 1979; Robinson et al., 1980; Weaver, 1984; Robinson and Bevins, 1986).

(ii)   *The Kübler index.* This index is defined as the half-height peak width of the first illite basal reflection (Figure 2.3b). This method of measurement was first proposed by Kübler (1964, p. 1107) and elaborated by Kübler (1967a, 1968) and Dunoyer de Segonzac et al. (1968). The numerical value of the Kübler index decreases with improving 'crystallinity'. In earlier studies the half-height peak width was expressed in mm, but is now generally given in $\Delta 2\theta°$. The measurement of the Kübler index depends on the adopted experimental conditions; illite 'crystallinity' standards must therefore be used. Kisch (1983, p. 348) has tabulated the experimental conditions used by various authors for the determination of the Kübler index, and the corresponding illite 'crystallinity' values equivalent to the limits of the anchizone. The limiting values used by Kübler (1984, p. 578) for the low-grade and high-grade limit of the anchizone are 0.42 and 0.25° $\Delta 2\theta$ CuK_α.

The error in determining the Kübler index decreases with improved 'crystallinity' (Kübler, 1968) and the method has been successfully applied by more than a hundred authors. A compilation of relevant references can be found in Kisch (1983, pp. 518–524).

(iii)   *The Weber index.* The technique of Weber (1972a) relates the illite peak width at half height ($Hb(001)$ illite = Kübler index) to the same property of the (100) peak of quartz ($Hb(100)$ quartz). The relative peak width at half height is consequently defined as

$$Hb_{rel} = \frac{Hb(001)\,illite}{Hb(100)\,quartz} \times 100$$

According to Weber (1972a) the use of an external quartz standard should 'enable comparison of reflection measurements from different X-ray equipment' (p. 268). However, this is true only if $Hb(100)$ quartz is standardized. The Weber index has been used mainly by German authors (e.g. Weber 1972b; Loeschke and Weber 1973; Ludwig 1972b, 1973; Teichmüller et al. 1979; Brauckmann 1984).

(iv)   *The Flehmig index.* Flehmig (1973) proposed an infrared method for determination of illite 'crystallinity' by measuring the extinction ratio of two selected absorption bands at $750\,cm^{-1}$ ($E_{Al-O-Si}$) and $3630\,cm^{-1}$ (E_{OH}), see Figure 2.3c. The extinction ratio ($E_{Al-O-Si}/E_{OH}$) of a well-crystallized muscovite was taken as 10. This leads to a variation of the Flehmig index from 0 to 10.

The advantage of the infrared method lies in the fact that contributions of all crystallographic directions to disorder are taken into account, while the X-ray methods favour disorder in the c-direction.

A linear relationship between the Kübler and Flehmig indices has been noted by Hunziker et al. (1986, Figure 4).

Because of the greater analytical expense, few studies have used the infrared method to determine illite 'crystallinity' (e.g. Flehmig, 1973; Flehmig and Langheinrich, 1974; Nyk, 1985; Hunziker et al., 1986).

(v)   The Weber–Dunoyer de Segonzac–Economou index. For this index the domain size $N \cdot d$ is calculated from the measured half-height width of the 10 Å illite peak B (which represents in fact the Kübler index) using the Scherrer equation (see above):

$$N \cdot d = \frac{K \cdot \lambda}{B \cdot \cos \theta}$$

Setting $K = 0.9$, using $CuK_{\alpha 1}$ radiation ($\lambda = 1.5405$ Å) and taking the Bragg angle for the illite 10 Å peak ($\cos \theta = 0.997$), the equation above reduces to

$$N \cdot d \simeq \frac{1.40}{B - B_0}$$

where B_0 represents the Kübler index of a mica standard (Weber, Dunoyer de Segonzac and Economou, 1976). This illite 'crystallinity' index has seen little application (e.g. Dunoyer de Segonzac and Bernoulli, 1976; Árkai and Toth, 1983).

2.5.2   Factors affecting illite 'crystallinity'

Illite 'crystallinity' is dependent on many variables and these are discussed briefly below.

(i)   Temperature. Temperature is believed to be the most important physical factor affecting illite 'crystallinity' (Kübler, 1967a, b, 1968). This is best illustrated in contact metamorphic aureoles (e.g. Schaer and Persoz, 1976) and is also supported by a small number of hydrothermal experiments (Smykatz-Kloss and Althaus, 1974; Krumm, 1984, p. 228).

(ii)   Fluid pressure. This variable is generally assumed to be of negligible importance, although direct evidence to prove or disprove this statement is lacking.

(iii)   Stress. The role of stress as a possible factor influencing illite 'crystallinity' is not yet clear. Kübler (1967a, b) described several regional-scale examples of anchimetamorphism without schistosity and of schistosity without anchimetamorphism. From this observation he concluded that the 'dynamic factor, one of the most important factors controlling schistosity, does not affect illite crystallinity in a perceptible manner' (Kübler, 1967b, p. 259). On the other hand, Kübler (1967a, Figure 12), Frey et al. (1973, Figure 6) and Aldahan and Morad (1986, p. 72) measured enhanced illite 'crystallinity' values in some tectonic shear zones. Several authors (Flehmig and Langheinrich, 1974; Gruner, 1976; Teichmüller et al., 1979, p. 211; Nyk, 1985) tried to establish a relationship between deformation and illite 'crystallinity' within a single fold on an outcrop scale. No positive correlation was found when using the Kübler index, but in two of these studies the Flehmig index furnished clear evidence for increasing illite 'crystallinity' with increasing tectonic strain (Flehmig and Langheinrich, 1974; Nyk, 1985), as shown in Figure 2.4. On the other hand, Roberts and Merriman (1985) were able to find enhanced illite 'crystallinity' values with the Kübler index in the hinge zone of a tight anticline, revealing a direct relationship between illite 'crystallinity' and strain on a regional scale (Figure 2.5).

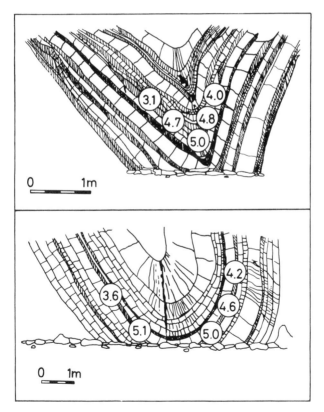

Figure 2.4 Variations of illite 'crystallinity' (Flehmig index) in two minor folds. (Modified from Nyk, 1985, Figure 4.)

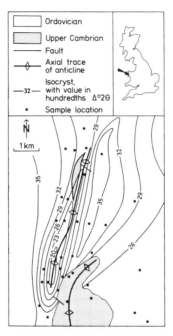

Figure 2.5 Isocrysts (lines of equal illite 'crystallinity' values) in Ordovician metapelites from North Wales. Isocrysts are symmetrically disposed in a narrow belt about the fold axial trace of an anticline, revealing a direct relationship between illite 'crystallinity' and strain. (Modified after Roberts and Merriman, 1985, Figure 2; and Roberts, pers. comm. 1986.)

(iv)   *Time.* Essene (1982, p. 190) pointed out that the attainment of a certain illite 'crystallinity' value is presumably rate-dependent. This would be typical behaviour of metastable material tending towards equilibrium. In other words, the longer an illite remained at constant temperature, the better crystallized it would be. However, at present no laboratory or field evidence exists to support this hypothesis.

(v)   *Lithology.* Lithology plays an important role in determining illite 'crystallinity'. Coarse-grained clastic sediments tend to have better-crystallized illites than fine-grained sediments for two reasons. First, the 'crystallinity' of detrital illites and muscovites will be higher. Secondly, coarse-grained sediments have higher porosities and permeabilities, allowing better circulation of interstitial solutions with concomitant aggradation of illite. Both factors may contribute to the presence of more crystalline illite in sandstones than in shales (Dunoyer de Segonzac, 1970, p. 316).

In metacarbonates the aggradation of illite may be retarded compared to 'normal' metaclastites due to a deficiency in potassium. In such extreme cases smectites may persist into the epizone (Wilson and Bain, 1970; Dunoyer de Segonzac and Abbas, 1976; Schaer and Persoz, 1976; Zingg *et al.*, 1976). A similar retarding effect can be found in sediments with a high organic matter content (Kübler, 1968, p. 391; Weber, 1972b, p. 337). In this case illite crystals are isolated from circulating ionic solutions by a mantle of hydrophobic organic material. In evaporites, on the other hand, the high potassium availability will enhance illite 'crystallinity' (Rumeau and Kulbicky, 1966; Kübler, 1968, p. 390).

According to Kübler (1968; 1984, p. 578) and Dunoyer de Segonzac (1970, p. 316) the influence of lithology on illite 'crystallinity' diminishes with increasing diagenetic or metamorphic grade and becomes negligible at the onset of the anchizone. Nevertheless, slight but systematic differences in illite 'crystallinity' have been determined in different rock types, even in the high-temperature part of the anchizone, by Árkai et al. (1981).

(vi) *Illite chemistry.* Increasing K content in illite leads to better illite 'crystallinity' (Riedel, 1966; Weaver and Beck, 1971, p. 21; Brauckmann, 1984; Hunziker et al., 1986, Figure 7). The required potassium may come from the destruction of clastic K-feldspar, muscovite or biotite.

Opinions concerning the influence, in illite, of Al content on 'crystallinity' are contradictory. Esquevin (1969) stated that 'crystallinity' values of only aluminous illites could be used to indicate metamorphic grade, magnesian illites being unsuitable. To distinguish between aluminous and magnesian illites, Esquevin assumed that the intensity ratio between the second and first basal reflection, $I(002)/I(001)$, could be used as a first approximation. However, as shown by Hunziker et al. (1986, Figure 15), little confidence is placed in the estimation of the chemical composition of illites from their intensity ratios at 10 and 5 Å.

(vii) *Interfering basal reflections.* The presence of other associated phyllosilicates with basal reflections close to 10 Å in the anchizone or epizone may cause a broadening of the illite (001) peak. Such minerals include pyrophyllite (basal reflection at 9.2 Å), paragonite (9.7 Å), mixed-layer paragonite/muscovite (9.8 Å), margarite (9.6 Å), and biotite (10.0 Å). In the presence of these phases, care should be taken in determining true illite 'crystallinity'.

(viii) *Experimental conditions.* Illite 'crystallinity' values determined on X-ray diffractograms are dependent on the shape of the illite 10 Å reflection. The peak shape is dependent on the instrumental conditions of the diffractometer (see Kisch, 1983, pp. 347–350) and on the sample preparation (see Appendix).

Comment. The concept of illite 'crystallinity' must be regarded as a pragmatic approach (Kübler, 1984, p. 575). As discussed above, illite 'crystallinity' is dependent on many factors, and several pitfalls must be avoided. If a large enough sample population is analysed, it may even be possible to map lines of equal 'crystallinity' values, called isocrysts by Roberts and Merriman (1985,

p. 617); see also Dunoyer de Segonzac *et al.* (1968, Figure 3), Weber (1972*b*, Figure 3), Krumm (1984, Figure 9). At the present state of knowledge it cannot be used for geothermometric purposes. Nevertheless, work carried out during the last two decades has shown that illite 'crystallinity' remains one of the most suitable and generally applicable monitors in very low-grade metamorphism of clastic sedimentary rocks.

2.5.3 Chlorite 'crystallinity'

Chlorite 7 Å peak width has recently been determined by several workers (Ludwig, 1973; Schamel, 1974; Le Corre, 1975; Schaer and Persoz, 1976; Deutloff *et al.*, 1980; Dandois, 1981; Duba and Williams-Jones, 1983*b*; Brauckmann, 1984). In general a good linear relation between chlorite 'crystallinity' and illite 'crystallinity' has been observed. In one case the absolute values of both were essentially the same (Duba and Williams-Jones, 1983*b*), but in three other studies chlorite showed an enhanced 'crystallinity' compared to that of illite from the same sample (Schaer and Persoz, 1976; Dandois, 1981; Brauckmann, 1984; see Figure 2.6). What causes the improved 'crystallinity' of chlorite with increasing metamorphic grade is not known; it may reflect an increased size of coherent domains of the chlorite crystal lattice, but it may also be a consequence of a more uniform chemical composition (note that chemically different chlorites have slightly different basal spacings causing peak broadening).

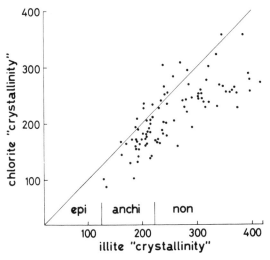

Figure 2.6 Illite 'crystallinity' (Weber index) *v.* chlorite 'crystallinity'. The line of equal illite and chlorite 'crystallinity' values is given for orientation purposes only. Note that chlorites are generally better crystallized than illites from the same sample. (Modified after Brauckmann, 1984, Figure 15.)

2.5.4   Kaolinite 'crystallinity'

Kaolinite 'crystallinity' has been derived by X-ray diffractometry either from the intensity ratio of some non-basal reflections (Hinckley, 1963) or from the 7 Å peak width (Brauckmann and Füchtbauer, 1983). The determination of the Hinckley index requires kaolinite-rich samples, and the kaolinite 7 Å peak width cannot be determined in the presence of chlorite. These restrictions may be responsible for the few examples where kaolinite 'crystallinity' has been measured in diagenetic and very-low grade metamorphic studies.

In a study of kaolinite–coal tonsteins, Eckhardt and von Gaertner (1962) and Eckhardt (1965) observed an increase in kaolinite 'crystallinity' (or stacking order) with increasing coal rank. At a temperature of 200 °C a maximum kaolinite 'crystallinity' was reached. Stadler (1971), on the other hand, noted varying kaolinite 'crystallinity' at a similar high coal rank corresponding to temperatures above 200 °C. He therefore concluded that no relationship existed between kaolinite 'crystallinity' and coal rank. Brauckmann and Füchtbauer (1983) found an exponential increase in kaolinite 'crystallinity' of siltstones towards the contact with a basaltic dyke.

An extensive literature dealing with order–disorder in kaolinite exists and for a review the reader is referred to Brindley (1980, pp. 146–152).

2.5.5   Pyrophyllite 'crystallinity'

Six indices to estimate pyrophyllite 'crystallinity' were proposed by Ianovici et al. (1981), five based on properties of X-ray basal reflections and one based on infrared spectroscopy. Pyrophyllite from the high-grade anchizone was shown to have a better 'crystallinity' than pyrophyllite from near the diagenesis/anchizone transition.

2.6   Polymorphism and polytypism

The characteristic of a chemical substance of crystallizing in more than one form is called polymorphism. Polytypism is a special type of polymorphism and is widespread among phyllosilicates, whereby different forms result from different stacking sequences of similar phyllosilicate structural units. In this section the utility of phyllosilicate polymorphism and polytypism as indicators of very low-grade metamorphism will be evaluated.

2.6.1   Illite polytypes

In illite the one-layer, monoclinic disordered polytype (1 Md) and the two-layer monoclinic polytype (2 M_1) are common, while the three-layer trigonal (3 T) and 2 M_2 structural forms are rare (see e.g. Yoder and Eugster, 1955, pp. 252–254; Dunoyer de Segonzac, 1970, p. 313). Opinions regarding the importance of the 1 M illite polytype are widely divergent (see e.g. Merriman and Roberts, 1985). Fine-grained illite shows the same polytypes as the coarser crystals of muscovite (Levinson, 1955). Diagnostic X-ray diffraction lines of muscovite polytypes are, for example, listed by Bailey (1980, p. 58; 1984, p. 9). The 1 Md → 2 M_1 illite polytype transformation has received most attention in studies of incipient metamorphism.

2.6.2 Determination of $2M_1/(2M_1 + 1\,Md)$ illite polytype ratio

This ratio is determined by measuring the peak areas of (*hkl*) reflections in randomly oriented samples. The proportion of 2 M_1 illite polytype has been estimated from the following peak intensity ratios:

$I(3.74\,\text{Å})/I(2.58\,\text{Å})$ (Velde and Hower, 1963)
$I(3.00\,\text{Å})/I(2.58\,\text{Å})$ (Reynolds, 1963)
$I(2.80\,\text{Å})/I(2.58\,\text{Å})$ (Maxwell and Hower, 1967)

The reflections at 3.74, 3.00 and 2.80 Å are unique to the 2 M_1 mica, while the 2.58 Å reflection is present for both polytypes. The lower limit of detection as well as the 1δ uncertainty are 10–20% 2 M, although some authors claim lower values. There are several reasons for the rather low precision of this method. They include the dependence of peak ratios on particle size (e.g. Velde and Hower, 1963), the possible peak interference with minerals other than illite, and the difficulty of reproducing randomly oriented samples. More experimental details are contained in the references mentioned above.

2.6.3 Experimental determination of muscovite $1\,M$–$2\,M$ polytype stabilities

Yoder and Eugster (1955) synthesized the 1 Md, 1 M, and 2 M_1 polytypes for the end-member muscovite composition. At the lowest temperatures or in runs of short duration the 1 Md mica was commonly obtained. Increasing temperature and run duration resulted in the growth of the ordered 1 M structure, which was later inverted to the 2 M polytype. Additional experiments were performed with natural muscovites of unknown composition. Again, at high temperatures the 1 M structural form converted to a 2 M polytype but it was not possible to demonstrate reversibility. From these observations the following conclusions were reached.

(i) The appearance of the 1 Md structure as the initial phase at all temperatures in some starting materials suggested that it is a metastable form

(ii) The subsequent 1 Md → 1 M ordering and persistence at low temperatures indicated that the 1 M muscovite is a stable form

(iii) At high temperatures the 2 M structure is the most stable form

(iv) On the basis of long runs the 1 M → 2 M transition appears to be between 200 and 350 °C at about 2 kbar water pressure.

Velde (1965) conducted hydrothermal experiments on the reaction 3 kaolinite $2 KOH \rightarrow 2$ muscovite $+ 5 H_2O$ at 1, 2, and 5 kbar water pressure. The sequence of polytype transformation was 1 Md → 1 M → 2 M, as for Yoder and Eugster (1955), and this conversion proceeded with an increase of either time, temperature, or pressure. Because the 1 M structural form was never produced from $2 M_1$ muscovite, Velde (1965, p. 442) concluded that $2 M_1$ is the only stable polytype of muscovite at low and moderate temperatures. However, field studies clearly indicate that 1 M muscovite is a low-temperature form (stable or metastable), and this explains why the $2 M_1 \rightarrow 1 M$ conversion failed in the 570–675 °C temperature range used by Velde (1965, Table 1). Another conclusion reached by Velde should be critically examined, namely the statement that 'the information from work at 4.5 kbar (Table 1) shows that $2 M_1$ is the stable polymorph at temperatures as low as 125 °C' (Velde, 1965, p. 441). It should be stressed that his conclusion is based on one single run using synthetic 1 M muscovite as starting material. In my opinion additional work is needed to support this result.

Mukhamet-Galeyev et al. (1986) investigated the stability of 1 M and $2 M_1$ muscovite polytypes from solubility data. At 300 °C and the pressure of saturated steam, the $2 M_1$ polytype was found to be the stable modification.

2.6.4 Field evidence

A progression from 1 Md toward $2 M_1$ illite with advanced diagenesis or incipient metamorphism has been noted by many authors (Reynolds, 1963; Maxwell and Hower, 1967; Gavrilov and Aleksandrova, 1968; Karpova, 1969; Artru et al., 1969; Dunoyer de Segonzac, 1969, 1970; Frey, 1969a, pp. 101–102; 1970; Gavish and Reynolds, 1970; van Moort, 1971; Eslinger and Savin, 1973; Ludwig, 1973; Foscolos and Kodama, 1974; Mitsui, 1975; Foscolos et al., 1976; Frank and Stettler, 1979; Brauckmann, 1984, pp. 44–50; Cloos, 1983; Lucas, 1984; Weaver and Broekstra, 1984; Merriman and Roberts, 1985; Hunziker et al., 1986). Several of these authors have therefore concluded that the ratio $2 M_1/(2 M_1 + 1 Md)$ constitutes a useful measure of metamorphic grade. However, as pointed out by Kübler (1967a, p. 108), inherited $2 M_1$ illite may obscure this relationship. Therefore, evidence is needed that only the 1 Md polytype was present before the onset of metamorphism.

The illite $1 Md \rightarrow 2 M_1$ polytype conversion with respect to metamorphic grade will be explored next. The relevant data are compiled in Table 2.3. It should be stressed that the information on 'metamorphic conditions' given by different authors is of unequal significance and ranges from 'lower grades of metamorphism' to specific $P-T$ values. Nevertheless a good correlation exists between the $1 Md \rightarrow 2 M_1$ progression and metamorphic grade or the few $P-T$

Table 2.3 Summary of information on the illite 1 Md → 2 M$_1$ polymorph conversion

No.	Author	Number of samples studied	Size fraction (μm)	Age and rock type	Range of % 2 M$_1$ studied	Metamorphic conditions at which the 1 Md → 2 M$_1$ conversion is completed or arrested
1	Reynolds (1963)	25	<2	Proterozoic carbonate rocks	5–79	'Lower grades of metamorphism'[1]
2	Maxwell and Hower (1967)	37	<2	Precambrian argillites	0–100	Biotite isograd
3	Gavrilov and Aleksandrova (1968)	?	?	Palaeozoic tuffaceous mudstones	?	Profound epigenesis to initial metamorphism (metagenesis) (?)
4	Karpova (1969)	?	?	Upper Palaeozoic terrigenous rocks	0–100(?)	Early metagenesis
5	Artru et al. (1969) Dunoyer de Segonzac (1969)	8	<2(?)	Jurassic black shales	0–90	Low-grade anchizone[2]
6	Frey (1969a, 1970) Hunziker et al. (1986)	70	0.1–2	Triassic shales and marls	0–100	~ Anchizone/epizone boundary
7	Gavish and Reynolds (1970)	11	0.5–2	Ordovician limestones	9–50	Biotite zone[3]
8	Eslinger and Savin (1973)	5	<0.5	Precambrian argillites and limestones	21–60	225 °C/1.7 kbar[4] at 20% 2 M$_1$ 310 °C/2.3 kbar[4] at 60% 2 M$_1$
9	Ludwig (1973)	?	<2 2–6	Ordovician slates	75–100 85–100	Low-grade epizone

10	Foscolos and Kodama (1974)	7	0.2–1	Cretaceous shales	20–65	65% 2 M₁ in late diagenetic stage (164 °C (?))
11	Foscolos et al. (1976)	13	<2	Cretaceous shales	20–50	50% 2 M₁ in middle diagenetic stage
12	Frank and Stettler (1979)	13	<1 to 6–20	Triassic shales and marls	45–100	350–380 °C/2–3 kb (?)
13	McMechan and Price (1982)	21	<2	Proterozoic argillites	60–100	Anchizone/epizone boundary
14	Cloos (1983)	44	?	Mesozoic shales and melange matrix	0–100	Beginning of greenschist facies
15a	Brauckmann (1984)	71	<2	Triassic biocalcarenites calcilutites	11–100	Low-grade epizone[5]
15b		19	<2		16–76	76% 2 M₁ at diagenesis/anchizone boundary[5]
16	Lucas (1984)	13	<2	Proterozoic marls	0–100	Anchizone/epizone boundary
17	Weaver and Broekstra (1984)	14	0.2–2	Cambrian shales	0–100	Anchizone/epizone boundary
18	Merriman and Roberts (1985)	314	<2	Precambrian to Silurian metapelites	0–100	Anchizone/epizone boundary

? = no data available

(?) = data somewhat uncertain due to insufficient information

[1] Carbonate rocks are reportedly interbedded with phyllites or slates

[2] Eventually high-grade anchizone; the unreported but possible presence of Na-bearing micas would lead to apparently low illite 'crystallinity'

[3] Doll et al. (1961); Thompson and Norton (1968)

[4] Temperature values based on illite-quartz oxygen isotope geothermometry. Pressure calculated with a geothermal gradient of ~ 36 °C/km (Eslinger and Savin, 1973) and assuming a rock density of 2.7 g/cm³.

[5] Contact metamorphic area

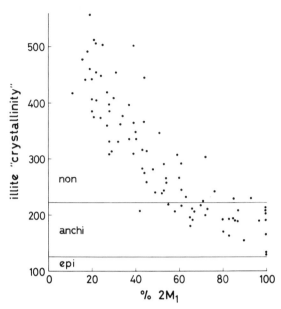

Figure 2.7 Illite 'crystallinity' (Weber index) $v.$ the amount of $2M_1$ polytype. (Modified after Brauckmann, 1984, Figure 11.)

values available. Most authors agree that the $1\,Md \rightarrow 2\,M_1$ conversion is completed approximately at the anchizone/epizone boundary as defined by illite 'crystallinity' data (see e.g. Figure 2.7) or the low-grade greenschist facies. This conclusion is at variance with that reached by Dunoyer de Segonzac (1969, Figure 105; 1970, Figure 12). According to this author the 1 Md illite polytype disappears at a lower grade, during deep (or late) diagenesis. However, the results presented by Dunoyer de Segonzac (1969, Figures 54, 63, 65) are not in favour of this conclusion.

2.6.5   Factors affecting the illite $1\,Md \rightarrow 2\,M_1$ polytype conversion

The determining factors for illite polytypism are essentially the same as those mentioned earlier for illite 'crystallinity'.

(i)   *Temperature.* The importance of this variable is verified both by field evidence and laboratory experiments as discussed above.

(ii)   *Fluid pressure.* The positive effect of fluid pressure was demonstrated in hydrothermal experiments by Velde (1965). The only field data from a high pressure terrane were provided by Cloos (1983), but in this case the $1\,Md \rightarrow 2\,M_1$ conversion was apparently not completed at a lower temperature (cf. Table 2.2).

(iii) *Stress.* It is to be expected that the degree of transformation of metastable 1 Md to 2 M_1 illite is a function of stress. However, this author is not aware of any relevant data.

(iv) *Time.* Yoder and Eugster (1955) pointed out the sluggish nature of the polytype transformations in their experiments, and Velde (1965) demonstrated the rate dependence of the 1 Md → 1 M and 1 M → 2 M_1 muscovite polytype transformations. However, there remains the question to what extent the results from short-run experiments are relevant to nature. An inspection of Table 2.3 does not confirm that time would be an important factor.

(v) *Lithology.* The relationship between host-rock lithology and illite polytypism has been reviewed recently by Kisch (1983, p. 333; see also Brauckmann, 1984, pp. 47–49). Hence sedimentological conditions (cf. amount of detrital 2 M_1 polytype), porosity and pore-fluid composition as well as the illite–carbonate ratio are important factors. See section 2.4.2, (v), for the development of corresponding arguments.

(vi) *Illite chemistry.* Radoslovich and Norrish (1962) have indicated that composition can significantly affect the mica structure and may influence the polytype. As an example, the Fe-rich dioctahedral micas glauconite and celadonite exhibit the 1 M or 1 Md polytype, but not the 2 M_1 structural form (see e.g. Bailey, 1984, Table 3). As illite may show considerable solid solution toward the celadonite end-member, it is conceivable that illite chemistry may influence the 1 Md → 2 M_1 transformation. The limited data presently available, however, do not allow any conclusion.

Conclusion. The illite 1 Md → 2 M_1 polytype conversion has seen much less application than illite 'crystallinity' to monitor incipient metamorphism. This is because the determination of the 2 M_1/(2 M_1 + 1 Md) ratio is much more time-consuming and relatively inaccurate, although the determining factors are essentially the same for both methods. If future work should show that time and chemical composition are important controlling factors, then one should be very cautious in applying the 1 Md → 2 M_1 transition to monitor the grade of metamorphism.

2.6.6 *Chlorite polymorphs and polytypes*

The 7 Å → 14 Å polymorph and the Ib → IIb (both 14 Å) polytype changes of trioctahedral chlorite are considered as follows.

(i) *The 7 Å → 14 Å polymorph transformation.* The use of the 7 Å → 14 Å polymorph change as a metamorphic indicator is hampered by two factors. First, the recognition of 7 Å chlorite (also termed 7 Å chamosite, berthierine or septechlorite) in the presence of 14 Å chlorite by X-ray methods is difficult due to peak coincidences. Secondly, not all 14 Å chlorite originates from 7 Å chlorite. Observations on natural samples have yielded the following information:

(i) Both 7 Å and 14 Å chlorites have been observed in the same sample, either by optical and XRD methods (Schoen, 1964; Delaloye, 1966; Tröhler, 1966; Frey, 1969a, 1970) or by the TEM/STEM technique (Lee *et al.*, 1984; Ahn and Peacor, 1985). In the two latter studies, 7 and 14 Å units were found to be transitional from one to the other along individual layers.

(ii) The 7 Å polymorph is nearly identical in chemical composition with the coexisting 14 Å polymorph (Ahn and Peacor, 1985).

(iii) Temperature information on the 7 Å → 14 Å polymorph transition is contained in only three studies. According to Karpova (1969), this transformation occurs during the initial catagenesis (zeolite facies, see Table 2.2), although no further details were given. Velde *et al.* (1974) inferred that 'the polymorphic transition could have occurred from 25 °C to somewhat above 100 °C'. Iijima and Matsumoto (1982) recorded a temperature of 160° or 190 °C for this transformation.

(iv) Some additional information on the temperature range of the 7 Å → 14 Å polymorph transformation is obtained from areas of incipient metamorphism where 7 Å chlorite is still present. Delaloye (1966), Tröhler (1966) and Frey (1969a, 1970) observed the 7 Å polymorph in the anchizone, that is in the temperature range of about 200 to 300 °C. These temperatures are definitively higher than the 100 °C proposed by Velde (1985a, p. 189) for the 7 Å → 14 Å chlorite transformation.

In summary, coexisting 7 Å and 14 Å chlorite have been found over a relatively large temperature range and the 7 Å phase is probably a metastable precursor of the 14 Å phase, as suggested by Schoen (1964), Ahn and Peacor (1985) and Cho and Fawcett (1986). It follows that the 7 Å → 14 Å chlorite polymorph change cannot be used as a reliable metamorphic indicator at the present time.

(ii) *The Ib → IIb polytype change.* During late diagenesis and incipient metamorphism a change in the polytype of 14 Å trioctahedral chlorite is observed. Bailey and Brown (1962) recognized four polytypes in a study of 303 natural chlorites, designated Ia, Ib ($\beta = 97°$), Ib ($\beta = 90°$), and IIb. For structural details and identification of the polytypes the reader is referred to Bailey (1980, pp. 86–96). Textural evidence from thin-section petrography and SEM studies led Hayes (1970) to the conclusion that type I chlorite structures are formed almost exclusively by diagenetic processes. The following diagenetic crystallization and stability sequence with increasing temperature was proposed, based on crystallographic considerations and on grinding experiments: Ibd (d = disordered) → Ib (β = 97°) → Ib (β = 90°). Conditions of incipient metamorphism finally cause the Ib (β = 90°) → IIb polytype conversion. Hence the IIb chlorite in unmetamorphosed sediments is most probably of detrital origin.

According to Hayes (1970), chemical composition of the chlorites appears to have little influence upon relative structural stabilities. Because Gibbs' free energy differences of polytypes are likely to be very small, chemical composition may be crucial, and this is supported by the following observations. Karpova (1969), Mitsui (1975) and Shirozu (1978, Figure 7.2) noted that the IIb polytype was usually more magnesium-rich than the Ib polytype, while Curtis *et al.* (1985) showed that metamorphic chlorites of the polytype IIb had generally higher tetrahedral Al and eventually also higher Mg contents than sedimentary chlorites of the polytype Ib.

Temperature information on the Ib → IIb polytype conversion is very limited. According to Hayes (1970, p. 301) this conversion occurs at 'the lowest grade of metamorphism, which is inferred to be about 150–200 °C'. Karpova (1970) observed the polytype conversion at the transition between the stages of initial and deep catagenesis (= epigenesis), that is within the zeolite facies (Table 2.1) and in a temperature range similar to that mentioned above. Mitsui (1975) found coexisting Ib and IIb chlorite polytypes at inferred temperatures of about 200–300 °C.

In summary, the view expressed by Hayes (1970) and Kisch (1983, p. 339) that 'chlorite polytypism is potentially a useful geothermometer' seems to be too optimistic at the present time. It should also be pointed out that only very chlorite-rich samples are suitable for the determination of chlorite polytypes on the basis of XRD data. The TEM method, on the other hand, is not hampered by this restriction.

2.6.7 Kaolinite polytypes

The occurrence and significance of dickite at the expense of kaolinite during advanced diagenesis and incipient metamorphism will now be discussed. Information on the occurrence of nacrite, yet another kaolinite polytype, is too limited to be included in the discussion.

The three kaolinite polytypes are distinguished by some non-basal reflections, and their diagnostic X-ray diffraction lines are listed by Bailey (1980, p. 30). Where kaolinite minerals occur as a main constituent, as for example in quartz–kaolinite sandstones, tonsteins or veins, the polytype is easily determined. In shales and slates, however, kaolinite minerals are often present as a minor constituent only, and the polytype determination by classical XRD methods is no longer possible. For this reason our knowledge on the distribution of kaolinite polytypes is limited, and much that is described in the literature as kaolinite may actually be dickite or nacrite.

A bibliographical search for dickite occurrences yielded about 25 references, excluding those dealing with dickite of obviously hydrothermal origin related to ore deposits and igneous intrusions. Shelton (1964), Bayliss *et al.* (1965), Dunoyer de Segonzac (1970), Shutov *et al.* (1970) and Kisch (1983) provide information on occurrences. The following observations are worth mentioning:

  (i) Authigenic dickite has been most commonly found in sandstones, and more rarely in limestones, ironstones, tonsteins, shales and slates.

 (ii) The common mode of occurrence appears to be in vugs, cracks, and cavities, some in replacement of faunal fragments, or along fractures or veins. This mode may reflect the easy way to identify dickite rather than its apparent absence as a matrix mineral in shales and slates.

(iii) In several cases dickite and kaolinite were found to coexist and it was inferred that dickite formed at the expense of kaolinite.

(iv) In other cases where dickite was found in sandstones, accompanying siltstones still contained kaolinite. It follows that high permeability appears to favour the kaolinite → dickite polytype transformation.

Information about diagenetic or metamorphic grade of dickite occurrences is as follows. Dickite has been reported from the diagenetic zone at present borehole temperatures of 100 °C (Ferrero and Kübler, 1964) and 80 °C (Dunoyer de Segonzac, 1969), from the low-grade part of the anchizone (Clauer and Lucas, 1970), and from the early and late epigenetic stage (Rodionova and Koval'skaya, 1974). A quartz–dickite facies is considered characteristic by Kossovskaya and Shutov (1965, 1970) for quartz–kaolinite rocks in the stage of deep epigenesis. However, the rather limited and divergent results do not favour the use of dickite as an index mineral.

2.6.8 *Pyrophyllite polytypes*

A triclinic and a monoclinic form of pyrophyllite can be distinguished (Brindley and Wardle, 1970), and both polytypes have been observed in nature (Frey, 1978; Ianovici *et al.*, 1981; Frey *et al.*, 1987). More data will be needed to establish whether a correlation exists between pyrophyllite polytype distribution and metamorphic grade. According to hydrothermal experiments of Eberl (1979), monoclinic pyrophyllite is the low-temperature form, a relation which is opposite to polymorphs of most silicates.

2.7 Mineral assemblages

Presumably more than 95% of very low-grade metaclastites contain the monotonous assemblage muscovite (or illite)–chlorite–quartz with or without feldspars, carbonates, haematite or organic material. This mineral assemblage is not diagnostic because it occurs in a large range of P–T conditions. This means that very low-grade metaclastite assemblages are generally not useful in determining metamorphic grade at subgreenschist facies conditions, as opposed to metabasic assemblages (see Chapter 3). For some metapelites and metamarls, however, the presence of or the first appearance of pyrophyllite, rectorite, mixed-layer paragonite/muscovite, paragonite, lawsonite, stilpnomelane, and Mg–Fe–carpholite may be indicative of very low-grade metamorphic conditions.

Before discussing specific mineral assemblages, some theoretical phase relations will be derived.

2.7.1 Phase relations

The chemical composition of pelites can be expressed by a complex system: K_2O–Na_2O–Al_2O_3–Fe_2O_3–MgO–FeO–SiO_2–H_2O. In order to express the chemical composition of marls, two additional components, CaO and CO_2, must be considered. The approach to an understanding of such complex systems involves the study of simpler subsystems. In the following, the subsystems CaO–Al_2O_3–SiO_2–H_2O (CASH) and CaO–Al_2O_3–SiO_2–H_2O–CO_2 (CASH–CO_2) will be considered. It should be emphasized that phase relations of medium- and high-grade metapelites are traditionally treated in the subsystem K_2O–Al_2O_3–MgO–FeO–SiO_2–H_2O. Because of the complexity of illitic and chloritic clay minerals, presentation of phase relations for this system is not as effective as those in higher-grade metapelites.

The petrogenetic grids presented below were produced through the courtesy of M. Engi and L. Baumgartner, using the thermodynamic database of Berman *et al.* (1985) Isobaric T–X_{CO_2} diagrams were calculated for ideal mixing of non-ideal H_2O and CO_2 using modified Redlich–Kwong equations of state given by Kerrick and Jacobs (1981). If non-ideal H_2O–CO_2 mixing was considered and/or electrolyte components (e.g. NaCl) included in the calculations, the predicted T–X_{CO_2} topologies would be altered primarily in

Table 2.4 Composition and abbreviations of phases in the system CaO–Al_2O_3–SiO_2–H_2O–CO_2

Name	CaO	Al_2O_3	SiO_2	H_2O	CO_2	Symbol
Calcite	1	—	—	—	1	Cc
Diaspore	—	1/2	—	1/2	—	Di
Grossularite	3	1	3	—	—	Gr
Kaolinite	—	1	2	2	—	Ka
Kyanite/andalusite	—	1	1	—	—	Ky/And
Laumontite	1	1	4	4	—	Lm
Lawsonite	1	1	2	2	—	Lw
Margarite	1	2	2	1	—	Ma
Prehnite	2	1	3	1	—	Pr
Pyrophyllite	—	1	4	1	—	Py
Quartz	—	—	1	—	—	Qz
Wairakite	1	1	4	2	—	Wr
Zoisite	2	$1\frac{1}{2}$	3	1/2	—	Zo

Table 2.5 Stoichiometries of stable reactions for the systems CASH and CASH–CO_2 shown in Figures 2.8 and 2.9, respectively. For phase compositions and abbreviations see Table 2.4

Reactions in Figure 2.8

(1) $And = Ky$

(2) $Py = And + 3Qz + H_2O$

(3) $Py = Ky + 3Qz + H_2O$

(4) $Ka + 2Qz = Py + H_2O$

(5) $Py + Wr = Ma + 6Qz + 2H_2O$

(6) $Ka + Wr = Ma + 4Qz + 3H_2O$

(7) $Ka + Lw = Ma + 2Qz + 3H_2O$

(8) $Py + Lw = Ma + 4Qz + 2H_2O$

(9) $Wr = Lw + 2Qz$

(10) $Ka + Lm = Ma + 4Qz + 5H_2O$

(11) $Lm = Lw + 2Qz + 2H_2O$

(12) $5Lw = Ma + 2Zo + 2Qz + 8H_2O$

(13) $5Wr = Ma + 2Zo + 12Qz + 8H_2O$

(14) $2Lw + Pr = 2Zo + Qz + 4H_2O$

(15) $2Lm + Pr = 2Zo + 5Qz + 8H_2O$

(16) $2Wr + Pr = 2Zo + 5Qz + 4H_2O$

(17) $Lm = Wr + 2H_2O$

(18) $Lm + Gr = 2Pr + Qz + 2H_2O$

(19) $2Pr = Lw + Gr + Qz$

(20) $5Lw + Gr = 4Zo + Qz + 8H_2O$

(21) $5Pr = 2Gr + 2Zo + 3Qz + 4H_2O$

Additional reactions in Figure 2.9a

(22) $Ka + Cc + 2Qz + 2H_2O = Lm + CO_2$

(23) $Ka + Cc + 2Qz = Wr + CO_2$

(24) $Py + Cc + H_2O = Wr + CO_2$

(25) $Ma + Cc + 6Qz + 3H_2O = 2Wr + CO_2$

(26) $2Py + Cc = Ma + 6Qz + H_2O + CO_2$

(27) $2And + Cc + H_2O = Ma + CO_2$

(28) $3Ma + 5Cc + 6Qz = 4Zo + H_2O + 5CO_2$

(29) $3Wr + Cc = 2Zo + 6Qz + 5H_2O + CO_2$

(30) $2Zo + 5Cc + 3Qz = 3Gr + H_2O + 5CO_2$

(31) $Wr + Cc = Pr + Qz + H_2O + CO_2$

Additional reactions in Figure 2.9b

(32) $2Ka + Cc = Ma + 2Qz + 3H_2O + CO_2$

(33) $2Ky + Cc + H_2O = Ma + CO_2$

Additional reactions in Figure 2.9c

(34) $Ka + Cc = Lw + CO_2$

(35) $Ma + Cc + 2Qz + 3H_2O = 2Lw + CO_2$

Additional reactions in Figure 2.9d

(36) $Ka = 2Di + 2Qz + H_2O$

(37) $2Di + 4Qz = Py$

(38) $2Di + Qz = Ky + H_2O$

(39) $Py + Cc + H_2O = Lw + 2Qz + CO_2$

Additional reactions in Figure 2.9e

(40) $3Lw + Cc = 2Zo + 5H_2O + CO_2$

(41) $3Ky + 4Cc + 3Qz + H_2O = 2Zo + 4CO_2$

the following ways relative to the diagrams presented here:

(i) H_2O-rich limbs of devolatilization reactions would be shifted towards the H_2O-axis, notably at low temperature and high salinity

(ii) Dehydration curves would be depressed to lower temperatures in proportion to the stoichiometric coefficient of H_2O.

Table 2.6 Comparison of P–T coordinates of some invariant points in the CASH petrogenetic grid from three different sources.

Invariant point (this study)	Perkins *et al.* (1980)	Chatterjee *et al.* (1984)	This study
I	340 °C/1.9 kbar	390 °C/2.6 kbar	390 °C/2.6 kbar
III	—	340 °C/6.5 kbar	310 °C/4.1 kbar
XI	240 °C/6.1 kbar	—	290 °C/4.7 kbar
XII	300 °C/6.0 kbar	440 °C/11.2 kbar*	440 °C/10.2 kbar

*Metastable invariant point

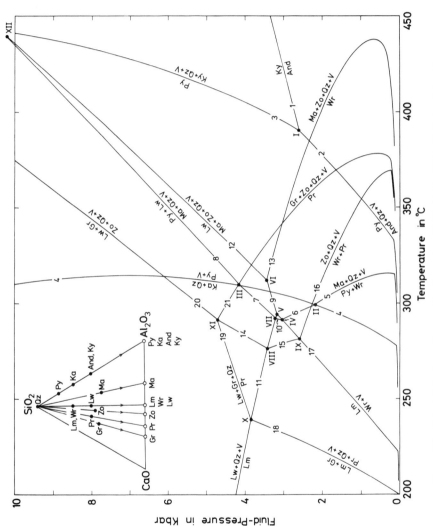

Figure 2.8 Calculated phase diagram for the system CaO–Al$_2$O$_3$–SiO$_2$–H$_2$O involving the phases And, Gr, Ka, Ky, Lm, Lw, Ma, Pr, Py, Wr, Zo with excess Qz and V. The inset shows the projection of phases from H$_2$O(V) and SiO$_2$(Qz) on CaO–Al$_2$O$_3$.

The combination of these effects might alter some of the univariant point topologies.

(i) *The $CaO-Al_2O_3-SiO_2-H_2O(CASH)$ system.* Minerals included in the calculations are andalusite, diaspore, kaolinite, kyanite, laumontite, lawsonite margarite, prehnite, pyrophyllite, quartz, wairakite, and zoisite. The composition of these phases is given in Table 2.4, and stoichiometric coefficients of stable univariant equilibria are presented in Table 2.5. The calculated phase relations with excess quartz and H_2O are shown in the $P-T$ diagram of Figure 2.8. Similar petrogenetic grids have been presented earlier by Perkins *et al.* (1980, Figure 9) and Chatterjee *et al.* (1984, Figure 4), but Perkins *et al.* ignore kaolinite, and both studies neglect laumontite and wairakite. Table 2.6 compares the $P-T$ coordinates of some invariant points encountered in the three phase diagrams mentioned above. While the agreement for invariant points I and XI is reasonable, it is less satisfactory for III and XII.

(ii) *The $CaO-Al_2O_3-SiO_2-H_2O-CO_2(CASH-CO_2)$ system.* Minerals included in the analyses are identical with those treated in the CASH system plus additional calcite. Laumontite, prehnite and wairakite are only rarely reported from very low-grade metaclastites (e.g. Schiffman *et al.*, 1985), but they have been included in the analyses in order to improve understanding of their rare occurrence. The calculated phase relations are depicted on isobaric temperature–fluid composition (X_{CO_2}) diagrams.

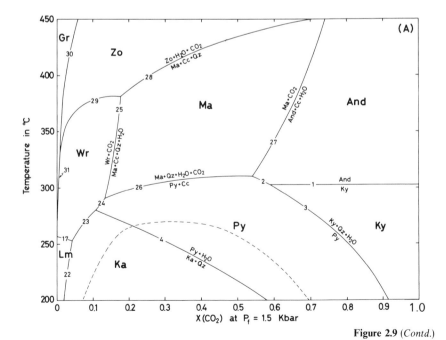

Figure 2.9 (*Contd.*)

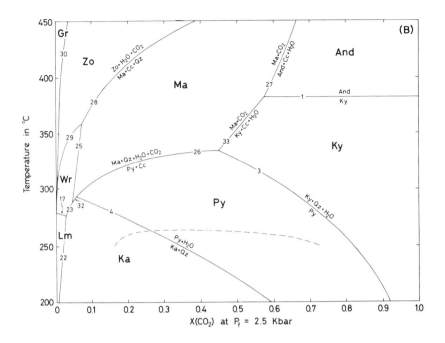

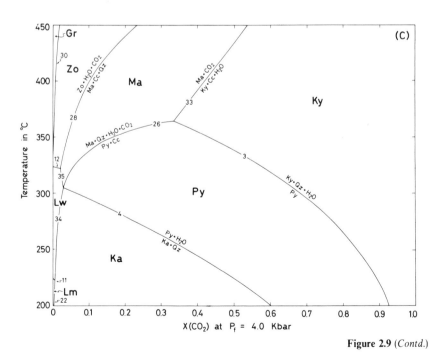

Figure 2.9 (*Contd.*)

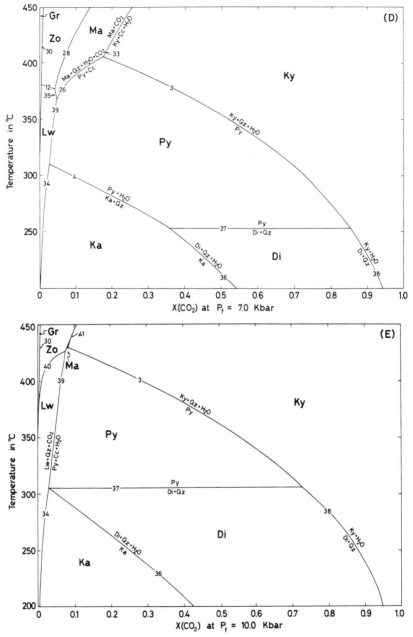

Figure 2.9 Calculated temperature–gas composition diagrams for the system CaO–Al$_2$O$_3$–SiO$_2$–H$_2$O–CO$_2$ involving the phases And, Di, Gr, Ka, Ky, Lm, Lw, Ma, Pr, Py, Wr, Zo with excess Qz and Cc at total fluid pressures of 1.5 kbar (A), 2.5 kbar (B), 4.0 kbar (C), 7.0 kbar (D), and 10.0 kbar (E). The dashed lines in (A) and (B) refer to the H$_2$O–CO$_2$ solvus. The calcite–aragonite transition in (D) and (E) has been ignored.

Many mineral assemblages belonging to the CASH–CO_2 system have excess quartz and calcite. Figure 2.9 shows the topologies of the isobaric T–X_{CO_2} diagrams with excess quartz and calcite. The pressures of 1.5 to 10.0 kbar were chosen in accordance with major changes of the topologies. The stoichiometries of the stable equilibria among the phases of Figure 2.9 are listed in Table 2.5.

It is evident from these diagrams that grossularite, laumontite, prehnite, and lawsonite in the presence of excess quartz + calcite are stable only in the presence of an extremely water-rich fluid composition, in accordance with conclusions reached in earlier studies (e.g. Coombs *et al.*, 1970, Figure 2; Seki and Liou, 1981, Figure 4).

2.7.2 *Pyrophyllite-bearing assemblages*

(i) *Occurrence*. Pyrophyllite has long been regarded as an essentially hydrothermal mineral (e.g. Deer *et al.*, 1962) but its occurrence during incipient regional metamorphism is now well established. More than sixty references referring to such pyrophyllite-bearing assemblages in shales, slates, phyllites, quartzites and conglomerates are listed in Table 2.8. One occurrence of apparent detrital origin described by Loughnan and Ward (1971) is also included because a 'diagenetic origin' should be reconsidered, as discussed by Day (1976, p. 1273). Illite 'crystallinity' data are available for 33 of these pyrophyllite occurrences. Provided that illite 'crystallinity' values were correctly determined (taking into account the possible peak broadening of the first illite basal reflection caused by the nearby first pyrophyllite basal reflection), 17 of these occurrences belong to the anchizone, 7 to the anchizone-epizone transition, and 9 to the low-grade epizone.

Table 2.7 Minerals coexisting with pyrophyllite in metaclastites

Very common	*Common*
Quartz	Kaolinite
Illite or muscovite	Paragonite
Chlorite	Paragonite/muscovite mixed-layer
Haematite	Calcite
Organic material	
Rare	*Very rare*
Rectorite	Diaspore
Chloritoid	Kyanite
Albite	Biotite
Dolomite	Margarite
	Clinozoisite
	Garnet
	Ferro-carpholite
	Magnesite
	Gypsum

Table 2.8 Pyrophyllite-bearing assemblages from metaclastites in five different chemical systems

$Al_2O_3-SiO_2-H_2O$	$CaO-Al_2O_3-SiO_2-H_2O$	$CaO-Al_2O_3-SiO_2-H_2O-CO_2$	$Al_2O_3-FeO-MgO\cdot SiO_2-H_2O$	$Al_2O_3-Na_2O\cdot K_2O-SiO_2-H_2O$
Py + Qz (1, 2, 4, 6, 7, 9–12, 14–16, 18–26, 28, 30, 31, 35–38, 42–49, 52–54, 56–59, 61, 62)	Py + Qz + Ma (49)	Py + Qz + Cc (2, 4, 6, 22–26, 42)	Py + Qz + Ch (1, 4, 6, 9, 10, 14, 20–26, 30, 36, 37, 43, 45, 49, 52–54, 56–59, 61)	Py + Qz + Il + Re (7, 26, 30, 43)
	Py + Qz + Cz (59, 62)	Py + Qz + Ka + Cc (6, 17, 54)		Py + Qz + Ka + Re (39)
Py + Qz + Ka (1, 3, 5, 6, 8, 11, 13, 17, 23, 27, 30–33, 39, 40, 43, 50, 51, 54, 55, 60)			Py + Qz + Ka + Ch (3, 5, 6, 8, 11, 13, 17, 23, 27, 30, 31, 33, 40, 43, 51, 54, 55, 60)	Py + Qz + Ka + Il + Re (3, 8, 13, 30, 43)
				Py + Qz + Il + Pa + Re (1)
Py + Qz + Ky (18, 22)			Py + Qz + Ct (2, 12, 23, 26, 49, 53, 56)	Py + Qz + Il + P/M (4, 19, 20, 25, 54, 56)
			Py + Qz + Ch + Ct (11, 15, 23, 26, 49, 52, 62)	Py + Qz + Il + Pa (16, 23, 28, 48, 53, 56)
			Py + Qz + Ka + Ch + Ct (1, 32)	Py + Qz + Il + P/M + Pa (10, 21, 23, 25, 36, 53, 61)
			Py + Qz + Ch + FeC (56)	
				Py + Qz + Ka + Il + P/M + Pa (33, 35)
Py + Ka + Di + Co (29)			Py + Qz + Ch + Ky (22)	Py + Qz + Il + Ab (9, 16, 22, 59)
				Py + Qz + Il + Re + Ab (15)
				Py + Qz + Ka + Il + Re + Ab (1)
				Py + Qz + Ka + Il + Ab (50, 54)
				Py + Qz + Ky + Il + Ab (22)

Mineral abbreviations
(in addition to those given in Table 2.4)

Ab	Albite	FeC	Ferro-carpholite
Ch	Chlorite	Il	Illite (also used for muscovite)
Co	Corundum	P/M	Paragonite/muscovite mixed-layer
Ct	Chloritoid	Re	Rectorite
Cz	Clinozoisite		

(1) Aparicio and Galán (1980)
(2) Ashworth and Evirgen (1984)
(3) Beuf et al. (1966)
(4) Breitschmid (1982)
(5) Brime and Perez-Estuan (1980), Brime (1985)
(6) Chennaux and Dunoyer de Segonzac (1967)
(7) Chennaux et al. (1970)
(8) Christensen (1975)
(9) Clauer and Lucas (1970)
(10) Davies (1983)
(11) De Swardt and Roswell (1975), Martini (written comm. 1976)
(12) Dobretsov et al. (1975, pp. 55–57)
(13) Dunoyer de Segonzac (1969)
(14) Dunoyer de Segonzac and Millot (1962)
(15) Dunoyer de Segonzac and Heddebaut (1971)
(16) Echle (1985)
(17) Ehlmann and Sand (1959)
(18) England (1972)
(19) Ferla and Lucido (1972)
(20) Fieremans and Bosmans (1982)
(21) Flehmig and Gehlken (1983)
(22) Franceschelli et al. (1984)
(23) Franceschelli et al. (1986)
(24) Frey (1970, 1978)
(25) Frey and Wieland (1975)
(26) Frey et al. (1987)
(27) Gill et al. (1977)
(28) Gomez-Pugnaire et al. (1978)
(29) Hayashi (1980)
(30) Henderson (1970, 1971)

(31) Hosterman et al. (1970)
(32) Ianovici et al. (1981)
(33) Juster and Brown (1984)
(34) Kazanskiy et al. (1972)
(35) Kossovskaya and Shutov (1965, 1970)
(36) Kramm (1978)
(37) Kubler (1967a), Dunoyer de Segonzac et al. (1968)
(38) Lécolle and Roger (1976)
(39) Loughnan and Ward (1971); see also Day (1976, p. 1273)
(40) Ludwig (1972a)
(41) Manby (1983)
(42) Melou and Plusquellec (1967)
(43) Paradis et al. (1983)
(44) Piqué (1982)
(45) Primmer (1985)
(46) Reed and Hemley (1966)
(47) Robinson et al. (1980)
(48) Rosenfeld (1961)
(49) Sagon (1965, 1970)
(50) Saupé et al. (1977)
(51) Scherp et al. (1968)
(52) Schramm (1974)
(53) Schramm (1977)
(54) Schramm (1978)
(55) Schramm et al. (1982)
(56) Seidel (1978), Viswanathan and Seidel (1979)
(57) Spackman and Moses (1961)
(58) Teichmüller et al. (1979, p. 220)
(59) Tobschall (1969)
(60) Weber (1972b)
(61) Wieland (1979)
(62) Zen (1961a, p.64)

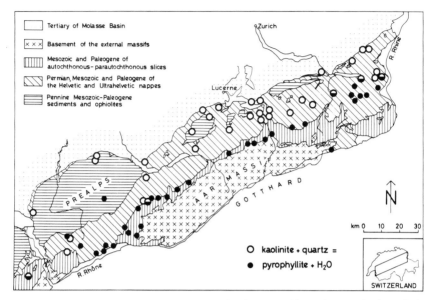

Figure 2.10 Distribution of Ka + Qz- and Py-bearing assemblages in metaclastites of the external zone of the Central Swiss Alps. (Modified from Frey, 1987.)

Pyrophyllite has also been reported from other slightly metamorphosed lithologies such as laterites (e.g. Goffé *et al.*, 1973; Goffé, 1979; Goffé and Saliot, 1977; Goffé, 1982; Jansen and Schuiling, 1976; Kamineni and Efthekhar-nezad, 1977) or siliceous volcanics (e.g. Wall and Kesson, 1969; Black, 1975). In a few cases pyrophyllite has also been found as a truly detrital mineral (e.g. Lafond, 1961, 1965; Biscaye, 1965; Dunoyer de Segonzac and Chamley, 1968).

(ii) *Mineral assemblages and mineral reactions.* Minerals which were reported to coexist with pyrophyllite in metaclastites are listed in Table 2.7, and corresponding mineral assemblages belonging to several model systems are given in Table 2.8. A few comments on some of these assemblages follow.

(i) *Py + Qz + Ka*: Field evidence indicates that pyrophyllite formed at the expense of dickite (Clauer and Lucas, 1970) or kaolinite (Frey, 1970, 1978; Schramm, 1978; Franceschelli *et al.*, 1986) according to the reaction

$$1 \text{ kaolinite} + 2 \text{ quartz} \rightarrow 1 \text{ pyrophyllite} + 1 \text{ H}_2\text{O} \qquad (4)$$

A corresponding reaction isograd has been mapped by Schramm (1978) and Frey (1987), see Figure 2.10. It is therefore somewhat surprising that many authors (Table 2.8) have encountered the assemblage Py + Qz + Ka, which is univariant in the system Al_2O_3–SiO_2–H_2O(ASH) at a fixed activity of water. This phenomenon might be explained by: the metastable persistence of kaolinite or the retrograde formation of the same phase; some solid solution in kaolinite and/or pyrophyllite involving additional components with respect to ASH; or the internal buffering of the fluid phase along reaction (4), whereby the assemblage Py + Qz + Ka would become divariant in P–T space.

(ii) $Py + Qz + Ky$: This assemblage represents the solid phases of the univariant reaction

$$1\,\text{pyrophyllite} = 1\,\text{kyanite} + 3\,\text{quartz} + 1\,H_2O \qquad (3)$$

in the system ASH. Arguments as mentioned for Py + Qz + Ka can be put forward to explain the presence of Py + Qz + Ky. Limited field evidence (England, 1972; Dobretsov et al., 1973, p. 205; Franceschelli et al., 1984, 1986) indicates that pyrophyllite is rarely consumed in nature by reaction (3).

(iii) $Py + Ab$: This assemblage has been reported in a few studies (Table 2.8), although Pa + Qz would be more stable. Textural evidence (e.g. through the use of backscattered electron images of SEM) would be needed to see whether pyrophyllite and albite occur in grain-to-grain contact.

(iv) $Py + Kf$: This assemblage was mentioned by Ferla and Lucido (1972) and Schramm (1978), but not entered in Table 2.8 since Mu + Qz should be more stable. Note that the strong X-ray reflection of K-feldspar at $d = 3.25\,\text{Å}$ coincides with a basal reflection of mixed-layer paragonite/muscovite (Frey, 1969b).

(v) $Py + Qz + Ch + Ct$: This assemblage represents the solid phases of the divariant reaction

$$\text{pyrophyllite} + \text{chlorite} = \text{chloritoid} + \text{quartz} + H_2O$$

in the system AFMSH. The rather common observation of this assemblage and attempts to map a corresponding reaction-isograd (e.g. Frey and Wieland, 1975) show that much pyrophyllite in metapelites is consumed by the above reaction.

(vi) $Py + Qz + Cc$: This assemblage is fairly common in very low-grade metamarls (Table 2.8) and phase relations presented in Figure 2.9 indicate that pyrophyllite in the presence of calcite may be consumed by one of the following reactions:

$$2\,\text{pyrophyllite} + 1\,\text{calcite} = 1\,\text{margarite} + 6\,\text{quartz} + 1\,H_2O + 1\,CO_2 \quad (26)$$

$$1\,\text{pyrophyllite} + 1\,\text{calcite} + 1\,H_2O = 1\,\text{lawsonite} + 2\,\text{quartz} + 1\,CO_2 \quad (39)$$

Field evidence in favour of reaction (26) during progressive regional metamorphism was presented by Frey et al. (1982) while the destabilization of lawsonite according to reaction (39) was described by Goffé (1979).

(iii) P–T–X stability. In this section the significance of pyrophyllite-bearing metaclastite assemblages of the ASH, CASH, and CASH–CO_2 systems with respect to pressure, temperature, and the composition of the fluid phase (X_{CO_2}) will be discussed.

Pyrophyllite has a rather narrow stability field with respect to temperature, but a much wider stability field with respect to pressure. In $P-T$ space, these limits are given by reaction (4), $Ka + Qz = Py + H_2O$, towards lower temperatures; by reactions (2, 3), $Py = Als + Qz + H_2O$, towards higher temperatures; and by reaction (37), $Py = Di + Qz$, towards higher pressures (Figures 2.8, 2.11). Under the condition that water pressure equals lithostatic pressure $(a_{H_2O} = 1)$, the presence of pyrophyllite indicates temperatures between 300–380 °C at 2 kbar, 310–440 °C at 5 kbar, and 350–420 °C at 10 kbar. In nature, however, it is to be expected that the water activity will be less than unity. Reactions involving carbonates will generate CO_2 and the maturation of organic material will produce CH_4 (see Chapters 4, 5). Under conditions of $a_{H_2O} < 1$, the univariant dehydration reactions (4) and (2, 3) will shift to lower temperatures, while the solid–solid reaction (37) is not affected (Figure 2.11). $P-T$ conditions at the kaolinite + quartz → pyrophyllite reaction isograd of Figure 2.10 have been estimated to be 2–3 kbar and 270 °C (Frey, 1987), which leads to a value of a_{H_2O} of about 0.7 (Figure 2.11). This rather low value of a_{H_2O} is corroborated by fluid inclusion studies from the area of Figure 2.10 (see Chapter 5).

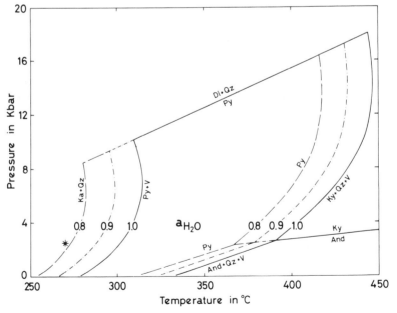

Figure 2.11 Calculated $P-T$ stability limits for pyrophyllite at different water activities. The $P-T$ conditions for the $Ka + Qz = Py + V$ reaction isograd from Figure 2.10 are indicated by a star.

From the T–X_{CO_2} sections of Figure 2.9 it can be seen that pyrophyllite in the presence of excess quartz + calcite is only stable for $X_{CO_2} > 0.06$ at 2.5 kbar pressure and for $X_{CO_2} > 0.03$ at 4.0 kbar. These values might even be lowered with calculations of non-ideal mixing of real gases, and with the possible presence of NaCl in the fluid phase (see for example Bowers and Helgeson, 1983).

(iv) *Conclusion.* Pyrophyllite is an index mineral of very low-grade metaclastites of the anchizone and the low-grade epizone. Kossovskaya and Shutov (1965, 1970) proposed the 'quartz-pyrophyllite facies' to fill the gap between the 'quartz-dickite facies' and the 'kyanite facies'. We shall desist from defining a 'pyrophyllite facies' here because (i) pyrophyllite is a relatively rare mineral, (ii) it has a rather large stability range with respect to pressure, and (iii) the temperature range of pyrophyllite stability is strongly dependent on a_{H_2O}. However, future work is needed to establish the stability field of pyrophyllite + chlorite versus those of chloritoid and Mg–Fe carpholite (Chopin and Schreyer, 1983), because the assemblage Py + Ch must have a more restricted P–T stability than pyrophyllite alone.

2.7.3 *Rectorite-bearing assemblages*

Rectorite is a 1:1 regular interstratification of a dioctahedral mica and a dioctahedral smectite. Details for its identification from X-ray diffractograms are given by Brown and Brindley (1980) and Kisch (1983, p. 366). Na^+ seems

Table 2.9 Minerals coexisting with rectorite in metaclastites

Common	Rare
Quartz	Paragonite
Illite or muscovite	Paragonite/muscovite mixed-layer
Chlorite	Dolomite
Pyrophyllite	Siderite
Kaolinite	Albite
Calcite	K-feldspar
Organic material	Chloritoid
	Haematite
	Anhydrite
	Gypsum

References used for compilation: Aparacio and Galán (1980), Aprahamian (1974), Artru *et al.* (1969), Beuf *et al.* (1966), Chennaux *et al.* (1970), Christensen (1975), Dunoyer de Segonzac (1969, p. 116), Dunoyer de Segonzac and Heddebaut (1971), Dunoyer de Segonzac and Abbas (1976), Eslinger and Savin (1973), Esquevin and Kulbicki (1963), Frey (1970, 1978), Frey *et al.* (1987), Füchtbauer and Goldschmidt (1959, p. 339), Gill *et al.* (1977), Henderson (1970, 1971), Hoffmann and Hower (1979), Hower *et al.* (1976), Logvinenko (1964), Loughnan and Ward (1971), Merriman and Roberts (1985), Paradis *et al.* (1983), Schmitz (1963, p. 134), Stadler (1971, p. 488), Stevaux (1967), Wieland (1979).

to be the dominant interlayer cation, but chemical data are generally not available from polyphase mixtures, the study of Paradis *et al.* (1983) being an exception. The name rectorite has priority over allevardite (Bailey, 1982).

Rectorite has been found mainly in shales and slates, and less frequently in limestones and dolomites. Minerals which were reported to coexist with rectorite in these lithologies are listed in Table 2.9. Those rectorite-bearing assemblages containing pyrophyllite and belonging to the system $Al_2O_3-Na_2O-K_2O-SiO_2-H_2O$ are given in Table 2.8.

Illite 'crystallinity' data are available for 13 rectorite occurrences, 11 indicating anchizonal and 2 low-grade epizonal conditions. This result confirms the conclusion reached earlier by Kisch (1983, p. 367) on a more limited database, that rectorite can be regarded as an anchizonal index mineral.

There are a few field studies providing direct or indirect temperature estimates on the occurrence of rectorite. Pevear *et al.* (1980) derived a minimum temperature of 145–160 °C from vitrinite reflectance data for K-rectorite in bentonites. Steiner (1968, Table 2 and Figure 3) and Eslinger and Savin (1973) found rectorite-like mixed-layer minerals in hydrothermal areas at measured borehole temperatures of 220 °C and 270 °C, respectively.

Hydrothermal experiments relevant to the stability of rectorite were conducted by Chatterjee (1973), Eberl and Hower (1977), and Eberl (1978). Chatterjee (1973) used natural low albite, kaolinite and quartz as starting materials. Rate studies with these mixes were performed at pressures of 2, 4 and 7 kbar, at temperatures between 300–600 °C and runs lasting up to 190 days. The phases to appear with increasing run durations or temperature were sodium–smectite, followed by rectorite, followed by various other regular mixed- layers with successively increasing paragonite components, ultimately giving way to paragonite + quartz. From these experiments Chatterjee (1973, p. 265) concluded that the various paragonite–smectite mixed-layer phases encountered were 'merely transient, metastable intermediate products'. Eberl and Hower (1977) used glasses of beidellite composition as starting materials. Temperatures ranged from 260–490 °C at 2 kbar pressure, and run times lasted up to 266 days. When sodium was the interlayer cation, increasing reaction produced the following series: glass → smectite → paragonite/smectite random mixed-layer + kaolinite + quartz at 260 °C; and glass → smectite → rectorite + pyrophyllite + quartz at 490 °C. As suggested by Eberl and Hower (1977) the temperatures given are probably higher than those found in nature, due to the short reaction times and the chemical purity of the synthetic system. Nevertheless, these experiments demonstrated the reaction series (although the final product was not reached): smectite → randomly interstratified paragonite/smectite → regularly interstratified paragonite/smectite (rectorite) → paragonite. Contrary to Chatterjee (1973), however, Eberl (1978, p. 337) was inclined to assume that Na-rectorite is a stable phase.

2.7.4 *Mixed-layer paragonite/muscovite-bearing assemblages*

Frey (1969b) noted the presence of a strong reflection at 3.25 Å and weaker reflections at 4.90 and 1.96 Å on diffractograms from anchimetamorphic slates. Since these reflections were located between basal reflections of paragonite and phengite, the presence of a mixed-layer paragonite/phengite was postulated. Moreover, a high-order reflection was observed in some samples, pointing to the existence of a regular interstratification. The position of (001) peaks at 1.96 and 3.25 Å indicated a paragonite:phengite ratio of about 6:4 for this mixed-layer mineral. In later studies, it was found that the coexisting K-white mica was not always of phengitic nature, and therefore the name paragonite/muscovite mixed-layer is now preferred.

When present in minor amounts, identification of mixed-layer paragonite/muscovite in the presence of discrete paragonite and muscovite can be a problem. Also, the main peak of K-feldspar near 3.25 Å coincides with the strongest basal reflection of paragonite/muscovite, and this led to mis-identifications in some earlier studies (e.g. Dunoyer de Segonzac and Heddebaut, 1971; Ferla and Lucido, 1972).

The presence of mixed-layer paragonite/muscovite in very low-grade metamorphic shales and slates is now well documented, and minerals which were reported to coexist with this phase are listed in Table 2.10. Some mineral assemblages of the system Al_2O_3–Na_2O–K_2O–SiO_2–H_2O (ANKSH) containing paragonite/muscovite and pyrophyllite are given in Table 2.8.

According to illite 'crystallinity' data of 21 of the above-mentioned studies, interstratified paragonite/muscovite is typically encountered in the anchizone and/or the low-grade epizone, but in two cases (Merriman and Roberts, 1985; Weaver and Broekstra, 1984) its appearance is already reported from the late diagenesis zone.

The origin of this mixed-layer paragonite/muscovite is not yet clear, but an irregular illite/smectite interstratification seems to be a reasonable candidate for a precursor phase (Frey, 1970). This suggestion is supported by Merriman and Roberts (1985, p. 310), who observed an irregular interstratification of Na-mica/K-mica with a minor smectite component based on XRD and electron-microprobe data. At higher grade, splitting up of the composite mixed-layer phase paragonite/muscovite into discrete paragonite and muscovite evidently occurs. Whether mixed-layer paragonite/muscovite is a stable or metastable precursor of paragonite (and muscovite) is not known, but it is hoped that a TEM study currently undertaken by D. Veblen will shed some light on this.

2.7.5 *Paragonite-bearing assemblages from the anchizone*

(i) *Occurrence.* Paragonite was formerly thought to appear only in the greenschist facies (e.g. Winkler, 1967, p. 96; Miyashiro, 1973, p. 200). However, its presence in anchimetamorphic shales and slates of pelitic or marly

Table 2.10 Minerals coexisting with paragonite/muscovite mixed-layer in metaclastites

Very common	Common	Rare
Quartz	Pyrophyllite	Kaolinite
Paragonite	Albite	Illite/smectite
Illite or muscovite	Calcite	random interstratification
Chlorite	Dolomite	Rectorite
	Chloritoid	Margarite
	Haematite	Magnesite
	Organic material	Siderite
		Andalusite
		Garnet (Mn-rich)

References used for compilation: Breitschmid (1982), Dandois (1981), Davies (1983), Dunoyer de Segonzac and Heddebaut (1971), Ferla and Lucido (1972), Fieremans and Bosmans (1982), Franceschelli *et al.* (1986), Flehmig and Gehlken (1983), Frey (1970, 1978), Frey and Wieland (1975), Frey *et al.* (1987), Gruner (1981), Gürler (1982), Kramm (1978, 1980), Ludwig (1973), Merriman and Roberts (1985), Schramm (1977, 1978, 1982b), Schramm *et al.* (1982), Thum and Nabholz (1972), Weaver and Broekstra (1984), Wieland (1979).

composition is now well documented (see Table 2.11 for references). Additional very low-grade paragonite occurrences in metaclastites (not documented by illite 'crystallinity' data) include: paragonitic hydromica of the 'stage of early metagenesis' (Karpova, 1966, 1969; see also Kisch, 1983, p. 364); paragonite crystallizing at lower grade than lawsonite (Black, 1975; 1977, Figure 2); an occurrence where associated metabasites contain pumpellyite (Kramm, 1978); paragonite from the kaolinite–pyrophyllite zone (Juster and Brown, 1984; Franceschelli, *et al.*, 1986); and from a blueschist facies terrain (Potdevin and Caron, 1986). On the other hand, paragonite is not known as an authigenic mineral from the diagenetic realm, but has been mentioned as a detrital mineral in a few cases (e.g. Hahn, 1969, p. 247; Herold, 1970).

Table 2.11 Minerals coexisting with paragonite in metaclastites from the anchizone

Very common	Common	Rare
Paragonite/muscovite mixed-layer	Pyrophyllite	Rectorite
	Kaolinite	Chloritoid
Illite or muscovite	Albite	Magnesite
Chlorite	Calcite	Siderite
Quartz	Dolomite	
	Haematite	
	Organic material	

References used for compilation: Brime (1985), Clauer and Lucas (1970), Davies (1983), Dunoyer de Segonzac and Heddebaut (1971), Frey (1970, 1978), Gruner (1981), Gürler (1982), Merriman and Roberts (1985), Schramm (1977, 1978, 1982b), Schramm *et al.* (1982), Thum and Nabholz (1972), Weaver and Broekstra (1984), Wieland (1979).

(ii) *Mineral assemblages and mineral reactions.* Minerals which were reported to coexist with paragonite in metaclastites from the anchizone are listed in Table 2.11. From this compilation some deductions regarding possible paragonite-forming reactions can be made. Zen (1960, p. 152) proposed the reaction 1 albite + 1 kaolinite = 1 paragonite + 2 quartz + 1 H_2O.

In the light of the reported coexistence of reactants and products (Gonzales Martinez *et al.*, 1970; Aparicio and Galán, 1980; Franceschelli *et al.*, 1986), this reaction seems to be a reasonable one. In order to obtain additional information some thermodynamic calculations in the system $Na_2O–Al_2O_3–SiO_2–H_2O$ were carried out. The solid phases involved were Ab, And, Ka, Ky, Pa, Py, and Qz and the database of Berman *et al.* (1985) was used. The results indicate that the reaction Ab + Ka = Pa + Qz + H_2O is metastable and is located at low temperatures ($T < 120\,°C$) and low pressures ($P < 500$ bar)! In nature, therefore, it is unlikely that paragonite + quartz is formed from albite + kaolinite.

The mixed-layer paragonite/muscovite is very often encountered with paragonite, and field evidence (Frey, 1978; Seidel, 1978; Fieremans and Bosmans, 1982; Merriman and Roberts, 1985) indicates that the modal content of the mixed-layered phase decreases with increasing metamorphic grade, promoting formation of paragonite plus muscovite. Therefore, a possible reaction series would be as follows (Frey, 1978): irregular mixed-layer illite/smectite → rectorite → regular mixed-layer paragonite/muscovite → discrete paragonite and muscovite.

Admittedly, the role of rectorite in this reaction series or as a direct precursor of paragonite is not well documented by field observations.

(iii) *Low-temperature stability limit of paragonite + quartz.* Hydrothermal experiments pertaining to the low-temperature stability limit of paragonite + quartz were conducted by Hemley and Jones (1964), Althaus and Johannes (1969), and Chatterjee (1973). Hemley and Jones (1964, p. 552 and Figure 2) determined a decomposition temperature of approximately 330–340 °C at 1 kbar total pressure for Na-montmorillonite + albite in a 0.5 M NaCl solution. The product consisted of mixed-layer montmorillonite/paragonite, and this phase was considered to be 'probably metastable' relative to paragonite. These experiments were not reversed and involved runs of short duration (3 to 4 weeks). For these reasons lower temperatures are suggested as equilibrium values. Althaus and Johannes (1969) performed experiments on the reaction montmorillonite + NaCl + quartz → paragonite + albite + muscovite + chlorite + salts + HCl + H_2O at 1 kbar total pressure, and the products were found to be stable at 350 °C, the lowest temperature used. Chatterjee's (1973) experimental study has already been discussed (see section 2.7.3). This author investigated the solid–solid reaction 3 Na-montmorillonite + 2 albite → 3 paragonite + 8 quartz. Despite non-reversal, the data of 335 °C at 2 kbar and of 315 °C at 7 kbar derived by linear extrapolation from rate studies were regarded as possible equilibrium conditions.

In conclusion, the experimental results presented above are broadly consistent with each other, although different reactions were investigated, and indicate a low-temperature stability limit of paragonite + quartz slightly above 300 °C. Taking into account the non-reversed experiments, a somewhat lower temperature is expected in nature.

2.7.6   Lawsonite-bearing assemblages

(i)   *Occurrence*. Lawsonite has rarely been reported from very low-grade metaclastites, contrary to its common occurrence in metabasites and grey-wackes of similar metamorphic grade (see Chapter 3). This phenomenon can be explained by considering a possible pressure–temperature–time path followed by many metamorphic rocks. After early crystallization of lawsonite in the blueschist facies (as defined in Chapter 3), subducted rocks generally undergo late recrystallization under greenschist facies conditions. During this process lawsonite in ductile metapelites and metamarls is partly to completely pseudomorphed by muscovite, paragonite, margarite, pyrophyllite, kaolinite, clinozoisite, zoisite, calcite, albite, and chlorite (see for example Ellenberger, 1960; Caron, 1977; Goffé, 1979; Sicard *et al.*, 1984), while lawsonite in less ductile metabasites and greywackes is much less affected. Therefore, the plurifacial history of lawsonite-bearing metaclastites renders the interpretation of mineral assemblages more difficult.

Minerals reported to coexist with lawsonite in very low-grade metaclastites are listed in Table 2.12. The most common mineral assemblage belonging to

Table 2.12 Minerals coexisting with lawsonite in metaclastites

Very common	Common	Rare
Quartz	Albite	Kaolinite
White mica	Na-amphibole	Pyrophyllite
Chlorite	Titanite	Diaspore
Calcite	Pyrite	Cookeite (Li-chlorite)
	Haematite	Chloritoid
	Organic material	Mg-carpholite
		Stilpnomelane
		Epidote
		Aragonite
		Dolomite
		Ankerite
		Rutile

References used for compilation: Black (1973), Bocquet-Desmons (1974, and pers. comm. 1986), Brothers (1970), Brothers and Yokoyama (1982), Caron *et al.* (1981), Davies (1983), Goffé (1979), Seidel (1977, 1978), Seki (1958), Sicard *et al.* (1984), Watanabe and Koboyashi (1984).

the model system CASH is Lw–Qz; while Lw–Py and Lw–Py–Qz have been reported by Goffé (1979) from metabauxites. In the CASH–CO_2 system the Lw–Qz–Cc assemblage is very common; while Lw–Qz–Cc–Ar, Lw–Py–Cc, and Lw–Py–Cc–Qz are rarely mentioned.

Scarce illite 'crystallinity' data indicate that lawsonite occurs in the low-grade anchizone and the epizone (Seidel, 1978).

(ii) *P–T–X stability*. With regard to the CASH system, and assuming that water pressure equals total pressure, the *P–T* stability limits of lawsonite are as follows (compare Figure 2.8). The low-pressure stability limit of about 3 to 4 kbar is given by reaction (11), Lm = Lw + 2 Qz + 2 H_2O, and reaction (9) Wr = Lw + 2 Qz; the low-temperature limit is provided by the reaction Hu* = Lw + Qz + H_2O at approximately 200 °C while the high-temperature limit is given by reaction (12), 5 Lw = Ma + 2 Zo + 2 Qz + 8 H_2O. Note that the stability field of lawsonite is enlarged with increasing pressure. At reduced activity of H_2O in the presence of CH_4, for example, but not of CO_2 (see below), these dehydration reactions would be shifted to lower temperatures in the case of (12), and to lower pressures in the case of (11).

Lawsonite is only stable at very low X_{CO_2}, as shown both experimentally (e.g. Nitsch, 1972) and theoretically (e.g. Thompson, 1971, Figure 4). The T–X_{CO_2} stability field of lawsonite in the presence of excess quartz + calcite increases with increasing total pressure, as outlined in Figure 2.9.

2.7.7 *Stilpnomelane-bearing assemblages of the anchizone*

(i) *Occurrence*. Under the microscope, stilpnomelane can easily be mistaken for vermiculite, biotite, oxychlorite, or even for sericite stained by secondary iron hydroxide! It is best identified by a strong basal reflection at 12 Å on X-ray diffractograms.

Stilpnomelane occurrences in the anchizone are well documented from the Central and Western Alps. Here stilpnomelane is reported from iron-rich lithologies, that is from glauconitic horizons (Niggli and Niggli, 1965; Martini and Vuagnat, 1970, p. 58; Martini, 1972, p. 262; Frey et al., 1973; Breitschmid, 1982; Gürler, 1982), from iron-oolitic horizons (Delaloye, 1966; Tröhler, 1966; Durney, 1974), and from iron ores and related limestones (Epprecht, 1946). In these areas stilpnomelane seems to be absent from more alumina-rich lithologies such as shales and slates.

Stilpnomelane has also been reported from very low-grade metaclastites and related lithologies where metamorphic grade has not been documented by illite 'crystallinity' data. Such occurrences include meta-ironstones and marbles from the blueschist facies of California (Coleman and Lee, 1962, 1963; Muir Wood, 1982); iron-rich metasediments and paraschists from the lawsonite and lawsonite-epidote transitional zone of New Caledonia

*Hu = heulandite

(Brothers, 1970; Black, 1975); impure limestones and quartz schists of the pumpellyite zone from several localities in Japan (Hashimoto and Kanehira, 1975); impure marbles from the blueschist facies terrain of Corsica (Caron et al., 1981); greywackes of the chlorite 1 zone from the Otago Schist belt of New Zealand (Turner, 1938); and quartzite-like sandstones of the 'stage of initial metagenesis' from the Russian Platform (Veselovskaya, 1967).

(ii) *Mineral assemblages and mineral reactions.* In this section the progressive metamorphism of glauconite-bearing formations from the Helvetic Alps will be described in some detail. Three metamorphic zones were distinguished by Frey *et al.* (1973) in the Glarus Alps (Figure 2.12). Zone I comprises unmetamorphosed sediments with the assemblage glauconite–calcite–quartz with or without chlorite and riebeckite. In zone II, stilpnomelane forms by the reaction glauconite + quartz $\pm$ chlorite = stilpnomelane + K-feldspar + $H_2O + O_2$. The characteristic mineral assemblage at the beginning of zone II is: stilpnomelane + glauconite + K-feldspar + calcite + quartz $\pm$ chlorite $\pm$ riebeckite, while at higher grades of this zone glauconite is absent. Evidently glauconite is consumed by sliding equilibria. The transition from zone I to zone II occurs approximately in the middle of the anchizone.

In zone III biotite appears, presumably by the reaction stilpnomelane + chlorite + K-feldspar = biotite + quartz + H_2O (Brown,   1975,   p. 269). Note that this biotite-forming reaction takes place at conditions where pumpellyite is still stable in greywackes. Frey *et al.* (1973) proposed that

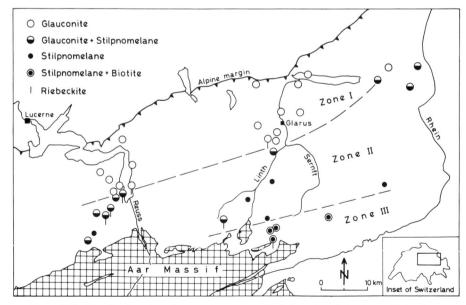

Figure 2.12 Distribution of very low-grade metamorphic index minerals in glauconitic horizons in the Helvetic zone of the Glarus Alps and the Lake Lucerne area, Switzerland. (Modified from Frey *et al.*, 1973, Figure 3; Breitschmid, 1982, Figure 18.)

stilpnomelane would be a product phase in the abovementioned reaction, but in the absence of chemical data on coexisting chlorite and biotite, this question cannot be answered. The typical mineral assemblage of zone III is biotite + stilpnomelane + K-feldspar + calcite + quartz + riebeckite. The transition from zone II to zone III is located at the beginning of the epizone.

Mineral zones I and II were also mapped further west in the Helvetic Zone of central Switzerland (Figure 2.12). In the Lake Lucerne area, Breitschmid (1982) observed stilpnomelane already in the high-grade diagenetic zone at conditions of $210\,^\circ C/1630$ bar, based on fluid inclusion data.

2.7.8 Mg–Fe–carpholite-bearing assemblages

Mg–Fe–carpholite, $(Mg, Fe^{2+})Al_2Si_2O_6(OH,F)_4$, has recently been described as an index mineral for relatively low-grade blueschist facies metapelites (Chopin and Schreyer, 1983). There exists a complete solid solution series between ferrocarpholite ($< 50\,mol\%\,Mg$ end member) and magnesiocarpholite ($> 50\,mol\%\,Mg$ end member).

Chopin and Schreyer (1983, Table 1) have compiled occurrences and mineral assemblages of Mg–Fe–carpholite, and some of their results and conclusions, complemented by those of other workers, are summarized below.

(i)   Mg–Fe–carpholite has been observed most frequently as vein fillings, segregations and as a rock-forming mineral in phyllites, schists, quartzites and metabauxites.

(ii) The common association of Mg–Fe–carpholite with chloritoid indicates a genetic relation between these two chemically similar minerals. According to field evidence Mg–Fe–carpholite is the low-grade equivalent of chloritoid. De Roever et al. (1967) have proposed for the relationship the reaction ferrocarpholite = chloritoid + quartz + H_2O.

(iii) Since in coexisting Mg–Fe–carpholite–chloritoid pairs Mg is preferentially incorporated into carpholite, the abovementioned reaction must be divariant, and, for a given water pressure, magnesiocarpholite must be stable at higher temperatures than ferrocarpholite (see also de Roever, 1977; Seidel, 1978, Figure 28).

(iv) Chemographic analysis in the system FMASH indicates that ferrocarpholite + quartz could be the high-pressure equivalents of pyrophyllite + chlorite + water (Viswanathan and Seidel, 1979).

(v)   A possible prograde sequence in the mineralogical evolution of blueschist facies metapelites involving Mg–Fe–carpholite-bearing assemblages is depicted in Figure 2.13.

(vi) Preliminary high-pressure experiments by Chopin and Schreyer (1983), combined with theoretical phase relations and field evidence, indicate that ferrocarpholite and magnesiocarpholite require minimum pressures of about 6 and 7 kbar, respectively. Ferrocarpholite would be expected in the temperature range of approximately 200–$400\,^\circ C$. Note that Seidel

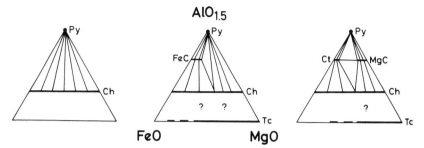

Figure 2.13 AFM-type diagrams for Mg–Fe–carpholite–bearing assemblages. (Modified from Chopin and Schreyer, 1983, Figure 2; Goffé and Velde, 1984, Figure 3.)

(1978, Figure 29) has observed ferrocarpholite to occur within the whole range of the anchizone.

2.7.9 *Does chloritoid exist in the anchizone?*

This question has been discussed by Kisch (1983, p. 370), who concluded that 'at the present-day state of knowledge, it is impossible to exclude the possibility of the local appearance of chloritoid at a metamorphic stage lower than the beginning of the greenschist facies', and that 'chloritoid cannot unequivocally be regarded as an indicator of the beginning of the epizone'. In the following, some of the apparent anchizonal chloritoid occurrences already discussed by Kisch will be examined again and some additional evidence from other occurrences will be presented.

According to Kübler (1967a, Figure 3; 1984, Figure 56) chloritoid is found in anchimetamorphic roofing slates, in which case Kubler (personal communication) was referring to the chloritoid in the 'Schistes d'Angers' described by Boudier and Nicolas (1968). This result was confirmed by Le Corre (1975, Figure 1) who reported chloritoid from the same formation with illite 'crystallinity' values of the medium and high-grade anchizone. However, whole-rock analyses from such chloritoid-bearing schists contain up to 1.5 wt% Na_2O (Lopez-Munoz, 1982, Figure 7), suggesting the presence of mixed-layer paragonite/mucovite and paragonite. As noted earlier (section 2.5.2) these layer silicates would lead to a peak broadening of the illite 10 Å basal reflection, resulting in apparent low illite 'crystallinities'. Therefore the possibility should be reconsidered that the chloritoid from the 'Schistes d'Angers' actually belongs to the epizone.

Anchimetamorphic chloritoid in black slates was also mentioned by Dunoyer de Segonzac and Heddebaut (1971). Using the same line of reasoning as above, this chloritoid may also be of epimetamorphic grade.

There remain, however, two other occurrences of anchizonal chloritoid. Seidel (1978, Figure 29) observed the appearance of chloritoid in a high pressure terrane in the middle of the anchizone, and Árkai *et al.* (1981, p. 276) ascribe the formation of chloritoid to the high-grade anchizone based on illite 'crystallinity' and vitrinite reflectance data.

In summary, the conclusions reached by Kisch (1983) and mentioned above are corroborated: chloritoid seems to occur rarely in the anchizone but is otherwise a typical epizonal index mineral.

2.8 Geothermometry and geobarometry

Few reliable geothermometers and geobarometers exist for very low-grade metaclastites. Data on 'crystallinity' and distribution of polytypes and polymorphs in sheet silicates cannot be used for quantitative temperature estimates as discussed in previous sections. The use of dehydration and mixed-volatile reactions to obtain $P-T$ estimates presupposes a knowledge of fluid composition (see below). In this author's opinion the most useful geo-thermometer and geobarometer for very low-grade metaclastites stems from the analysis of fluid inclusions, as discussed in Chapter 5. Other helpful geological thermometers may be derived from coal rank and oxygen-isotope data, and the former is treated in Chapters 4 and 7.

(i) *Calcite–dolomite geothermometry.* This thermometer has been discussed by Turner (1981, pp. 131–134) in some detail, and therefore only a short summary will be given here.

In the system $CaCO_3-MgCO_3$, a solvus exists between calcite and dolomite, and the temperature-dependence of the amount of $MgCO_3$ in calcite in equilibrium with dolomite can be used for estimating metamorphic temperatures. The effect of pressure on temperature estimates is small. Most natural carbonates, however, contain additional components like $FeCO_3$ or $MnCO_3$. A thermodynamic evaluation of the iron correction to the calcite–dolomite geothermometer was made by Bickle and Powell (1977). This thermometer was further improved through an experimental study of the system $CaCO_3-MgCO_3-FeCO_3$ in the temperature range 300–450 °C and at pressures of 3 and 5 kbar, combined with a theoretical analysis by Powell *et al.* (1984). The determination by electron microprobe of the mole fractions $X_{Fe,cc}-X_{Mg,cc}$ and $X_{Fe,dol}-X_{Mg,cc}$ leads to two independent temperature estimates which become increasingly uncertain towards lower temperatures, and particularly towards higher $X_{Fe,cc}$. According to Powell *et al.* (1984, p. 40) 'uncertainties of less than 30 °C should be attainable at low $X_{Fe,cc}$ but these will certainly be greater at higher $X_{Fe,cc}$.'

The difficulty of temperature estimation is exacerbated by the considerable range of carbonate compositions usually found even on a microscopic scale in rocks. Fairchild (1985, p. 177), for example, observed extreme inhomogeneity of Fe and Mg in calcites coexisting with dolomite; re-equilibration under high-pressure greenschist facies conditions was insufficient to allow the derivation of temperatures by calcite–dolomite geothermometry. McDowell and Paces

C

(1985), on the other hand, found equilibrium coexistence of calcite with dolomite and ankerite near 200 °C in a geothermal system. This allowed construction of an isothermal section in the Ca–Mg–(Fe + Mn) carbonate phase diagram and provided a low-temperature constraint on the calcite limb of the calcite–dolomite solvus. McDowell and Paces suggest that this calcite solvus be used for low-temperature, low-pressure geothermometry.

(ii) *Phase equilibria.* The stability relations of some mineral assemblages involving pyrophyllite, paragonite, lawsonite, and Mg–Fe–carpholite may provide *P–T* estimates for very low-grade metaclastites as discussed in previous sections. However, it should be stressed again that the presence of organic material will result in lowered water activities, thus limiting the direct application of petrogenetic grids calculated for unit water activity (Figure 2.8). As an example, the effect of reduced water activity on the stability field of pyrophyllite is demonstrated in Figure 2.11.

(iii) *Muscovite b_0-geobarometry.* Sassi (1972) and Sassi and Scolari (1974) first attempted to calibrate the b_0 muscovite cell parameter as a geobarometer for greenschist facies metapelites. Later Padan *et al.* (1982) extended the method to shales and slates of the high-grade anchizone as well; see also Robinson and Bevins (1986).

This geobarometer is based on two facts: firstly, the celadonite content of potassic white mica in specific mineral assemblages (or suitable bulk composition) increases with increasing pressure if temperature is held constant; secondly, there exists a positive correlation between the b_0 parameter and celadonite content (see Guidotti, 1984, for a review). In the following we shall comment on (i) the choice of the mineral assemblage and (ii) on the method of obtaining b_0 values.

With regard to (i), Guidotti and Sassi (1976, pp. 107–108, 124–126) provide a detailed discussion. These authors advocated the use of the assemblage muscovite + albite + quartz ± chlorite ± carbonates ± graphite because of its very common occurrence. It was realized that the assemblages Mu–Pa–Py–Qz, Mu–Pa–Ab–Qz or Mu–Ab–Kf–Qz (which are limiting assemblages in the model system KNASH) would be preferable to the proposed non-limiting assemblage Mu–Ab–Qz. However, the abovementioned four-phase assemblages are not common, which dictates the use of Mu–Ab–Qz, an assemblage in which the celadonite content of muscovite must depend on rock composition. In spite of that, empirical testing of the b_0-geobarometer by Sassi, Guidotti and co-workers has shown that the disadvantage of using the Mu–Ab–Qz assemblage may be overcome if several precautions are considered. Accordingly, suitable metapelites should contain abundant muscovite; samples with abundant quartz and/or carbonates should be avoided; and rocks containing one or more of the following minerals should be excluded:

paragonite, margarite, pyrophyllite, haematite, and magnetite (see Guidotti and Sassi, 1976, for details).

With regard to (ii), the experimental approach in obtaining b_0 through the measurement of the $d(060)$ spacing should be critically considered. Conventional powder diffractometry cannot be applied, because, for 2 M_1 muscovite, the (060) reflection is overlapped by the strong (331) reflection (Frey *et al.*, 1983). This problem is largely eliminated by using thin rock slices cut perpendicular to the foliation (Sassi and Scolari, 1974) or by using the Guinier camera technique. Still another problem should be mentioned here. Very low-grade metaclastites often contain detrital and authigenic potassic white mica with a large range of chemical composition (see for example Hunziker *et al.*, 1986, Figure 9). This may result in very broad '(060)' reflections without any clear peak maximum (Frey, unpublished results).

In practice the mean b_0 muscovite spacing from about 30 or more samples from an area of similar metamorphic grade is determined and the data presented in the form of cumulative curves. The standard deviation of an acceptable group of samples should be < 0.01 Å. The facies series (Miyashiro, 1961) may then be determined as follows (Sassi *et al.*, 1976; Guidotti and Sassi, 1986:

$b_0 < 9.000$ Å: low-pressure facies series

$9.000 < b_0 < 9.040$: intermediate-pressure facies series

$b_0 > 9.040$ Å: high-pressure facies series

Guidotti and Sassi (1986, Figure 1) went even further and constructed an empirical P–T diagram contoured for different muscovite b_0 values. As pointed out by these authors, this plot should be regarded as qualitative because extensive data are available only for the lower temperature range; but it is naturally attractive to use this diagram for quantitative purposes. However, one should be aware that mineral equilibria controlling the celadonite content in muscovite (and hence also b_0) are dehydration reactions which are dependent on the activity of water in the fluid phase. Higher b_0 values are therefore expected for very low-grade metapelites containing organic material.

In summary, this author still has some reservations regarding the muscovite b_0 geobarometer. Further testing should clarify the matter.

2.9 Conclusions

(i) The determination of illite 'crystallinity' is a rapid and convenient method to broadly define metamorphic grade of very low-grade metaclastites, provided that numerous precautions are respected. This method allows the recognition of the anchizone, which covers an approximate temperature range between 200 and 300 °C based on fluid inclusion data (see Chapter 5).

(ii) The determination of the illite polytype ratio $2 M_1/(2 M_1 + 1 Md)$ is another useful method for recognizing incipient metamorphism of clastic sedimentary rocks.

(iii) The mineralogy of very low-grade metaclastites is dominated by the non-diagnostic assemblage illite + chlorite + quartz ± feldspars ± carbonates. Nevertheless, the occasional presence of or the first appearance of pyrophyllite, rectorite, mixed-layer paragonite/muscovite, paragonite, lawsonite, stilpnomelane, and Mg–Fe–carpholite may be helpful for mineral zoning purposes. At the present state of knowledge, however, it is not possible to define different metamorphic facies with mineral assemblages of very low-grade metaclastites in contrast to metabasites, as described in Chapter 3.

(iv) Whenever possible, the mineralogical investigation of very low-grade metaclastites described in this chapter should be combined with coal rank (Chapter 4) and fluid inclusion studies (Chapter 5).

Acknowledgements

Written communications by Hanan J. Kisch helped to clarify several points. Martin Engi and Lukas Baumgartner performed the calculation of phase diagrams. Steve Ayrton improved the English. Thoughtful reviews were provided by Steve Ayrton, Niranjan D. Chatterjee, and J.G. Liou. To all these friends and colleagues my sincere thanks.

3 Very low-grade metamorphism of volcanic and volcaniclastic rocks–mineral assemblages and mineral facies

JUHN G. LIOU, SHIGENORI MARUYAMA and MOONSUP CHO

3.1 Introduction

In the upper portions of the Earth's crust and at the surface, rocks interact with aqueous solutions of diverse origins in various ways determined by the prevailing physical and chemical conditions, which include temperature, pressure, mineralogy, petrology, bulk composition, texture, permeability, and residence time of the fluids. Rocks are rearranged in terms of their mineralogy, texture, trace element and isotopic compositions, and locally bulk composition. Recrystallization proceeds chiefly according to dissolution and precipitation reactions, both disequilibrium and equilibrium in nature. In this chapter, we confine our discussion mainly to basaltic rocks interacting with dominantly aqueous fluids at temperatures below about 400 °C.

Basaltic rocks constitute the major portion of the oceanic crust and most appear to have been subjected to mid-oceanic hydrothermal metamorphism immediately after formation at a spreading ridge. When transported to the continental margin, the oceanic mafic rocks are again recrystallized at or near convergent plate junctions; the alteration in mineralogy and readjustment in texture and structure depend on whether the oceanic crust participated in a collision with the continent or was subducted or overridden by a continental plate. On the other hand, andesitic rocks and their related calc-alkaline volcanics are the dominant lithologies within island arcs and Pacific-type continental margins. These volcanics are repeatedly subjected to alteration by hydrothermal fluids as evidenced by present-day geothermal activity in many island arc terranes. Recrystallization of andesitic rocks by fluids with moderate to high CO_2 and SO_4^{2-} contents results in very different mineral assemblages. The processes of low-T metamorphism are, therefore, a function of tectonic setting; each environment is characterized by certain lithologies, metamorphic gradient, fluid composition and other intensive parameters.

This chapter describes (i) characteristic metabasaltic–andesitic assemblages; (ii) critical reactions involving facies transitions; (iii) $P_{fluid}–X_{CO_2}$ stabilities for index mineral assemblages; and (iv) a quantitative petrogenetic grid for low-T metamorphism of basaltic rocks. We hope to better understand the physico-chemical and tectonic processes for low-T recrystallization of basaltic-andesitic rocks in various tectonic environments.

3.2   Characteristics of very low-grade metavolcanics

3.2.1   *Types of very low-grade metamorphism*

Volcanic rocks of both basaltic and andesitic compositions occur as pillow lavas, massive flows, hyaloclastic breccias, and tuffs. Their thickness varies depending on tectonic setting. Some volcanic sequences, such as those from ophiolites and those inferred from spreading centres, are very thin. On the other hand, the basaltic to andesitic volcaniclastics in island arcs such as those of Japan and New Zealand can be over 20 km thick. Because of differences in tectonic setting, and in physical and chemical conditions for the formation of secondary minerals, several types of very low-grade metamorphism have been described (see Table 3.1 for summary): (1) *Ocean ridge + ocean-floor metamorphism* of oceanic crust at spreading ridges or off-axis; (2) *burial metamorphism* of thick volcanic piles; (3) *subduction-zone metamorphism* of oceanic crust and its overlying clastic wedges; (4) *thermal metamorphism* of volcanic rocks, adjacent to plutons; and (5) *hydrothermal metamorphism* of volcaniclastic rocks in active and fossil geothermal systems. These terms will be used to describe metamorphic mineral assemblages formed at various tectonic–geologic settings in later parts of this chapter. However, it should be noted that, in any tectonic setting, more than one type of metamorphic recrystallization may proceed simultaneously. For example, hydrothermal, thermal and burial metamorphism may occur concurrently at an oceanic ridge. Therefore, classification according to tectonic–geologic setting is purely for descriptive purposes.

3.2.2.   *General characteristics*

With the exception of recrystallization in deep subduction zones, all these 5 types of metamorphism are characterized by lack of strong shear stress and of widespread penetrative deformation. Moreover, many of these volcanics may not show significant change in bulk composition. Hence, most of the very low-grade metavolcanics retain remarkably well their original igneous structures and compositions; in the field, these volcanics may appear to be unmetamorphosed.

Other common features of very low-grade metavolcanics include extensive to incipient development, sporadic distribution, and selective growth of secondary minerals in vesicles and fractures, and the topotaxic growth of these minerals after primary plagioclase, clinopyroxene, olivine, hornblende, opaques, and volcanic glass. In a prograde sequence, although no systematic pattern of association among secondary minerals is apparent, the degree of recrystallization and extent of replacement may increase with increasing grade. Because of local domains with varying effective compositions, different associations of secondary minerals may develop in vesicles, veins and after primary phases even within a single thin section. The persistence of igneous

Table 3.1 Schematic representation of facies series and zonation of Ca–Al silicates for various types of very low-grade metamorphism

Type of metamorphism	Representative region	Facies series and zonal distribution	Metamorphic gradient*	References
I. Ocean-floor	(1) Dredged and drilled oceanic basalts	ZEO–(PP)–PrA–GS Lm–(Pm)–Ep	100–500 °C km^{-1}	Humphris and Thompson (1978) Alt (1985)
	(2) Del Puerto Ophiolite, California	ZEO–PP–PrA–GS Hu–Lm–Pm–Ep	100–200 °C km^{-1}	Evarts and Schiffman (1983)
	(3) Horokanai Ophiolite, Japan	ZEO–(PP)–PrA–GS Chabazite–Lm–Wr–Ep	100–200 °C km^{-1}	Ishizuka (1985)
II. Hydrothermal	(1) Iceland (IRDP hole)	ZEO–PP–PrA–GS Lm–Pm–Act	80–90 °C km^{-1}	Viereck et al. (1982); Mehegan et al. (1982); Exley (1982)
	(2) Onikobe, Japan	ZEO–PrA–GS Mo–Lm–Yu–Wr	80–90 °C km^{-1}	Seki et al. (1983) and Liou et al. (1985b)
III. Thermal	(1) Tanzawa Mt., Japan	ZEO–PP–PrA–GS St–Lm–Wr–Pm–Act	50 °C km^{-1}	Seki et al. (1969)
	(2) Karmutsen volcanics, British Columbia	ZEO–PP–PrA–GS	50–60 °C km^{-1}	Cho et al. (1986); Cho and Liou (1987)
	(3) Takitimu volcanics, New Zealand	ZEO–PP–PrA–GS	40–50 °C km^{-1}	Houghton (1982)
IV. Burial	(1) Wakatipu belt, New Zealand	ZEO–PP–PA–GS	25–35 °C km^{-1}	Coombs (1954); Kawachi Cho and Liou (1987)
	(2) Greentuff Frm, Japan	ZEO–PP–(PA)–GS	25–35 °C km^{-1}	Utada (1965); Seki (1976)
V. Subduction	(1) Franciscan, California	ZEO–(PP)–BS Lm–Lw–Pm–Ep	10–15 °C km^{-1}	Coleman and Lee (1963); Maruyama and Liou (1987)
	(2) Sanbagawa belt, Japan	ZEO–PP–PA–(BS)–GS Lm–Pm–Ep	15–20 °C km^{-1}	Nakajima et al. (1977)

*Metamorphic gradients for these types of metamorphism are not necessarily constant throughout the facies series.

phases and differences in domain mineral assemblages make it difficult to define the equilibrium compositions and compatibilities of the coexisting minerals in very low-grade metamorphic rocks.

3.2.3 Persistence of primary phases

In very low-grade metamorphic terranes, relict phases such as plagioclase and clinopyroxene are common. In some occurrences, plagioclase phenocrysts are completely replaced by albite together with Ca–Al silicates or Ca carbonate while clinopyroxene remains virtually unattacked. In other terranes, plagio- clase tends to be unaltered whereas clinopyroxene is chloritized (e.g. Coleman, 1977). The presence of abundant relict igneous phases in metabasites and meta-andesites must be considered when interpreting stable assemblages. The contrast in the alteration of plagioclase and pyroxene may be due to differences in fluid composition, fluid/rock ratio and other physicochemical parameters. Clearly, the relict igneous phases are metastable during low-T recrystallization. The presence of unaltered clinopyroxene, for instance, may indicate that the effective composition is more aluminous than the original bulk rock composition. Such variations in effective bulk composition would influence the phase proportions, mineral compositions, and perhaps even mineral assemblages in very low-grade metamorphic rocks.

3.2.4 Domains of equilibrium

As described by many previous investigators (e.g. Smith, 1968), most secondary minerals in very low-grade metavolcanics vary significantly in their compositions and assemblages even within a single thin section. Such variations can be attributed to local differences in effective bulk composition as well as metastable equilibria, evidenced by compositional zoning and persistence of primary phases. Therefore, the common assumption that the assemblage at thin section scale represents equilibrium (e.g. Nakajima et al., 1977) cannot be applied to rocks of very low-grade metamorphism. In this chapter, minerals in physical contact are assumed to represent equilibrium as advocated by Zen (1974) and Kawachi (1975). In practice, however, the analysed compositions for a group of minerals within one amygdule a few millimetres to one centimetre in diameter, or veins 10–50 mm in thickness, or minerals replacing relict phases in a single thin section, are considered together to approximate the equilibrium compositions and their trends with increasing grade. In fact, it is extremely difficult to find 2 or 3 Ca–Al silicates together with quartz, albite and chlorite in contact relationship in an assumed equilibrium domain.

3.2.5 Chemical changes

Very low-grade metamorphism of basaltic to andesitic rocks generally results in only minor chemical changes. However, recent studies on the mobilities of elements during very low-grade metamorphism suggest that abundances of many major and minor elements are somewhat susceptible to change by secondary processes (e.g. Smith and Smith, 1976; Humphris and Thompson, 1978a). The occurrence of quartz, chlorite, albite and some Ca–Al silicates filling amygdules and fractures provides petrographic evidence for at least microdomain-scale mobility of most major elements. Replacement of olivine by chlorite or pumpellyite, and of plagioclase by albite and Ca–Al silicates, together with minor K-feldspar or sericite, suggest substantial transfer of K and Fe from solution into pre-existing feldspar and of Al, Fe, and/or Ca into earlier olivine. Concentrations of SiO_2, MgO, Na_2O, K_2O, Rb, Sr and the light REEs may be variably modified and may not reflect original igneous abundances. However, relatively immobile elements including Nb, La, Ce, P, Zr, Ti and Y have been suggested to be useful in evaluating the petrogenetic affinity and possible tectonic settings (e.g. Pearce and Cann, 1973; Wood et al., 1979). Such conclusions have applied even to some high-grade metabasites (e.g. Sheraton, 1985). Apparently, element mobility and extent of chemical changes are dependent on effective water/rock ratio (e.g. Seyfried et al., 1979), solution composition, alteration intensity, and P–T conditions. The most significant chemical changes include hydration–dehydration, carbonation–decarbonation, oxidation–reduction, and limited domain migration of selective elements (e.g. Smith, 1968). Except for spilites and some metasomatized rocks such as rodingites, very low-grade metamorphism described in this chapter refers to processes that do not markedly change the bulk-rock chemistry.

3.3   Model basaltic and andesitic systems

Mineral assemblages of low-T metamorphism of basaltic and andesitic rocks commonly consist of one or two Ca-Al hydrosilicates together with albite, white mica, quartz, chlorite (or smectite), sphene, carbonates, pyrite, haematite and/or magnetite. Ca- and Na-amphiboles are characteristic of greenschists and blueschists, respectively. Common low-T Ca-Al index minerals include Ca-zeolites, lawsonite, prehnite, pumpellyite, and epidote. Andradite–grossular garnet has been recorded in low-T recrystallization and in geothermal systems (Coombs et al., 1977; Malley et al., 1983; Schiffman et al., 1985). Smectite, illite, corrensite, and other mixed-layer clay minerals described in Chapter 2 are also common. Jadeitic and omphacitic pyroxenes are restricted to high-P blueschists. Compositions and abbreviations of these phases are listed in Table 3.2.

Compositions of these low-T metamorphic minerals may be described by 13 oxide components: SiO_2, Al_2O_3, Fe_2O_3, CaO, FeO, MgO, MnO, Na_2O, K_2O, TiO_2, SO_2, H_2O and CO_2 (Kuniyoshi and Liou, 1976a). Ignoring chain silicates, Na_2O, K_2O, and TiO_2 are assumed to be retained only in albite, white mica and sphene, respectively. Except for considerable substitutions of FeO or MnO for MgO in amphibole, mica, chlorite and clay minerals, and

Table 3.2 Compositions and abbreviations of phases and facies used in this chapter

Zeolites
Ch = Chabazite, $CaAl_2Si_6O_{16}$ $4H_2O$
Lm = Laumontite, $CaAl_2Si_4O_{12}$ $4H_2O$
Wr = Wairakite, $CaAl_2Si_4O_{12}$ $2H_2O$
Hu = Heulandite, $CaAl_2Si_7O_{18}$ $6H_2O$
St = Stilbite, $CaAl_2Si_7O_{18}$ $7H_2O$

Yu = Yugawaralite, $CaAl_2Si_6O_{16}$ $4H_2O$
Na = Natrolite, $Na_2Al_2Si_{13}O_{12}$ $2H_2O$
Th = Thomsonite $NaCa_2Al_5Si_5O_{20}$ $6H_2O$
Am = Analcime, $NaAlSi_2O_6$ H_2O
Mo = Mordenite, $(Na_2,Ca)Al_2Si_{10}O_{24}$ $7H_2O$

Other Ca–Al silicates
Pr = Prehnite, $Ca_2(Al, Fe)$ $AlSi_3O_{10}(OH)_2$
Lw = Lawsonite, $CaAl_2Si_2O_7(OH)_2$ H_2O
Gr = Grossular, $Ca_3Al_2Si_3O_{12}$
And = Andradite, $Ca_3Fe_2Si_3O_{12}$
Grd = Grandite, $Ca_3(Al, Fe)_2Si_3O_{12}$

Zo = Zoisite, $Ca_2Al_3Si_3O_{12}(OH)$
Cz = Clinozoisite, $Ca_2Al_3Si_3O_{12}(OH)$
Ep = Epidote, $Ca_2(Al, Fe)_3Si_3O_{12}(OH)$
Pm = Pumpellyite, $Ca_4(Al, Fe)_5MgSi_6O_{21}(OH)_7$

Amphibole (Amp) and clinopyroxenes (Cpx)
Di = Diopside, $CaMgSi_2O_6$
Jd = Jadeite, $NaAlSi_2O_6$
Ac = Acmite, $NaFeSi_2O_6$
Aug = Augite
Cr = Crossite

Tr = Tremolite, $Ca_2Mg_5Si_8O_{22}(OH)_2$
Act = Actinolite, $Ca_2(Mg, Fe)_5Si_8O_{22}(OH)_2$
Gl = Glaucophane, $Na_2(Mg, Fe)_3Al_2Si_8O_{22}(OH)_2$
MRi = Magnesioriebeckite, $Na_2(Mg, Fe)_3Fe_3Si_8O_{22}(OH)_2$
Hb = Hornblende, $NaCa_2(Mg, Fe, Al)_5Si_6Al_2O_{22}(OH)_2$

Other phases
Ab = Albite, $NaAlSi_3O_8$
An = Anorthite, $CaAl_2Si_2O_8$
Ol = Oligoclase
Pl = Plagioclase, $(CaAl, NaSi)AlSi_2O_8$
Mt = Magnetite, Fe_3O_4
Ht = Haematite, Fe_2O_3
F = fluid

Cc = Calcite, $CaCO_3$
Ar = Aragonite, $CaCO_3$
Do = Dolomite, $(CaMg)CO_3$
Qz = Quartz, SiO_2
Py = Pyrope, $Mg_3Al_2Si_3O_{12}$
Sph = Sphene, $CaTiSiO_5$
Chl = Chlorite, $(Fe, Mg)_{6-x}Al_x(Si_{4-x}Al_x)O_{10}(OH)_8$

Metamorphic facies
ZEO = Zeolite facies
BS = Blueschist facies
GS = Greenschist facies
AM = Amphibolite facies

PA = Pumpellyite–actinolite facies
PrA = Prehnite–actinolite facies
EA = Epidote–amphibolite facies
PP = Prehnite–pumpellyite facies

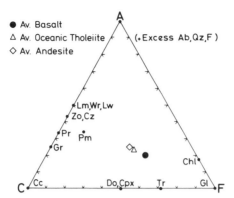

Figure 3.1   Chemographic relations of some Ca–Al silicates, carbonates, chlorite, pyroxene and amphibole in the ACF diagram in the presence of excess albite, quartz, and fluid phase. Compositional fields of basaltic and andesitic volcanics are also shown. Abbreviations and compositions of phases are listed in Table 3.2.

Fe_2O_3 for Al_2O_3 in most of the Ca-Al hydrosilicates, compositions of these phases can be modelled by the system NCMASH ($Na_2O–CaO–MgO–Al_2O_3–SiO_2–H_2O$). With the assumption that quartz, albite, phengite, sphene and fluid are present in excess, the NCMASH system is further simplified on to the ternary ACF diagram as shown in Figure 3.1. Andesitic rocks contain higher SiO_2, Al_2O_3, and alkalis and lower (MgO + FeO) and CaO than basaltic compositions (e.g. Le Maitre, 1976), but basaltic and andesitic rocks have rather similar ACF compositions. With the assumptions stated above, the NCMASH system can be extended to andesitic rocks. The present model first considers the case of unit activity of H_2O, despite the common presence of CO_2, CH_4, NaCl and other solutes in the fluid phase during recrystallization. The effects of CO_2 on the model system will be explained later to demonstrate the effect of variable fluid composition.

3.4   Graphic representation of basaltic–andesitic mineral parageneses

Many diagrammatic presentations of mineral assemblages for basaltic to andesitic rocks have been proposed to correlate specific assemblages with variations in chemical composition and in physical conditions according to the phase rule. Mathematic treatments for graphic representation of minerals have been discussed (e.g. Korzhinskii, 1959; Greenwood, 1975; Spear *et al.*, 1982). The three graphical projections described below are useful for low-*T* basaltic and andesitic assemblages.

3.4.1   *ACF diagram*

The conventional ACF diagram of Eskola (1915) has been used to show broad paragenetic features and mutual relations among common metabasites.

However, because of the assumptions described in previous sections, the ACF plot cannot fully account for the compositional variations of phases, particularly in terms of Fe^{3+}–Al substitution. In order to illustrate such substitutions, the two projections described in the following sections are employed. However, for the model NCMASH system with quartz, albite, white mica, sphene and fluid present in excess, the ACF diagram provides a conceptually simple method for showing the variation of mineral parageneses as a function of compositional variables and intensive properties.

3.4.2   The Al–Ca–Fe^{3+} projection

Introduction of FeO and Fe_2O_3 components into the model system allows the possibility of significant chemical substitution of FeO for MgO in chlorite and amphibole and of Fe_2O_3 for Al_2O_3 in most Ca–Al silicates present in low-T metamorphism. In order to evaluate the effect of the Fe^{3+}–Al substitution on compositions of the coexisting phases in low-grade metabasites, a simplified chlorite projection has been proposed (Liou *et al.*, 1985a). Compositions of the common low-T minerals are plotted in a tetrahedron Al–Ca–Fe^{3+}–(Fe + Mg) in Figure 3.2. The ranges of solid solution for epidote, prehnite, grandite garnet, pumpellyite, amphibole and chlorite are shown. Complete solid solution exists in the grandite garnet, epidote solid solution is continuous at least in the range of Ps 10 to Ps 33, and prehnite extends from $X_{Fe^{3+}} = 0.00$ up to 0.30 without compositional discontinuity. Other low-T minerals are either fixed in composition (e.g. laumontite, lawsonite, iron oxides) or present

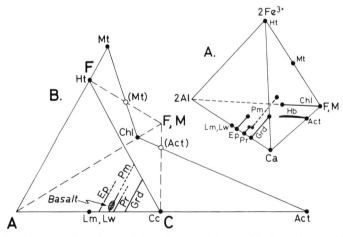

Figure 3.2   *A:* Compositions of some Ca–Al silicates, calcite, amphiboles, and iron oxides in the tetrahedron 2Al–Ca–2Fe^{3+} – (Fe^{2+}, Mg). Compositional ranges of these phases are shown. *B:* Projection of these phases from a chlorite composition on to a 2Al–Ca–2Fe^{3+} diagram. Note that the compositional variations of prehnite, epidote, pumpellyite, and grandite garnet are parallel to each other. Basaltic compositions are also shown.

in excess (e.g. quartz, albite, sphene), or both. For subgreenschist facies assemblages, Ca-amphibole is confined to actinolite composition.

The $Al-Ca-Fe^{3+}$ plane was used as the projection plane and chlorite of fixed composition as the projection point. This is justified by a number of factors: (i) compositional variation of the Ca–Al silicates is defined mainly by $Fe^{3+}-Al$ substitution; (ii) chlorite is ubiquitous in subgreenschist facies assemblages; and (iii) compositions of these phases except for actinolite lie on or very close to the projection plane, hence a slight change in chlorite composition does not significantly affect their dispositions on the ternary diagram.

Depending on the oxidation state and sulphur fugacity during metamorphism, haematite, magnetite and/or pyrite may occur in low-grade assemblages. The effect of f_{O_2} on compositional shifts has been discussed by Liou et al. (1983). Also shown in Figure 3.2B is the chemical composition of an average basalt in terms of $Al-Ca-Fe^{3+}$, which plots very close to the composition of Al-pumpellyite. This projection has been used by Coombs et al. (1977) to discuss variations in mineral compositions and grandite-bearing assemblages in low-grade metamorphic rocks from New Zealand.

3.4.3 Projection for blueschist parageneses

For high-P facies series of very low-grade metamorphism, where glaucophane–crossite amphibole is diagnostic and jadeitic pyroxene may

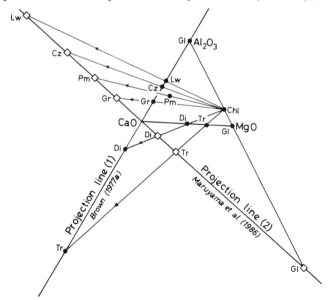

Figure 3.3 Compositions of common blueschist–greenschist facies minerals in an ACF diagram and their projection from chlorite on to a projection line (1) by Brown (1977a) and a projection line (2) by Maruyama et al. (1986).

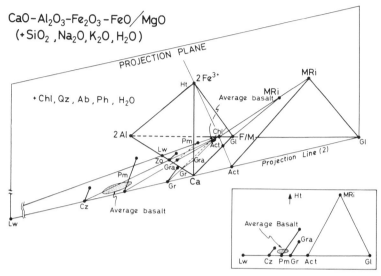

Figure 3.4 Introduction of Fe_2O_3 and FeO into the model system to show disposition of blueschist–greenschist facies minerals from the chlorite projection on to a plane normal to the Ca–2Al–F/M base. Also shown are the average composition of basalts and compositions of sodic amphiboles in terms of actinolite–glaucophane–Mg–riebeckite components.

occur in basaltic assemblages, a modified ACF projection should be used. Compositions of lawsonite, chlorite, clinozoisite, pumpellyite, grossular-pyrope garnet, diopside, glaucophane, jadeitic pyroxene, tremolite and chlorite are plotted in the $CaO–Al_2O_3–MgO$ diagram (Figure 3.3). They are projected from excess chlorite of constant composition on to two projection lines: one along the $Al_2O_3–CaO$ join used by Brown (1977a) and the other on a projection line which is inclined about 80° to the $Al_2O_3–CaO$ join (Maruyama *et al.*, 1986). Note that the latter projection can better illustrate the coexistence of the sodic and calcic amphiboles in blueschist facies metamorphism.

For the system NCMASH–Fe_2O_3, the projection plane is perpendicular to the $CaO–Al_2O_3–MgO$ base (Figure 3.4). The Fe^{3+}–Al substitution for epidote, pumpellyite, and sodic amphiboles in progressive high P–T facies series can be shown. Compared with the projection used by Brown (1977a), the present plot has three advantages: (i) all blueschist facies minerals are plotted on the positive side, hence, phase relations are easily depicted; (ii) composition of clinopyroxene in the acmite–jadeite–diopside system can be correlated with variation in bulk rock composition and in P–T conditions; and (iii) compositional gaps for Na–Ca amphiboles and pyroxenes can be illustrated. Also shown in this diagram is the composition of an average basalt.

3.5   Mineral assemblages for very low-grade metamorphic facies

Subdivision of the various subgreenschist facies has been mainly based on metabasaltic and meta-andesitic assemblages, inasmuch as most pelitic and other rock types described in Chapter 2 do not contain the index Ca–Al silicates. Although equilibrium is difficult to demonstrate for very low-grade assemblages, the systematic and predictable changes in mineral parageneses and compositions with increasing grade indicate that equilibrium has been approached sufficiently for the stable mineral assemblage to be identified.

3.5.1   Zeolite (ZEO) facies

The zeolite facies defined by Coombs et al. (1959, p. 91) included '... at least all those assemblages produced under physical conditions in which the following are commonly formed: quartz–analcime, quartz–heulandite, and quartz–laumontite.' The index assemblages are Am + Hu (or St) + Qz + Chl/Sm and Lm + Ab + Qz + Chl/Sm ($\pm$ Pm $\pm$ Pr $\pm$ Ep), respectively, for low- and high-T zeolite facies. The boundary for such a subdivision probably occurs at about 180 °C, where analcime + quartz give way to albite, and heulandite decomposes to laumontite + quartz (Liou, 1971a; Cho et al., 1987). Occurrence of wairakite is not definitive for the zeolite facies.

The zeolite parageneses, including three-phase assemblages such as thomsonite + laumontite + analcime, thomsonite + analcime + natrolite, and thomsonite + analcime + chabazite, have been described in dredged oceanic basalts, in ophiolites, and in metabasites from other tectonic settings as described in later sections. Depending on the size of the domain of equilibrium and effective bulk composition, zeolite facies assemblages vary quite substantially. However, occurrence of laumontite together with Fe-pumpellyite + chlorite is common, and assemblages of Lm + Pm + Ep, and Lm + Pm + Pr have been documented in zeolite facies metabasites (e.g. Cho et al., 1986).

3.5.2   Prehnite–pumpellyite (PP) facies

Coombs (1960) and Seki (1961) proposed the prehnite–pumpellyite facies to bridge the gap between the zeolite and greenschist facies. Characteristic basaltic assemblages of this facies include Pm + Pr, Pm + Ep, Pr + Pm + Ht, and Pm + Pr + Ep together with (+ Chl + Qz + Ab). In PP facies rocks of New Zealand, andradite + Pr + Ht and grandite + Ep + Ht have been described by Coombs et al. (1977) and Sivell (1984). Prehnite–pumpellyite facies metabasites are common in all tectonic settings and characterized by the stable association of Pm + Ep $\pm$ Pr and the absence of laumontite and actinolite. Partly because of high CO_2 activity and partly because of the bulk compositions of basaltic and andesitic rocks, prehnite is not a common phase and usually is restricted to Ca-rich domains such as veins and vesicles.

3.5.3   Prehnite–actinolite (PrA) facies

Liou et al. (1985a) proposed this facies to describe metabasite assemblages of Pr + Act + Ep (+ Chl + Ab + Qz + Sph) which occur at P–T conditions transitional from the zeolite or prehnite–pumpellyite to greenschist facies at low pressures. Other characteristic assemblages include epidote + prehnite + magnetite in active geothermal systems (Bird *et al.*, 1984; Schiffman *et al.*, 1985). Assemblages of this facies are characterized by the stable coexistence of prehnite and actinolite and by the absence of pumpellyite; some examples are described in later sections of this chapter. This facies was not previously recognized because of the relatively rare occurrence of prehnite + actinolite for basaltic compositions, partly due to high CO_2 activity. However, from the experimental study of a modal basaltic system by Liou *et al.* (1985*a*), the index assemblages for this facies can be seen to occupy a large $P–T$ field at low pressures. We therefore suggested a new facies for the Pr + Act + Ep (+ Chl + Ab + Qz + Sph) assemblages which have been generally assumed to belong to the greenschist facies.

3.5.4   *Pumpellyite–actinolite (PA) facies*

In many regionally metamorphosed terranes, pumpellyite commonly coexists with actinolite. The assemblage Pm + Act + Ep (+ Chl + Ab + Qz) occurs in deeply buried volcanics of considerable thickness and has been used as an index assemblage for the pumpellyite–actinolite facies (Hashimoto, 1966). Many previous descriptions of the field relations and mineral parageneses, chemistries and textures for such rocks (e.g. Bishop, 1972; Zen, 1974; Kawachi, 1975; Coombs *et al.*, 1976) indicate that the pumpellyite–actinolite facies occupies a $P–T$ field intermediate between the prehnite–pumpellyite, blueschist, and greenschist facies. Some Fe^{3+}-rich metabasites of the pumpellyite–actinolite facies may contain sodic amphibole (Nakajima *et al.*, 1977; Maruyama and Liou, 1985).

3.5.5   *Blueschist (BS) facies*

Deeply subducted rock sections display blueschist facies assemblages. A broad $P–T$ range for the blueschist facies from the appearance of lawsonite at about 3 kbar and 150–200 °C to the coexistence of omphacite + Ab + Qz + Ep at about 15 kbar and 450 °C has been recognized. Three $P–T$ fields characterize the stabilities of the assemblages Lw + sodic amphibole + Chl (lawsonite zone), Pm + Gl + Chl (pumpellyite zone), and Ep + Gl + Chl + Act (epidote zone). Aragonite and jadeitic to omphacitic pyroxene are commonly associated with these assemblages, and garnet and winchite may appear in epidote zone blueschists.

Winkler (1967) subdivided the blueschist facies into the lawsonite–albite and lawsonite–glaucophane–jadeite facies on the basis of the Jd + Qz = Ab reaction. Such subdivision was found to be useful for metagreywackes in New Zealand (e.g. Kawachi, 1975), and in New Caledonia (Brothers, 1970). However, recent investigations of blueschist facies rocks indicate that sodic amphiboles and pyroxenes may occur at lower pressures and lawsonite–albite–chlorite associations are ubiquitous in blueschists (e.g. Maruyama and Liou, 1987). Therefore, in this chapter, lawsonite–albite facies rocks are included within the blueschist facies.

3.6 Petrogenetic grid

Considerable progress has been made in evaluating the P–T conditions for various very low-grade metamorphic facies during the last two decades by both field and experimental studies. Three experimental approaches have been taken: (i) determination of mineral stability for its own simple bulk composition; (ii) investigation of phase relations and compositions using natural basaltic assemblages; and (iii) investigation of phase relations in the model basaltic system.

The first approach, which provides direct evidence regarding mineral stability, involves the reversible and systematic growth and disappearance of phases from mineral mixtures in the appropriate P_{fluid}–T fields. Some of these data are summarized in Figure 3.5. Reactions shown in this and other figures are listed in Table 3.3. Applications of equilibria derived from this approach to natural parageneses are restricted due to the complexity of natural systems. Nevertheless, stability studies of pure end-member phases permit retrieval calculations of internally consistent thermodynamic properties (Zen, 1972; Helgeson et al., 1978; Berman et al., 1985; Chatterjee et al., 1984; Holland and Powell, 1985; Powell and Holland, 1985) and delineation of maximum P–T conditions for their occurrences. From these thermodynamic bases, phase relations involving some index minerals in very low-grade metamorphism have been calculated (e.g. Kerrick and Ghent, 1979; Perkins et al., 1980; Bird and Helgeson, 1981; Perchuk and Aranovich, 1981; see also Chapter 2 of this book).

The second approach, using natural basaltic assemblages, encountered several difficulties because of the complexity of the rock composition, slow reaction kinetics, persistence of metastable phases, and the control of f_{O_2} in low-T metamorphism. Hence, except for water–basalt interaction experiments at 100–400 °C and P_{fluid} of 500 bars, this approach has been adopted only for facies transitions at higher temperatures (e.g. greenschist–amphibolite by Liou et al., 1974, and Moody et al., 1983; greenschist–epidote amphibolite by Apted and Liou, 1983).

The third approach is very useful for modelling parageneses and compositions of minerals for basaltic systems for very low-grade metamorphic conditions. Having justified the use of the model system, we constructed a petrogenetic grid for the model system, and identified and experimentally located several key reactions defining the facies transitions. We then evaluated the effects of FeO and Fe_2O_3 on compositions and phase relations in the model system using the principle of sliding equilibria and available com-

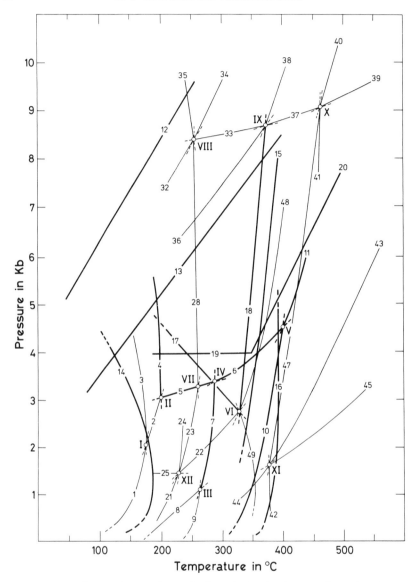

Figure 3.5 *P–T* diagram for experimentally determined univariant reactions relevant to very low-grade metamorphism of basaltic volcanics.

positions of natural Ca–Al silicates from classical regions (see next section). T–$X_{Fe^{3+}}$ relations were established for parageneses in the zeolite, prehnite–pumpellyite, pumpellyite–actinolite, prehnite–actinolite and blueschist facies. The results of these procedures are described below.

P–T relations of Ca-zeolites (St, Hu, Lm, Yu, and Wr), prehnite, pumpel-

Table 3.3 Reactions for both the NCMASH and NCMASH–Fe$_2$O$_3$ system described in this chapter. For phase compositions and abbreviations see Table 3.2.

Reaction No.	Reaction	Reaction No.	Reaction
1	St = Lm + H$_2$O	2	Hu = Lm + Qz + H$_2$O
3	St = Hu + H$_2$O	4	Hu = Lw + Qz + H$_2$O
5	Lm = Lw + Qz + H$_2$O	6	Lw + Qz = Wr
7	Lm = Wr + H$_2$O	8	Lm + Qz = Yu
9	Yu = Wr + Qz + H$_2$O	10	Wr = An + Qz + H$_2$O
11	Lw = An + Qz + H$_2$O	12	Jd + Qz = Ab
13	Ar = Cc	14	Am + Qz = Ab + H$_2$O
15	Pm = Cz + Gr + Chl + Qz + H$_2$O	16	Pr = Cz + Gr + Qz + H$_2$O
17A	Pr + Chl(Mg$_5$Al$_2$Si$_3$O$_{10}$(OH)$_8$) + H$_2$O = Pm + Tr + Qz		
B	Pr + Chl(Mg$_6$Si$_4$O$_{10}$(OH)$_8$) + Qz = Pm + Tr + H$_2$O		
19,20	Stability limits of glaucophane (Maresch, 1977)		
18	Pm + Chl + Qz = Cz + Tr + H$_2$O	21, 24	Lm + Pr = Cz + Qz + H$_2$O
22	Pm + Qz = Cz + Pr + Chl + H$_2$O	23	Lm + Pm = Cz + Chl + Qz + H$_2$O
25	Pr + Chl + Lm = Pm + Qz + H$_2$O	26	Pr + Chl + Lw = Pm + Qz + H$_2$O
27	Pr + Lw = Cz + Qz + H$_2$O	28	Lw + Pm = Cz + Chl + Qz + H$_2$O
29	Pm + Chl + Qz + H$_2$O = Lw + Tr	30	Lw + Pr + Chl = Pm + Qz + H$_2$O
31	Lw + Tr = Pr + Chl + Qz + H$_2$O	32	Lw + Gl = Pm + Chl + Ab + Qz + H$_2$O
33	Pm + Chl + Ab = Cz + Gl + H$_2$O	34	Lw + Gl = Cz + Chl + Ab + Qz + H$_2$O
35	Lw + Pm + Ab = Cz + Gl + Qz + H$_2$O	36	Pm + Gl + Qz + H$_2$O = Tr + Chl + Ab
37	Cz + Gl + Qz + H$_2$O = Tr + Chl + Ab	38	Pm + Gl + Qz = Cz + Tr + Ab + H$_2$O
39	Cz + Gl + Qz + H$_2$O = Hb + Chl + Ab	40	Cz + Gl + Tr + Qz = Hb + Ab + H$_2$O
41	Tr + Chl + Ab = Hb + Gl + H$_2$O	42	Ol + Tr + Chl = Hb + Ab + H$_2$O
43	Cz + Chl + Ab + Qz = Ol + Hb + H$_2$O	44	Cz + Chl + Ab + Qz = Ol + Tr + H$_2$O
45	Cz + Hb + Ab + Qz = Ol + Tr + H$_2$O	46	Cz + Chl + Qz = Wr + Tr + H$_2$O
47	Cz + Chl + Tr + Qz = Hb + Tr + H$_2$O	48	Pm + Qz = Cz + Pr + Tr + H$_2$O
49	Pr + Chl + Qz = Cz + Tr + H$_2$O	50	Ep + MRi + Chl + Qz = Tr + Ab + Ht + H$_2$O
51	Pm + Ht + Ab + Qz = Ep + MRi + Chl + H$_2$O	52	Pm + MRi + Qz = Ep + Tr + Ab + Chl + H$_2$O
53	Pm + MRi + Chl + Qz = Tr + Ht + Ab + H$_2$O	54	Pm + Ht + Qz = Ep + Tr + Chl + H$_2$O
55	Cc + Chl + Qz = Cz + Tr + H$_2$O + CO$_2$	56	Pr + CO$_2$ = Cz + Cc + Qz + H$_2$O
57	Cc + Chl + Qz = Pr + Tr + H$_2$O + CO$_2$	58	Do + Chl + Qz = Cz + Tr + H$_2$O + CO$_2$
59	Cc + Chl + CO$_2$ = Cz + Do + Qz + H$_2$O	60	Do + Qz + H$_2$O = Cc + Tr + CO$_2$
61	Pm + CO$_2$ = Cz + Cc + Chl + Qz + H$_2$O	62	Pm + Cc + Qz = Pr + Chl + H$_2$O + CO$_2$
63	Lm + Cc = Cz + Qz + H$_2$O + CO$_2$	64	Lm + Cc + Chl = Pm + Qz + H$_2$O + CO$_2$
65	Lm + Do = Cz + Chl + Qz + H$_2$O + CO$_2$	66	Lm + Do = Cc + Chl + Qz + CO$_2$
67	Pm + Qz + CO$_2$ = Cz + Pr + Cc + Chl + H$_2$O		

17A   Pr + Chl (Mg$_5$Al$_2$Si$_3$O$_{10}$(OH)$_8$) + H$_2$O = Pm + Tr + Qz

B   Pr + Chl (Mg$_6$Si$_4$O$_{10}$(OH)$_8$) + Qz = Pm + Tr + H$_2$O

19,20   Stability limits of glaucophane (Maresch, 1977)

Lm + MRi + Chl + Qz = Tr + Ht + Ab + H$_2$O

Lm + Cc + Chl = Pm + Qz + H$_2$O + CO$_2$

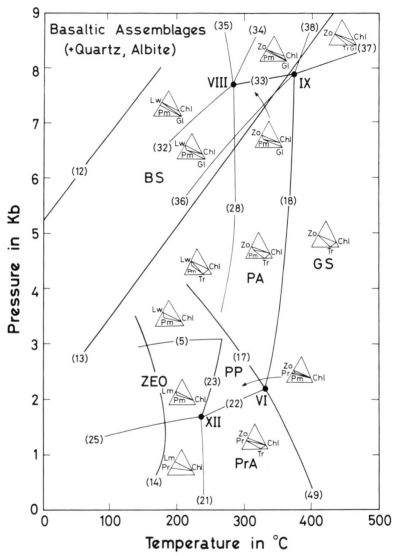

Figure 3.6 *P–T* diagram showing a petrogenetic grid for various low-grade metamorphic facies (uppercase letters, see Table 3.2 for abbreviations) and their basaltic assemblages (+ Chl + Qz + Ab) in the model basaltic system.

lyite, zoisite-epidote, plagioclase, actinolite, hornblende, Na–Ca pyroxene, lawsonite, glaucophane and chlorite were constructed for the model NCMASH system. With the assumptions described above, five univariant lines radiate from one invariant point in a pseudoternary system. If we further assume that chlorite is present in excess, the chlorite-absent reactions become metastable and many invariant points shown in Figure 3.5 have four radiating univariant lines. The $P-T$ relations summarized in Figure 3.5 are compiled from available experimental data mainly on stabilities of minerals for their own bulk compositions (solid, thick lines), and from natural parageneses, together with geometrically derived equilibria. The resulting curves were bound together by means of the Schreinemakers rule. In some places, the relations are well constrained; in others, they hinge heavily on calculated slopes and deduced data from natural parageneses. Considerable care was taken to make the grid internally consistent.

Figure 3.6 summarizes some experimentally-determined key reactions defining facies boundaries for various very low-grade metamorphic facies and their stable mineral assemblages for basaltic compositions (Liou *et al.*, 1985a). Although there is some uncertainty on $P-T$ positions for some of the inferred reactions, this diagram is useful for determining mineral parageneses for various metamorphic facies series. In order to make the petrogenetic grid as complete as possible, stabilities of Jd + Qz, calcite-aragonite and Am + Qz are also included in Figures 3.5 and 3.6.

3.7   Continuous *v.* discontinuous reactions for very low-grade metabasites

3.7.1   *Continuous and discontinuous reactions*

Continuous and discontinuous reactions have been differentiated and commonly used in metapelites to illustrate the detailed behaviour of mineral assemblages and compositions during progressive metamorphism (for details, see e.g. Thompson, 1976). The terms *continuous* and *discontinuous* are synonymous with *simple* and *complex* reactions as used by Thompson and Norton (1968). The same principle can be applied for metabasites.

A continuous reaction is *divariant* in the model system which involves exchange components such as $AlFe_{-1}$. Reactant and product phases involved in the continuous reaction define a *continuous reaction assemblage*. These phases may vary their compositions and proportions *continuously* following the constraints of the Gibbs' phase rule and mass balance. Accordingly, their relative abundances change for a given bulk composition as T or P varies. For example, at a constant P, the continuous reaction assemblage becomes univariant. With increasing or decreasing T, all the participating phases change their compositions and modal abundances continuously until one of the reactant phases is finally consumed. In nature, however, the consumption of one phase may be governed by a discontinuous reaction as described below.

A discontinuous reaction is *univariant* in the model basaltic system, and its constituent phases define a *discontinuous reaction assemblage*, which becomes invariant at constant P or T. Thus, in isobaric $T-X$ sections, a discontinuous reaction occurs at constant T and the compositions of all the phases are fixed. It should be pointed out that a discontinuous reaction results in a distinct change in topologic relations for ternary phase diagrams. There are two types of discontinuous reactions depending on the phase relations among the reactant and product phases: *terminal* and *non-terminal* types. The former defines disappearance or appearance of a phase regardless of bulk composition. This characteristic feature of disappearance of a reactant phase for a wide range of compositions makes the terminal-type discontinuous reaction most suitable in defining a metamorphic isograd. On the other hand, the non-terminal reaction represents a crossing tie-line relation in the projected phase diagrams. Above and below the $P-T$ conditions of a non-terminal reaction, two reactant phases become incompatible and one of them disappears. The disappearance or survival of a particular phase is dependent on the bulk composition. Nevertheless, non-terminal reactions are also useful for isograds, because they generally yield a change in mineral assemblages.

Both continuous and discontinuous reaction assemblages are also called low-variance *reaction assemblages* or simply low-variance assemblages. They are often referred to as *buffered assemblages*, because they are capable of buffering the compositions of minerals involved in the reactions. In the study of natural assemblages, it is important to identify low-variance assemblages for mineral analyses in order to delineate the $P-T-X$ relations described below. Note also that, by decreasing one component, or in the end-member subsystems, a discontinuous reaction becomes an invariant point, and continuous reactions become discontinuous reactions.

3.7.2   *Model basaltic–Fe_2O_3 system*

The experimental results for facies transitions, summarized in Figure 3.6, are for the model basaltic system. With the introduction of Fe_2O_3 into the model system (for convenience, we use 'model-Fe_2O_3 system'), the three Ca-silicates for each univariant line now define a three-phase field on the $Al-Ca-Fe^{3+}$ diagram described in the previous section. These three phases are stable over the whole $P-T$ field but have fixed compositions in terms of $X_{Fe^{3+}}$ at constant P and T. They may systematically change their $X_{Fe^{3+}}$ in response to a change in either T, P or both; hence the specific three-phase triangle may be continuously displaced within the field with changing P or T. These three phases define a continuous reaction for the model-Fe_2O_3 system and their compositions may vary systematically with increasing grade, as can be shown in a series of the $Al-Ca-Fe^{3+}$ diagrams. On the other hand, an invariant point of the model system with four Ca-silicates ($+ Chl + Qz + Ab$) becomes a univariant line in the model-Fe_2O_3 system. These four phases shown on the chlorite-projection diagram display either a cross-tie-line relationship or mark the disappearance or appearance of a phase. These four phases define a discontinuous reaction which reflects a distinct change in the topology of the $Al-Ca-Fe^{3+}$ diagrams.

Because the discontinuous reactions are independent both of the bulk composition in the model-Fe_2O_3 system and of the proportions of the participating phases, they are easily detected in a progressive metamorphic sequence in metabasalts and useful as isograds. It should be pointed out that most experimentally-determined reactions (Figure 3.6) are discontinuous reactions in the end-member model system and become continuous reactions in the model-Fe_2O_3 system. Moreover, through the introduction of an additional component such as FeO into the model-Fe_2O_3 system, the discontinuous reactions described above also become continuous. Therefore, most reactions in a complex natural system are continuous in nature; parageneses and compositions of Ca–Al silicates stable in a prograde sequence

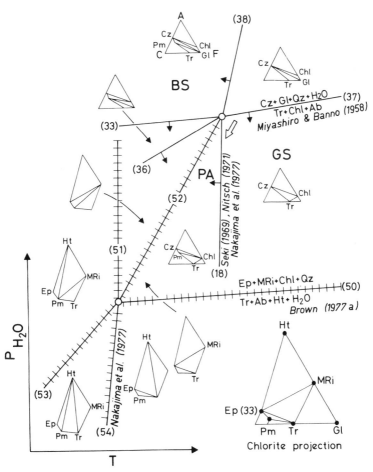

Figure 3.7 $P-T$ diagram showing phase relations among BS, GS and PA facies assemblages around the invariant point IX of Figure 3.6 for the model basaltic system (as shown in simplified ACF plots) and for the model–Fe_2O_3 system (lines with hatch marks and assemblages shown in the chlorite-projection plots). Arrows refer to the shifts of univariant lines and invariant point due to the introduction of Fe_2O_3 into the model system.

in low-T metamorphism will be extremely complicated. Using the assumptions in the previous sections, the phases defined both continuous and discontinuous reactions in the model-Fe_2O_3 system have compositions systematically vary with T or P. Hence, these phases constitute a *reaction* (or *buffered*) assemblage.

Another characteristic feature of discontinuous reactions as described by Thompson (1976) is that a discontinuous reaction terminates at the end-member invariant points *without a metastable extension*. The location of discontinuous reactions relative to the stable and metastable univariant curves in the model system is determined by the Schreinemakers principle (see Zen, 1966). An example is illustrated in Figure 3.7. For the model system, the phase assemblages of the BS, PA and GS facies can be shown by a series of ACF diagrams. They are related by five univariant lines on a $P-T$ diagram; at the invariant point, the assemblage $Cz + Pm + Tr + Gl$ ($+ Chl + Qz + Ab$) occurs. In the model-Fe_2O_3 system, this assemblage defines a discontinuous reaction and its $P-T$ position is shown with a hatch mark. This discontinuous reaction starts from the invariant point of the model system and terminates at another invariant point from which four other discontinuous reactions radiate. The assemblages in the model-Fe_2O_3 system can be illustrated using an $Al-Ca-Fe^{3+}$ diagram. The phase relations shown in Figure 3.7 will be discussed later in terms of $T-X$ and $P-X$ diagrams relevant to the facies transitions between the PA and GS and between the BS and GS facies.

3.8 $P-T-X_{(Al-Fe^{3+})}$ Diagrams: sliding equilibria in low-T metamorphism

The behaviour of three-phase assemblages ($+ Chl + Ab + Qz$) in the model-Fe_2O_3 system can be best represented by pseudo-binary $T-X_{(Al-Fe^{3+})}$ and $P-X_{(Al-Fe^{3+})}$ diagrams following the procedures outlined by Thompson (1976) and Spear *et al.* (1982). Such diagrams explain at least three features: (i) both continuous and discontinuous reactions are easily recognized; (ii) the relative $Al-Fe^{3+}$ distribution between coexisting Ca–Al silicates can be estimated and compositions of Ca–Al silicates can be seen to be highly dependent on the overall mineral assemblages; and (iii) complex compositional trends for Ca–Al silicates in a progressive sequence can be satisfactorily shown. To construct such diagrams, we need to know the $P-T$ positions of the univariant lines in the model system and the distribution coefficients (K_D) between coexisting pairs of minerals to be contoured in a $P-T$ diagram. Because of the lack of K_D data as a function of P, T, and chemical composition the $P-X$ and $T-X$ diagrams are at best qualitative. Three examples are described below to illustrate the effect of Fe_2O_3 on the model systems.

3.8.1 $T-X$ plot for the transition from the zeolite to prehnite–pumpellyite facies

Figure 3.8 illustrates the $P-T$ positions of both univariant lines and invariant points for the model system and their systematic displacements if the Fe_2O_3

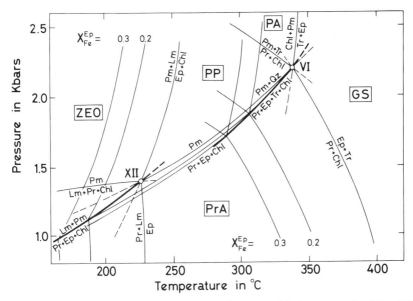

Figure 3.8 $P-T$ diagram showing continuous reactions around the invariant points VI and XII and displacement of the invariant point along a discontinuous reaction at the introduction of Fe_2O_3 into the model system. Series of isopleths were designated as $X_{Fe^{3+}}$ of epidote for these continuous reactions (see Liou *et al.*, 1985*b*, for details).

component is introduced. For example, four continuous reactions radiating from the invariant point XII are used for discussion of transition equilibria between the ZEO and PP facies. These reactions balanced for Al end-member phases are:

$$\text{(Pm)} \; 2\,Lm + Pr = 2\,Ep + 5\,Qz + 8\,H_2O \qquad (21)$$

$$\text{(Lm)} \; 5\,Pm + 3\,Qz = 3\,Ep + 7\,Pr + Chl + 5\,H_2O \qquad (22)$$

$$\text{(Pr)} \; 14\,Lm + 5\,Pm = 17\,Ep + Chl + 32\,Qz + 6\,H_2O \qquad (23)$$

$$\text{(Ep)} \; 6\,Lm + 17\,Pr + 2\,Chl = 10\,Pm + 21\,Qz + 14\,H_2O \qquad (25)$$

The $P-T$ positions of these continuous reactions were determined based on the experimental data and predicted $P-T$ grid in the model basaltic system (Liou *et al.*, 1985*a*; Cho *et al.*, 1986). The invariant point generated by the intersection of these Al–end-member reactions is located at $P = 1.4 \pm 0.5$ kbar and $T = 228 \pm 30\,°C$.

The effect of Fe^{3+} substitution for Al in Ca–Al silicates on the four continuous reactions may be estimated from the determined $X_{Fe^{3+}}$ values of coexisting phases. The microprobe data from very low-grade Karmutsen metabasites (Cho *et al.*, 1986) indicate that the $X_{Fe^{3+}}$ values increase in the order of prehnite, through pumpellyite to epidote. This order of relative $X_{Fe^{3+}}$ values will not change for the case of ideal solutions of prehnite, pumpellyite, and epidote, when a limited range of $X_{Fe^{3+}}$ values is considered. The displacements of the Al end-member reactions by the increase in Fe^{3+} content are schematically shown in Figure 3.8, as isopleths of fixed $X_{Fe^{3+}}$ values of epidote. These isopleth curves are calculated using both simple activity–composition relations and the observed Fe^{3+} partitioning between epidote and pumpellyite in the Karmutsen metabasites (see Cho et al., 1986, for details). These isopleths for the continuous reactions intersect each other, producing a discontinuous reaction: $Lm + Pm + Qz = Pr + Ep + Chl + H_2O$.

A schematic pseudobinary $T-X_{(Fe^{3+}-Al)}$ diagram at constant P_{fluid} of 1.1 kbar is constructed in Figure 3.9. Each mineral in the assemblage of a continuous reaction has its own compositional loop, and its composition is highly dependent on the reaction assemblage in which the mineral is found. For the chlorite-excess system described for metabasites, four continuous reaction loops intersect at one discontinuous reaction, where the compositions of four phases ($+ Chl + Ab + Qz + F$) and temperature are fixed at constant P. Note that the slopes of reactions (22) and (25) are positive in the $T-X_{Fe^{3+}}$ diagram; thus, as reactions proceed with increasing T, prehnite, pumpellyite, and epidote of these reaction assemblages become more Fe-rich at a constant P. The same information on the Fe-enrichment of epidote can be obtained from Figure 3.8, when the isopleths of epidote are taken into account at constant P. On the other hand, the reactions (21) and (23) have negative $T-X_{Fe^{3+}}$ slopes, suggesting that the epidote of the (Pr) and (Pm) assemblages decreases its Fe^{3+} content with increasing T.

Figure 3.9 shows the $T-X$ relations for the discontinuous reaction for the ZEO to PP facies transition: $Lm + Pm + Qz = Pr + Ep + Chl + H_2O$, which occurs at 185 °C and 1.1 kbar. Compositional data indicate that the $X_{Fe^{3+}}$ values of prehnite, epidote and pumpellyite vary systematically with increasing T. For example, both prehnite and epidote of the ZEO facies reaction assemblage $Lm + Ep + Pr$ ($+ Chl + Ab + Qz$) display increasing Al content with increasing temperature. On the other hand, prehnite, epidote and pumpellyite of the PP facies metabasites decrease their Al contents with increasing grade. Hence, compositional trends of these Ca–Al silicates in progressive metamorphic sequences may become very complicated, depending on the bulk rock composition and mineral assemblage.

Three divariant loops are stable for the ZEO facies assemblages: $Lm + Pr + Pm$, $Lm + Pm + Ep$, and $Lm + Pr + Ep$. However, only one loop representing the prehnite–pumpellyite-epidote assemblage is stable in the PP facies. As shown in Figure 3.9, disappearance of laumontite in the model-Fe_2O_3 system is governed by the (Pm) loop defined by the reaction (21) or by the discontinuous reaction. Thus, at a temperature interval between T of the discontinuous reaction (185 °C) and the T_{21} at 227 °C, laumontite can be stable only for the aluminous bulk composition. But for a common basaltic rock with a significant Fe_2O_3 component, disappearance of laumontite is governed by the discontinuous reaction and occurs at a significantly lower

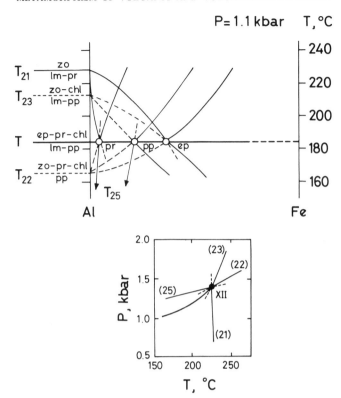

Figure 3.9 Schematic isobaric $T-X_{Fe^{3+}}$ diagrams showing compositional variations of buffered assemblages for continuous and discontinuous reactions to illustrate the relationship for the zeolite and prehnite-pumpellyite facies transition. Temperature estimates and compositions of Ca–Al silicates are from Cho *et al.* (1986).

temperature. Therefore, the stability of laumontite or the zeolite facies assemblage is highly dependent on the bulk composition and that the transition between the ZEO and PP facies assemblages is multivariant.

3.8.2 *T–X plot for the transition from the pumpellyite–actinolite to greenschist facies*

Transition from the PA to GS facies in the model-Fe_2O_3 system is defined by two reactions: (18) $Pm + Chl + Qz = Cz + Tr + F$, and (54) $Pm + Ht + Qz = Ep + Tr + Chl + F$. Their $P–T$ positions are qualitatively shown in Figure 3.7 and an isobaric $T–X_{(Fe-Al)}$ diagram is modified from that of Nakajima *et al.* (1977) in Figure 3.10. The $T–X$ diagram was constructed based on the experimental data of Liou *et al.* (1985a) and on the epidote isopleths for a reaction assemblage $Ep + Act + Pm + Chl + Ab + Qz$ from

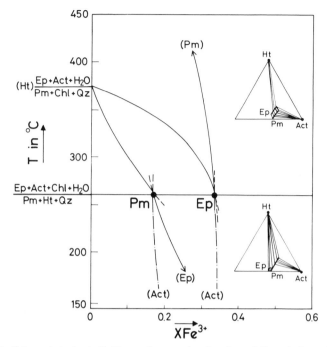

Figure 3.10 Schematic isobaric $T-X_{Fe^{3+}}$ diagram showing the stability relations of pumpellyite, epidote, and haematite in a model–Fe_2O_3 system with excess chlorite, actinolite, albite and quartz (see Nakajima *et al.*, 1977, for details).

Nakajima *et al.* (1977). Four reaction loops representing the four univariant lines for the model system intersect at a discontinuous reaction. However, for clarity in illustrating the facies transition, only two loops are shown in Figure 3.10. At P_{fluid} of 5 kbar and f_{O_2} higher than the HM buffer, Mg-pumpellyite breaks down at about 370 °C, whereas the discontinuous reaction for the appearance of Fe-epidote occurs at about 260 °C (see Schiffman and Liou, 1983). In between, both pumpellyite and epidote are stable, and their compositional relationship is defined by a continuous reaction (18). With increasing temperature, the coexisting pumpellyite and epidote decrease in $X_{Fe^{3+}}$ values. At constant T and P, Fe^{3+} preferentially concentrates in the epidote. The epidote isopleths of Nakajima *et al.* (1977) for the PA to GS facies transition are very sensitive to temperature change, hence are good geothermometers (see below).

3.8.3 *P–X plot for the transition from the blueschist to greenschist facies*

As shown in Figure 3.7, the BS–GS facies transition equilibria are bounded by two reactions in the model–Fe_2O_3 system: (37) $Cz + Gl + Qz + F = Tr +$

Chl + Ab, and (50) Ep + MRi + Chl + Qz = Tr + Ab + Ht + F. The $P-T$ locations of these two reactions are estimated by Brown (1974, 1977a) based on natural parageneses and experimentally determined by Maruyama $et\ al.$ (1986). Both reactions possess similar $P-T$ slopes, but reaction (37) occurs at much higher pressures than reaction (50). Therefore, a GS/BS transitional assemblage of sodic amphibole + Act + Ep (+ Ab + Chl + Qz) occurs in the $P-T$ region between these two reactions. Compositions and proportions of these phases for a given bulk composition change systematically along a continuous reaction (37) as a function of P and T. Hence, the compositions of blue amphibole and epidote in the reaction assemblage can be used to estimate the pressure of metamorphism.

Figure 3.11 schematically illustrates the inferred compositional variations of sodic amphibole and epidote in the reaction assemblage as a function of pressure at constant temperature (300 °C). At f_{O2} conditions higher than that defined by the HM buffer, sodic amphiboles may vary their compositions along the join glaucophane (Gl)–magnesioriebeckite (MRi), tremolite may have a very limited FeO substitution, and the stable iron oxide is haematite. P_3 of Figure 3.11 refers to the equilibrium pressure at 300 °C for the reaction (37) in the model system. With gradual introduction of Fe_2O_3 into the model system, this reaction is displaced continuously toward lower pressure, and both epidote and sodic amphibole become more Fe^{3+}-rich. The Fe^{3+}/Al partitioning between epidote and sodic amphibole for the reaction assemblage

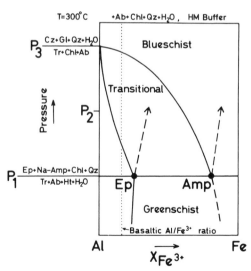

Figure 3.11 $P-X_{Fe^{3+}}$ plot at 300 °C showing changes in compositions of sodic amphiboles and epidote for buffered assemblages in the blueschist–greenschist transitional zone: (1) the solid lines are for Ep + Tr + sodic amphibole, and (2) the dashed lines for Ep + Ht + sodic amphibole. P_1 and P_3 are, respectively, the equilibrium pressure for the discontinuous and continuous reactions for the facies transitions (modified from Maruyama $et\ al.$, 1986).

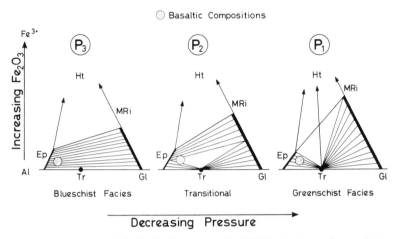

Figure 3.12 Three schematic isobaric diagrams at $T = 300\,°C$ showing variation of mineral assemblages and compositions of sodic amphiboles as a function of pressure (modified from Maruyama *et al.*, 1986).

is complex, as discussed in detail in Maruyama *et al.* (1986). In the Fe^{3+}-saturated system, the discontinuous reaction (50) occurs at about 4 kbar. So long as f_{O_2} is maintained above that defined by HM, the discontinuous reaction remains at fixed pressure isothermally and has fixed mineral compositions for a given bulk composition. Hence, the $P–X$ relation for the discontinuous reaction appears as a horizontal line in Figure 3.11. If compositional variations of sodic amphibole of the reaction assemblage were calibrated with pressure, its composition can be used as a geobarometer (see below for more details).

Phase assemblages and compositions of sodic amphiboles at P_1, P_2, and P_3 at $300\,°C$ are illustrated in three chlorite-projection diagrams of Figure 3.12. Tie-lines were schematically drawn for the coexisting phases. Complete solid solution is assumed for the Gl–Mri and Cz–Ep (Ps33) joins. These three diagrams illustrate the paragenetic and compositional variations of BS–GS reaction assemblage as a function of not only pressure, described above, but also of bulk composition. For example, basaltic rocks have compositions between those of Ep and Tr whereas ironstones may be very oxidized, containing abundant Fe_2O_3. At $P = P_3$ where reaction (37) occurs, basaltic rocks may contain the typical BS assemblage Ep + Gl (+ Ab + Chl + Qz), whereas ironstones contain Ep + MRi + Ht. At intermediate pressures (e.g. $P = P_2$), metabasites, depending on their bulk composition, may contain (i) the BS assemblage, (ii) the GS assemblage, or (iii) the reaction assemblage Ep + Act + sodic amphibole where all three phases are of fixed composition.

If pressure continuously decreases, both epidote and sodic amphibole of the reaction assemblage may systematically vary their Fe^{3+}/Al ratio, and the compositional change of sodic amphibole is most evident. When pressure is lowered to P_1, a discontinuous reaction occurs and compositions of the participating phases are fixed. At pressures lower than P_1, basaltic rocks with high Fe^{3+}/Al ratios contain the GS assemblage Ep + Tr + Ht whereas ironstones may have Ep + Tr + Ht or Tr + MRi + Ht depending on their bulk composition. Compositions of sodic amphiboles from the basaltic three-phase reaction assemblage Ep + Act + sodic amphibole may differ significantly from those of ironstones. The composition of the former is shown as a solid line in Figure 3.11 for the basaltic blue amphiboles, and the latter compositions (ironstone amphiboles) as a broken line.

3.9   The effects of CO_2 and $P–T–X_{CO_2}$ diagrams

The role of CO_2 in diagenesis and very low-grade metamorphism has been well documented from experimental, theoretical, and petrographic studies (see Seki and Liou, 1981, for review and references). For example, over a narrow range of temperature at constant pressure, different assemblages are possible in an isochemical system as a function of $\mu CO_2/\mu H_2O$ ratio. In addition to influencing mineral assemblages, the contamination of the fluid phase by CO_2 would drastically affect reaction temperature, kinetics, diffusion and transport processes. Laumontite- and heulandite-bearing assemblages of the ZEO facies can be obtained isothermally and isobarically from the kaolinite-calcite-quartz assemblage by increasing μH_2O relative to μCO_2. Ca-zeolites and lawsonite are stable with respect to calcite–quartz–clay minerals only if the X_{CO_2} of the fluid phase is less than about 0.1 (see Chapter 2 for calculated phase diagram). A similar effect of CO_2 on prehnite, pumpellyite, epidote, and other Ca–Al silicates and on BS facies assemblages has also been deduced. The absence of Ca-zeolites or prehnite has been noted in some active geothermal areas in which P_{CO_2} in the fluid phase was estimated to be just a few bars. The most significant effects of CO_2 on very low-grade metamorphic parageneses include (i) decreasing the dehydration temperatures of Ca–Al and other hydrosilicates; and (ii) stabilizing carbonates at the expense of Ca–Al silicates.

Many very low-grade metabasaltic rocks may lack index Ca–Al silicates as a direct consequence of pervasive high CO_2 activity. An example in Taiwan has been described by Liou (1981), and formation of propylite assemblages in the Greentuff Formation of Japan (Seki, 1973) is another. In such terranes, the physical conditions of very low-grade metamorphism have been extremely difficult to assess; hence, some of these areas have been considered as unmetamorphosed in many geological studies.

Previous studies on $T–X_{CO_2}$ relations for simple systems indicate that most characteristic Ca-hydrosilicates in very low-grade metamorphism are stable only at X_{CO_2} values less than 0.1. In the model basaltic system, the effects of CO_2 are evaluated based on available experimental and thermodynamic data. Phases considered are only those which are commonly encountered in metabasalts; dolomite is also included, even though its occurrence is rare (Coombs *et al.*, 1970). The results with index assemblages of ZEO, PP, PrA, and GS facies are illustrated at 2 kbar in Figure 3.13. Although the fluid phase may have an immiscibility gap at low temperatures, two-phase regions in the H_2O-CO_2 mixture are not shown because we are dealing with very low X_{CO_2} conditions. It is evident from this diagram that these index assemblages are stable only at low X_{CO_2} values, probably less than 0.15, and the carbonate–chlorite assemblage occupies quite a large $T-X_{CO_2}$ space. Therefore, the clay–carbonate assemblages described by Zen (1961b) and Coombs *et al.* (1970) may be common in natural metabasite parageneses. For example, with increasing X_{CO_2} in the metamorphic fluid at constant temperature and P_{fluid}, successive transitions from ZEO facies assemblage through $Lm + Cc + Chl$, and

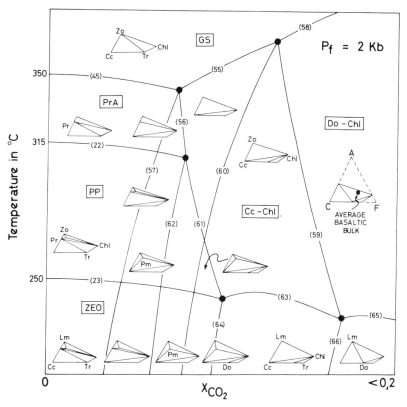

Figure 3.13 Schematic $T-X_{CO_2}$ diagram at $P_{fluid} = 2$ kbar. Dehydration temperatures for reactions 22, 23 and 49 are from Liou *et al.* (1985a). Mineral assemblages for basaltic rocks in each $T-X_{CO_2}$ space are shown on a series of ACF diagrams. See Table 3.3 for reactions.

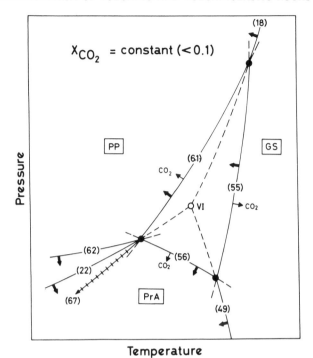

Figure 3.14 Schematic $P-T$ diagram showing relations of mineral assemblages among the PP, PrA and GS facies (for simplicity, the PA assemblage is not considered) with constant X_{CO_2} less than 0.1. Compared with Figures 3.6 and 3.8, the effect of introduction of very low CO_2 content into the model system is to generate 3 invariant points and several univariant lines. $P-T$ positions of the univariant lines and invariant point (open circle) are from Figure 3.8. The univariant line with hatch marks is for a discontinuous reaction (67). Large solid arrows show the shift of the univariant lines with introduction of Fe_2O_3 into the model–CO_2 system and small arrows for increasing CO_2 activity. See Cho and Liou (1987) for details.

Lm + dolomite + Chl, to kaolinite + dolomite + Chl assemblages may occur at low temperatures. Similarly, transition of the PP (or PrA, PA) facies assemblage through Ep + Cc + Chl, and Ep + dolomite + Chl, to pyrophyllite + dolomite + Chl assemblages may occur at higher temperatures.

Figure 3.14 shows $P-T$ relations illustrating the effect of CO_2 on the facies transitions among the PP, PrA and GS facies for the model system (compare this diagram to Figure 3.8). Introduction of CO_2 causes the invariant point VI of the model system to become metastable and generates several CO_2-bearing continuous and discontinuous reactions. It is apparent from this diagram that, with increasing CO_2 content, the stability field of the calcite–chlorite assemblages encroaches upon the PP and PrA facies at low P and low T and upon the GS facies at high T. Therefore, for a given metamorphic gradient, a prograde paragenetic sequence from the PP facies through the PrA to carbonate–clay and GS facies occurs in low CO_2 environments, whereas at

D

high CO_2 the PrA facies assemblage would not be encountered. This relationship explains why PrA facies rocks are not common in several of the classical metamorphic terranes described in later sections.

3.10 Geothermometry and geobarometry

3.10.1 *General statements*

Geological thermometry and barometry and their applications to various metamorphic systems have been extensively described (e.g. Essene, 1982). For very low-grade metamorphic rocks, most proposed geothermometers and geobarometers are for metapelites. These methods employ the degree of illite crystallinity, the nature of illite/smectite mixed-layer clays, colours of conodonts, vitrinite reflectance, fluid inclusions, oxygen isotopes and others; some of these methods are described in various chapters of this book (see also Kisch, 1983). In addition to those precautions and limitations summarized by Essene (1982), application of these methods to low-grade rocks is also plagued by metastable equilibrium, persistence of relict phases and most importantly, slow reaction kinetics. For metabasites, we rely on phase equilibria and compositions of minerals from reaction assemblages for $P-T$ estimates; their compositions vary systematically as a function of intensive properties as described in previous sections. Several examples are illustrated below. In practice, the following steps must be followed: (i) to identify the minerals in low-variance assemblages within a particular metamorphic domain; (ii) to analyse the compositions of these phases and determine their equilibrium compositions; (iii) to apply the appropriate $P-T-X$ diagram or equations; and (iv) to use at least two independent methods of thermometry and barometry.

3.10.2 *Zeolite equilibria*

The stability relations of a number of Ca-zeolites (Hu, St, Lm, Yu, and Wr), analcime, and lawsonite in the presence of excess quartz and fluid are summarized in Figure 3.15. These relations, together with lawsonite stability relations, to some extent model the transitions among ZEO, PP, GS and BS facies mineral assemblages; the attending plagioclase in nature much more closely approaches albite than anorthite. At pressures above 3 kbar, laumontite breaks down to $Lw + Qz + F$, and wairakite reacts with fluid to form $Pr + smectite + Qz$. Under conditions where fluid pressure is less than total pressure, the stability fields of the Ca-zeolites are encroached upon by those of BS and PP facies minerals at higher pressures, and by those of PrA facies minerals at higher temperatures. At low H_2O activities caused by dissolved solutes or presence of CH_4, these zeolites would be restricted to lower $P-T$ conditions than those determined by experiments. Nevertheless, this diagram

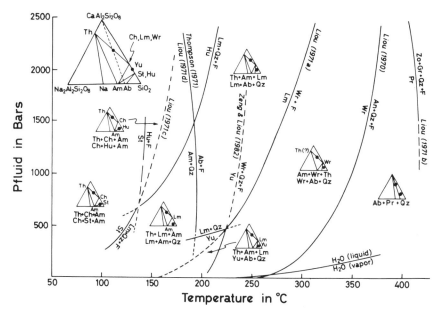

Figure 3.15 Experimentally determined $P-T$ relations among various Ca–zeolites (St, Hu, Lm, Yu, Wr), analcime, prehnite, albite and anorthite in the presence of excess quartz and fluid. Parageneses of Ca- and Na-zeolites for two bulk compositions (X = plagioclase of An 50; and Y = basaltic composition) are shown in terms of $CaAl_2Si_2O_8$, $Na_2Al_2Si_2O_8$, SiO_2 and excess H_2O components.

indicates that stilbite, heulandite and laumontite are confined to low-T and low-P conditions, wairakite and yugawaralite are stable only along high-T metamorphic gradients, and the minimum depth for formation of lawsonite in quartzose rocks is about 10 km. The assemblage Am + Qz is restricted to diagenetic and low-T ZEO facies rocks. At temperatures above 200 °C, Na-bearing wairakite is common in high-T ZEO facies and in PrA facies rocks in active geothermal systems. Of course, changes in composition and substitution of other components within these end-member phases would modify the thermal stability ranges for these minerals.

The paragenetic depth sequence of the Ca-zeolites is highly dependent on the imposed metamorphic gradient, on the P_{H_2O}/P_{total} ratio, and on other factors such as fluid composition. Two invariant points among zeolite equilibria related to such sequences need to be discussed. One is the invariant point for the stable coexistence of stilbite, heulandite and laumontite at pressures lower than 1 kbar and temperatures less than 150 °C (see Cho *et al.*, 1987). Heulandite, which is stable between the stability fields of stilbite and laumontite, can occur only at pressures higher than that of the invariant point. These data are consistent with natural parageneses in low-grade metamorphic rocks recrystallized under conditions of low X_{CO_2} in the fluid. The direct zonal transition from stilbite to laumontite without heulandite may be favoured by lower-pressure conditions, such as at Tanzawa Mt., Japan, whereas regional distribution of heulandite and its transformation to laumontite, as is found in New Zealand, may be typical in a higher pressure burial sequence. Certainly, variations in heulandite compositions including substitution of alkalis for Ca and difference in SiO_2 and H_2O contents significantly affect the heulandite stability. Hence, heulandite occurrence has been reported in some submarine metabasalts (e.g. Malley et al., 1983) and in active geothermal systems (e.g. Mehegan et al., 1982).

The other invariant point at 230 °C and 0.5 kbar is that for the coexistence of laumontite, yugawaralite and wairakite. Zeng and Liou (1982) showed that yugawaralite has a very restricted $P-T$ stability field; in geothermal systems where P_{fluid}/P_{total} ratio is about 0.3, its occurrence is restricted to depths shallower than 500 m. In a geothermal system with a relatively high metamorphic gradient and a high P_{H_2O}/P_{total} ratio, yugawaralite may be stable and the sequence of zeolites with depth could be mordenite–laumontite–yugawaralite–wairakite. On the other hand, in regions with a lower metamorphic gradient and a lower P_{H_2O}/P_{total} ratio, yugawaralite is not stable and the observed zonation of Ca-zeolites would be mordenite–laumontite–wairakite. Because many geologic, geochemical and hydrologic conditions may control both the P_{H_2O}/P_{total} ratio and the metamorphic gradient, different depth zonation patterns of Ca-zeolites may occur even within a single geothermal system or burial metamorphic sequence. Such variations have been recorded in the Onikobe geothermal area by Liou et al. (1985b).

3.10.3 Cpx–Ab–Qz equilibria

The jadeite–quartz–albite equilibrium defined by the reaction Jd + Qz = Ab is the single most important reaction for delineating $P-T$ conditions of metamorphism in various BS terranes. However, application of the jadeitic pyroxene equilibria to natural parageneses involves two fundamental difficulties: one is the choice of the experimental calibration of the Jd–Ab–Qz curve and the other is the lack of data on activity-composition relations among the Ca–Na pyroxenes.

Because of its petrological significance, the Jd–Ab–Qz equilibrium for the jadeite end-member composition has attracted a great deal of attention among experimental petrologists (e.g. Birch and LeComte, 1960; Newton and Smith, 1967; Holland, 1980). All these experiments have been performed at temperatures above 500 °C, using high albite for starting material. Although the equilibrium pressures at 500–700 °C are reasonably consistent among different investigators, considerable uncertainty remains as to the slope of the equilibrium curve and its $P-T$ position at higher temperatures. These uncertainties lead to a large discrepancy in the equilibrium pressures extrapolated to BS facies metamorphic temperatures of 200–350 °C. For example, the Holland's curve is extrapolated to 5.8 kbar at 200 °C, whereas the Birch–LeComte curve requires 8.2 kbar at the same temperature. Many petrologists have used the latest determination by Holland for their P–T estimates (e.g. Maruyama and Liou, 1985; Yokayama et al., 1986; Jayko et al., 1986). However, as described below, Holland's data are not consistent with some observed field petrological data.

Holland's extrapolated curve crosses the aragonite–calcite transition of Carlson (1983) at 4.3 kbar and 170 °C. At temperatures below 170 °C, calcite can coexist with jadeite + quartz. To our knowledge, such an association has never been recorded. Moreover, the X_{Jd} isopleths calculated by Holland (1983) indicate that pyroxenes of $Jd_{50}Ac_{50}$ and $Jd_{50}Di_{50}$ compositions cannot coexist with aragonite at temperatures lower than 350 °C and 370 °C, respectively. However, such assemblages are common in Franciscan meta-basites (Maruyama and Liou, 1987) which were recrystallized at tempera-tures lower than 300 °C (Taylor and Coleman, 1968).

The X_{Jd} isopleths for clinopyroxenes coexisting with albite and quartz were calculated based on the Jd–Ab–Qz stability relations recommended by Popp and Gilbert (1972) and using the equation given by Essene and Fyfe (1967). The calculated isopleths of X_{Jd} in both the Jd–Di and the Jd–Ac series are shown with the calcite–aragonite transition equilibrium in Figure 3.16. This diagram provides reasonable pressure estimates for BS facies metamorphism in

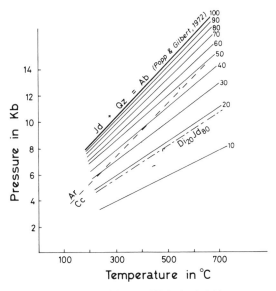

Figure 3.16 P–T diagram showing the sliding equilibria for jadeitic pyroxene + albite + quartz in the system Jd–Ac (or Di)–Qz. The P–T line for the end–member jadeite is from Popp and Gilbert (1972) and for the calcite–aragonite transition is from Carlson (1983). The Jd isopleths in both Jd–Ac and Jd–Di joins are calculated following the methods of Essene and Fyfe (1966) and Holland (1983).

Franciscan, Sanbagawa and New Caledonian terranes. It is apparent that aragonite can be stable with pyroxenes even with X_{Jd} as low as 20 mole percent at temperatures less than 200 °C.

3.10.4 $X_{Fe^{3+}}$ of epidote

Compositions of epidote in the buffered assemblage Ep + Pm + Chl + Act (+ Ab + Chl + Qz) are sensitive to temperature changes in the PA to GS facies transition. However, the assemblage is trivariant in the model-Fe_2O_3 system, and the compositions of the phases are dependent on P, T and host rock Fe^{2+}/Mg ratio. At given P and host rock Fe^{2+}/Mg ratio as represented by the constant Fe*/Mg of chlorite, the compositions of epidote and pumpellyite are a function of temperature. This isobaric $T-X_{Fe^{3+}}$ relation was first proposed by Nakajima et al. (1977), and slightly modified by our unpublished experimental data as shown in Figure 3.10. At 5 kbar, epidote of the buffered assemblage changes from Ps 33 at 260 °C to nearly pure clinozoisite at about 370 °C. The coexisting pumpellyite decreases its $X_{Fe^{3+}}$ with increasing T, but to a lesser extent than epidote. The exact position of the compositional loop in Figure 3.10 remains uncertain due to a lack of crystallographic-thermodynamic data on the effect of $Fe^{3+} = Al$ substitution in pumpellyite.

Using the same principle, the $X_{Fe^{3+}}$ of epidote for various low-T reactions discussed in the previous sections may also be used as a geothermometer (see Figures 3.8 and 3.9). However, because these continuous reactions involve H_2O, variations in P_{fluid}/P_{total} ratio or in fluid composition could significantly affect the T estimates. Previous applications of this geothermometer have assumed $P_{H_2O} = P_{total}$. Combination of this method with the pyroxene geobarometer and oxygen isotope geothermometer may allow us to estimate P, T and P_{H_2O}/P_{total} during very low-grade metamorphism.

3.10.5 Al_2O_3 content of sodic amphibole

The $P-X_{Fe^{3+}}$ relations shown in Figure 3.11 indicate that the composition of sodic amphibole coexisting with Ep + Act + Chl + Ab + Qz decreases systematically with decreasing pressure and can be used as a pressure indicator for metamorphism. Several isopleths have been drawn for constant compositions for both epidote and sodic amphibole in the BS–GS facies transition by Maruyama et al. (1986). The results indicate a rapid decrease in equilibrium pressures for increasing Fe^{3+}–content, especially at low pressures. The slopes of the calculated isopleths vary from very gently positive to gently negative with decreasing pressure. The gentle, gradual nature of the slope for these isopleths confirms their suitability for use as a geobarometer for the BS–GS facies transition.

This transition equilibrium is also trivariant, dependent on P, T and host-rock FeO/MgO ratio. The effect of FeO on the transition equilibria was evaluated to be of minor importance. An empirical diagram using the Al_2O_3 content of sodic amphibole in the buffered assemblage for common metabasites with X_{Fe*} in chlorite $= 0.4-0.5$ is shown in Figure 3.17A for the proposed geobarometer. Sodic amphibole composition is expressed as wt% Al_2O_3 in order to avoid the uncertainty of Fe^{3+} estimates from microprobe analysis. Pressure estimates using this method for selected BS terranes in

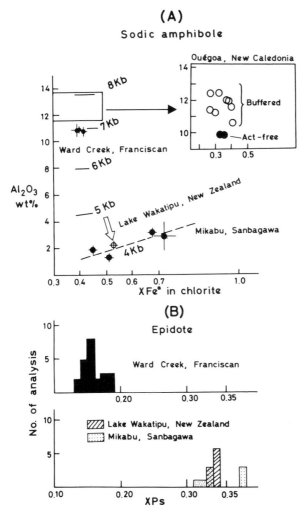

Figure 3.17 A: Al_2O_3 wt% of sodic amphibole of the buffered assemblage sodic amphibole + Act + Ep + Ab + Chl + Qz plotted against the X_{Fe*} of chlorite from the Sanbagawa, the Franciscan, New Zealand and New Caledonia. Metamorphic pressures for these blueschist terranes are estimated from pyroxene geobarometry. See Maruyama *et al.* (1986) for details. B: Compositions of epidotes (as expressed in X_{Ps} values) from three blueschist terranes showing the pressure dependence of the epidote composition.

California, Japan, New Zealand, and New Caledonia are in good agreement with those derived from sodic pyroxene geobarometry.

It should be noted that epidote minerals are common in BS facies rocks, and those coexisting with sodic amphibole and actinolite (+ Ab + Chl + Qz) can be used as a geobarometer. For example, epidotes in Franciscan metabasites have Ps values of 13 to 19, while for those in Sanbagawa and New Zealand, Ps = 31–38 as shown in Figure 3.17B. This relation indicates that glaucophane and clinozoisite were stable at higher pressures in Franciscan metabasites than were crossitic amphibole and epidote in the Sanbagawa and New Zealand metabasites.

3.10.6 $Na-M_4$ content of calcic amphibole

Brown (1977b) suggested that the crossite content of Ca-amphibole coexisting with albite + iron oxide + chlorite could be used as a potential geobarometer. Because of high variance in this assemblage, the $Na-M_4$ content of Ca-amphibole may be highly dependent on variations in Fe^{2+}/Mg, and Fe^{3+}/Al ratios of the coexisting minerals. However, based on the estimated pressures for metamorphism in the Sierra Nevada of California, the Otago district of New Zealand, and the Shuksan area of Washington, an empirical relationship between the $Na-M_4$ content of Ca-amphibole with pressure was established (see Figure 3.10 of Brown, 1977b). This geobarometer can be applied to rocks with metamorphic grades higher than the crest temperature of the compositional gap between the Ca- and Na-amphiboles, and therefore, for metamorphism in the GS and lower amphibolite facies.

3.10.7 Sphalerite geobarometry

The iron content of sphalerite in equilibrium with hexagonal pyrrhotite and pyrite provides a good pressure estimate in the temperature range 300–650 °C (Scott, 1973, 1976). Reasonable pressure estimates have been obtained by this geobarometer for disseminated, unexsolved, coexisting three-phase assemblages in GS or higher-grade rocks (e.g. Brown et al., 1978). However, at temperatures less than 300 °C, extensive resetting of sulphide compositions and transformation of hexagonal to monoclinic pyrrhotite may introduce considerable uncertainty to the pressure estimates. Hence, the validity of this geobarometer in very low-grade metamorphism remains to be investigated.

3.11 Examples of natural parageneses

3.11.1 Ocean-floor metamorphism

3.11.1.1 *Alteration and metamorphism of oceanic basalts.* Ocean-floor basaltic rocks near the sea bottom have suffered hydrothermal alteration through interaction with circulating hot sea water. In the axial zone of a spreading ridge where injection of basaltic magma and circulation of hot sea water dominate, intense recrystallization of basaltic rocks by hydrothermal, thermal, and local dynamic metamorphism occurs. Such hydrothermal metamorphism is characterized by incomplete and static recrystallization and occurrence of non-equilibrium assemblages; hence, original igneous textures and primary phases, particularly clinopyroxene, are well preserved. Metamorphic gradients from 500 to 1400 °C/km have been postulated for the ridge axis, and sea water is heated to temperatures as high as 500 °C and circulated through the ridge system (Humphris and Thompson, 1978b). The distribution of secondary minerals and extent of downward prograde recrystallization are strongly controlled by the volume of circulating sea water, that is, by the water–rock ratio.

Two types of ocean-floor hydrothermal alteration have been described by Thompson (1983). *Low-temperature alteration* of basaltic rocks at temperatures to about 100 °C is a ubiquitous process at the sea floor. The majority of dredged oceanic basalts generally referred to as 'weathered' basalts, are characterized by only small changes in their mineralogy from the original igneous precursor. Low-temperature alteration proceeds in four stages: I, formation of palagonite; II, formation of smectites; III, formation of carbonates; and IV, compaction and dehydration. The diagenetic 'Brownstone facies' alteration of Cann (1979) does not produce unequivocally metamorphic minerals such as laumontite and chlorite. *High-temperature alteration* of basaltic rocks at $P-T$ conditions ranging from zeolite to amphibolite facies is widespread in dredged samples. These metabasites have been recovered predominantly from the axial valley of spreading mid-ocean ridges and the large fault scarps of transform faults. By far the most common metabasalts recovered in dredge hauls are greenstones.

Under zeolite facies conditions, variable zeolite assemblages with depth as shown in Figure 3.18 have been described (e.g. Miyashiro *et al.*, 1971; Malley *et al.*, 1983). Haematite, smectite, calcite, Fe-rich pumpellyite and prehnite are also common. Wairakite has been reported by Jehl *et al.* (1976) but apparently is not common. 'Spilitic' metamorphism is used to describe metasomatized basalts resulting from interaction of hot circulating H_2O with the upper part of the oceanic ridge (e.g. Coleman, 1977). Comparison of bulk analyses of spilitic basalts with unaltered tholeiites indicates that the former generally are enriched in Na_2O and depleted in CaO relative to the latter.

Humphris and Thompson (1978b) subdivided the 'greenschist facies' metabasites into epidote-rich and chlorite-rich assemblages. The chlorite-rich assemblage shows significant chemical changes, whereas the epidote-rich samples show very little change from basaltic compositions. Examples of these two assemblages are often found within a single pillow: the pillow rims are characterized by a chlorite-rich assemblage and by a higher MgO content whereas the pillow cores contain the epidote-rich assemblage and are high in

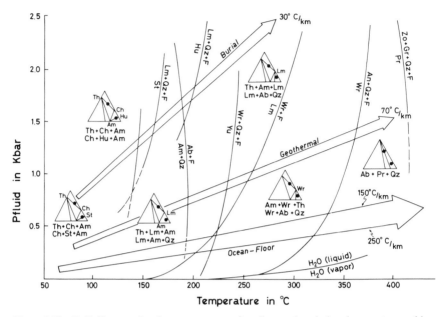

Figure 3.18 *P–T* diagram showing parageneses of zeolites and prehnite along metamorphic gradients suggested for ocean-floor, thermal-geothermal, and burial metamorphism.

Na_2O, CaO, and SiO_2. Such variation may be attributed to alteration at different sea water/rock ratios as demonstrated by many sea water–basalt experiments (see Mottle, 1983a, b for summary). Because of the inherent permeability of pillow basalt piles, the pillow core has a limited access to sea water and alters at a low water/rock ratio, whereas the glassy rim is apparently exposed to much greater volumes of sea water and consequently alters at a high water/rock ratio.

Recent deep-sea drilling has penetrated pillow lava into the underlying sheeted dyke complex and recovered in-situ basalts containing greenschist facies assemblages at depths up to 1350 m below the sea floor (e.g. Alt and Honnorez, 1984). Many descriptions of altered basaltic samples can be found in every issue of *Initial Reports of the Deep Sea Drilling Project*. Alt (1985) identified three stages of alteration in a transition pillow dyke complex and its underlying sheeted dyke section. The first stage is characterized by crystallization of Chl + Act + Pr + Ab + Qz at temperatures of at least 200–250 °C, the second stage by Qz + Ep + sulfide at up to 380 °C, and the third stage by formation of laumontite, heulandite, and prehnite at up to 250 °C. A transition zone with PrA facies assemblage exists between zeolite facies alteration in the overlying pillow section and greenschist facies parageneses in the underlying dyke complex.

The characteristic features for subgreenschist facies assemblages recovered from the dredged oceanic metabasites are summarized as follows.

(i) The index assemblage for the PP facies is not common in the recovered samples. Mevel (1981) described dredged metabasites with abundant pumpellyite from the Vema fracture zone, and attributed the rare occurrence of pumpellyite elsewhere to its being stable at lower f_{O_2} than epidote. However, as shown in Figure 3.8, with the characteristically high metamorphic gradient for the ocean-floor metamorphism, the PP facies assemblages may not be stable.

(ii) Instead, the assemblage Pr + Act + Ep + Chl + Ab of the PrA facies should be common inasmuch as this facies occupies a large $P-T$ space for very low P/T metamorphism. Lack of positive identification of minor prehnite in altered basalts may have caused difficulty in assignment of this facies. Moreover, as shown in Figure 3.14, minor amounts of CO_2 would significantly expand the stability field of the Cc + Chl + Ep + Qz assemblage at the expense of the PrA facies assemblages.

(iii) The statement by Thompson (1983, p. 235) that 'greenstones are metabasalts of the greenschist facies' needs to be modified. Greenstone is a general term used to describe a fine-grained metabasite which is greenish in colour and contains abundant smectite, corrensite, or chlorite. Greenstone has been generally used for both subgreenschist and greenschist (e.g. Archaean greenstones) facies rocks.

(iv) Depending on the $P-T$ conditions, oceanic crust of layers 1 and 2 may have been subjected to metamorphism of the ZEO and minor PP facies, through PrA and GS facies, to transitional and amphibolite facies with increasing depths. Because of differences in water–rock ratio and intensity of hydrothermal alteration, some basaltic rocks may have been significantly modified in their bulk composition as well as their mineralogy. Hence, index assemblages for the zeolite and PrA facies may not be well developed, and depth variations for mineral assemblages may not be apparent from dredged samples. In this regard, petrological studies of on-land ophiolite sequences have shed significant light on ocean-floor metamorphism.

3.11.1.2 *On-land ophiolites.* Parageneses of secondary minerals resulting from ocean-floor metamorphism have been documented in many on-land ophiolites (e.g. Liou, 1979; Evarts and Schiffman, 1983). The secondary minerals show a typical ophiolite pattern of increasing grade, but generally decreasing intensity, of metamorphism with depth as shown, for example, in Figure 3.19. The zeolite and minor PP facies assemblages are present in the upper part of the pillow sequence, passing downward through PrA and GS facies assemblages for lower volcanic sequences and to low amphibolite facies for the plutonic sequences. The total vertical distance may be up to about 3–4 km, and suggests a metamorphic gradient of $100-200\,°C\,km^{-1}$.

Four metamorphic zones, as shown in Figure 3.19, are delineated for the Del Puerto ophiolite in California (Evarts and Schiffman, 1983). The topmost zone is identified by the presence of heulandite or laumontite in volcaniclastic sediments and in the uppermost volcanic member of the ophiolite. Beneath it

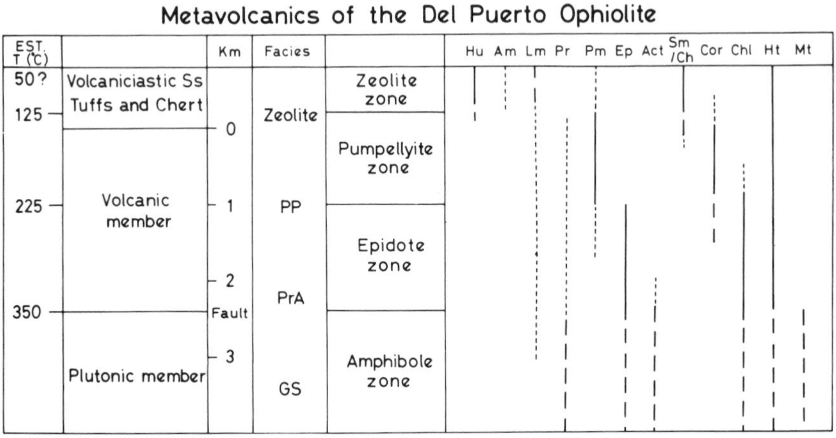

Metavolcanics of the Del Puerto Ophiolite

EST T(°C)		Km	Facies		Hu Am Lm Pr Pm Ep Act Sm/Ch Cor Chl Ht Mt
50?	Volcaniclastic Ss Tuffs and Chert	0	Zeolite	Zeolite zone	
125					
				Pumpellyite zone	
225	Volcanic member	1	PP		
		2	PrA	Epidote zone	
350		Fault			
	Plutonic member	3	GS	Amphibole zone	

Figure 3.19 Parageneses of secondary minerals in the Del Puerto ophiolite, California (modified after Evarts and Schiffman, 1983).

are the pumpellyite and epidote zones, respectively, for the upper and lower members of the volcanic sequences. The pumpellyite zone is characterized by the occurrence of PP facies assemblages. The epidote zone may be equivalent to that of the PrA facies, except for the virtual absence of actinolite. The amphibole zone is coincident with the plutonic member and is separated from other zones by a fault. Dominant greenschist facies assemblages were removed during the fault juxtaposition between the volcanic and plutonic sequences.

The Horokanai ophiolite in Hokkaido, Japan, (Ishizuka, 1985) exhibits a complete metamorphic gradation over the temperature interval 100–750 °C, ranging through the zeolite, greenschist, greenschist–amphibolite, amphibolite, and pyroxene hornfels facies. The zeolite zone is subdivided into three subzones by the sequential appearance of chabazite (stilbite), laumontite and wairakite; their characteristic assemblages are listed below:

Chabazite subzone: Chl + Ch + Am + Th; Chl + Ch + Am + St

Laumontite subzone: Chl + Lm + Am + Th; Chl + Lm + Th + Ab

Wairakite subzone: Chl + Wr + Am + Th; Chl + Wr + Th + Ab;
                      Chl + Wr + Pm + Ab; Chl + Wr + Ab + Qz.

Some of the chlorite minerals in the chabazite and laumontite subzones are mixed-layer smectite/chlorite clays. In the wairakite subzone, zeolite-free assemblages such as Pr + Pm + Chl, Pr + Act + Chl occur in the interpillow matrix, whereas the wairakite subzone assemblages occur in nearby pillow cores. With increasing grade (depth), the greenschist facies assemblage Act + Ab + Ep + Chl becomes ubiquitous in the middle of the extrusive sequence. The observed zeolite parageneses in this and other oceanic metabasites together with the probable metamorphic gradient for ocean-floor metamorphism are summarized in Figure 3.18.

3.11.2 Hydrothermal metamorphism in geothermal systems

Many previous studies on low-T metamorphism and metasomatism in active geothermal systems have documented systematic zonation of temperature-dependent hydrothermal minerals (e.g. Browne, 1978). These mineral zonations have been correlated with those produced by progressive low-T metamorphism observed in fossil hydrothermal systems. Parageneses of Ca–Al silicates in geothermal systems suggest a very low-pressure facies series ranging from the ZEO through PrA to GS facies (see Figure 3.18 for zeolite parageneses). Authigenic mineral zonations in Iceland and Onikobe (Japan) geothermal fields are described, respectively, for hydrothermal metamorphism of basaltic and andesitic rocks. Other recent petrologic studies on this subject include Schiffman et al. (1985) for the Cerro Prieto geothermal field of Mexico, Cavarretta et al. (1982) for the Larderello field of Italy, and Bird et al. (1984) for the Salton Sea geothermal field of the USA (see also Zen and Thompson, 1974, for the Wairakei area and other geothermal areas in New Zealand).

3.11.2.1 *Icelandic geothermal systems.* Iceland presents a very well-exposed example of oceanic crust formed at the Mid-Atlantic Ridge. The Icelandic crust has been extensively drilled in connection with geothermal research and utilization, with over 120 wells deeper than 1000 m and the deepest well 3085 m. Extensive studies of hydrothermal metamorphism from available drill cuttings have been summarized by Kristmannsdottir (e.g. 1983). Icelandic geothermal areas have been subdivided into two types according to subsurface temperatures at 1 km depth. The high-T geothermal areas, with temperatures exceeding 200 °C, are confined to zones of active volcanism and rifting, and the low-T zones, with temperatures less than 150 °C, are located in Quaternary rocks on the flanks of the active zones.

A simple correlation between temperature and alteration zones has been made for interaction of basaltic crust and meteoric water as summarized in Figure 3.20. A downward transition from low-T zeolite zones into a laumontite zone followed by an epidote zone occurs in the Tertiary section of northern Iceland. Successive zeolite depth zones (chabazite–thomsonite, mesolite–scolecite), described by Walker (1960) for the exposed Tertiary basaltic sequences, are also observed in active, low-T geothermal areas. However, laumontite is the most abundant zeolite in recovered samples, and first appears at about 100 °C. Wairakite occurs locally in deeper sections, and the general absence of wairakite in Tertiary sections has been explained by (i) regionally low thermal gradients during the main alteration events, and (ii) retrograde replacement of wairakite by laumontite. Calcite, quartz, and pyrite are common throughout. Smectite, corrensite, and chlorite are success-

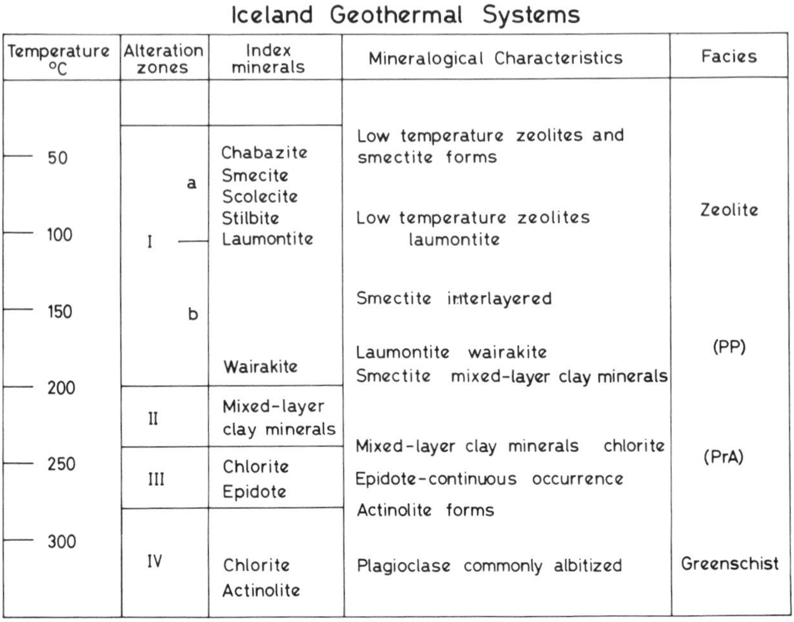

Iceland Geothermal Systems

Temperature °C	Alteration zones	Index minerals	Mineralogical Characteristics	Facies
— 50	a	Chabazite Smecite Scolecite Stilbite Laumontite	Low temperature zeolites and smectite forms	Zeolite
— 100	I		Low temperature zeolites laumontite	
— 150	b		Smectite interlayered	
— 200		Wairakite	Laumontite wairakite Smectite mixed-layer clay minerals	(PP)
	II	Mixed-layer clay minerals	Mixed-layer clay minerals chlorite	
— 250	III	Chlorite Epidote	Epidote-continuous occurrence	(PrA)
			Actinolite forms	
— 300	IV	Chlorite Actinolite	Plagioclase commonly albitized	Greenschist

Figure 3.20 Correlation of alteration zones in tholeiite basalt with increasing temperature from active geothermal areas in Iceland (after Kristmannsdottir, 1982).

ively the dominant sheet-silicate with depth. Epidote is the common Ca–Al silicate and becomes abundant with depth. Actinolite appears at 275 °C. Prehnite occurs sporadically at similar depth levels to epidote; pumpellyite has not been positively identified in drill cuttings. The fossil 'geothermal gradient' obtained from the depths of appearance of laumontite and epidote is about 75 °C km^{-1} for the Eyjafjordur field where the present gradient is measured to be 65 °C km^{-1}.

The IRDP borehole, 1919 m deep, near Reydarfjordur, eastern Iceland, together with a 1-km thick exposed section nearby, provides a nearly complete alteration sequence for the upper 3 km of Icelandic crust. Detailed petrological and mineralogical studies of secondary minerals (e.g. Mehegan et al., 1982; Exley, 1982) yield three distinct alteration zones. The low-T zeolite zone described by Walker (1960) occurs down to 1.6 km crustal depth with a maximum palaeotemperature of 100–120 °C. The high-T zeolite zone from 1.6 to 2.8 km crustal depth is dominated by laumontite, chlorite, calcite, and quartz, with minor epidote and prehnite in the lower part where the maximum temperature of about 230 °C is reached. Starting at about 1200 m and extending to the base of the core (2.8 to 3.5 km crustal depth), higher-temperature assemblages with abundant epidote, quartz, prehnite, chlorite and albite and minor amounts of wairakite, pumpellyite and actinolite occur at temperatures below 300 °C. These data suggest an average fossil thermal gradient of 80–90 °C km^{-1} for the upper 3.5 km of crust. Contact metamorphism, indicated by formation of andradite and wollastonite in other Icelandic geothermal areas (Kristmannsdottir, 1982), appears to be superimposed on the pre-existing assemblages.

Using the estimated thermal gradients of 80 to 90 °C km^{-1}, Robinson et al. (1982) suggested that the onset of the GS facies may occur at about 4 km, and of the amphibolite facies at about 6 km crustal depth. At depths from 3 to 4 km (about 270–350 °C), the basaltic rocks may contain the characteristic assemblages of the PrA facies, whereas at shallower depths, the ZEO facies with minor PP facies assemblages predominate.

3.11.2.2 *Onikobe geothermal area of Japan.* The Japanese island arc has experienced repeated volcanism, both terrestrial and submarine. Intensive volcanism since the early Miocene has been intimately connected with generation of fumaroles and hot springs, evolution of the Greentuff tectonic belt, and intensive alteration of rocks at both the surface and in the subsurface at numerous existing geothermal areas. Many exploratory and production wells up to 2000 m depth have been drilled in more than 30 geothermal areas in Japan. Studies of rocks from drill holes show pervasive alteration of volcanic and sedimentary host rocks by hot fluids and gases.

Parageneses of secondary minerals of the Onikobe geothermal system (Seki et al., 1983; Liou et al., 1985b) are used to exemplify the hydrothermal metamorphism of andesitic–dacitic rocks in the Japanese island arc. The zonal distribution of clay and zeolite minerals with increasing depth can be simplified as follows: alkaline smectite or smectite–corrensite–chlorite for clay minerals and mordenite–laumontite–yugawaralite–wairakite for zeolites. In general, smectite occurs with mordenite, corrensite with laumontite, and chlorite with wairakite. Both prehnite and epidote become increasingly abundant with depth. Characteristic features of the Iceland geothermal systems are also observed in Onikobe. However, there are some differences: (i) pumpellyite was not positively identified in any Onikobe core samples; (ii) yugawaralite occurs in this and other geothermal areas in Japan; (iii) wairakite is the most abundant Ca-zeolite as opposed to laumontite in most Icelandic geothermal areas; (iv) epidote appears at very shallow depths (e.g. about 160 m); (v) calcite and Ca-zeolites are abundant at depth; and (vi) minor actinolite was identified at depths of about 1 km. These features suggest that the Onikobe geothermal area may possess a higher fossil 'geothermal gradient' than the Iceland areas.

3.11.3 *Burial metamorphism*

Greywackes and volcaniclastic rocks of mafic to felsic compositions are the most common rock types in marginal basin, trench, and trench-slope deposits.

Their thicknesses may amount to more than 20 km in continental margins such as the Torlesse greywacke basement of New Zealand. Progressive burial metamorphism of these rocks has attracted a great deal of attention (see Zen and Thompson, 1974, for review). Other than those described below, recent studies of burial metamorphism of basaltic rocks include PP, PA, and GS facies rocks from Maine, eastern USA (Richter and Roy, 1976), from Central Sweden (Nyström, 1983), and from the Hamersley Basin, western Australia (Smith *et al.*, 1982), and ZEO and PP facies rocks from the Welsh basin, UK (Bevins and Rowbotham, 1983), and from Central Chile (e.g. Levi *et al.*, 1982).

3.11.3.1.   *New Zealand.* In New Zealand, metamorphic rocks of the ZEO, PP, PA, and GS facies occur within the basement greywacke sequence of Permian to Cretaceous age. The mineral parageneses developed within the basement rocks in South Island have been the topic of study for many years (e.g. Landis and Coombs, 1967). These include ZEO to PP facies rocks in the Taringatura area and PP through PA to GS facies rocks in the Upper Wakatipu district. The paragenetic sequence with depth is summarized in Figure 3.21. Apparent metamorphic gradients and mineral compatibilities for rocks of basaltic composition are shown in Figure 3.22. Both ZEO and PP facies assemblages are well developed in Taringatura, which is the type locality for the ZEO and PP facies rocks (Coombs *et al.*, 1959; Coombs, 1960). However, Boles and

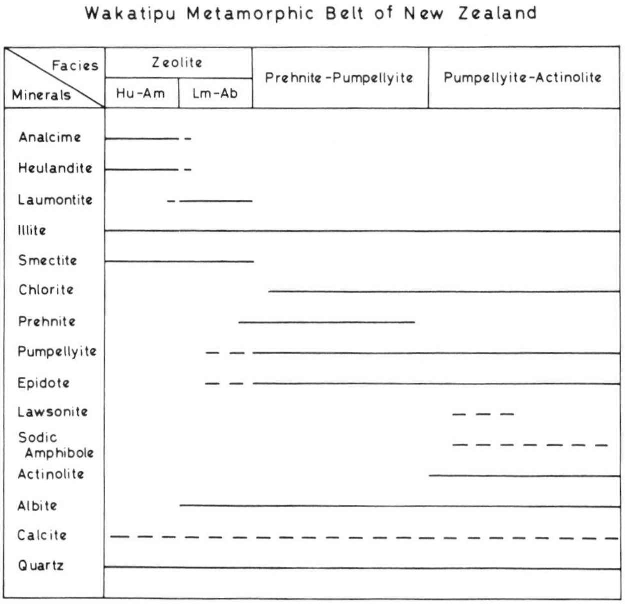

Wakatipu Metamorphic Belt of New Zealand

Facies / Minerals	Zeolite		Prehnite-Pumpellyite	Pumpellyite-Actinolite
	Hu–Am	Lm–Ab		
Analcime	———	–		
Heulandite	———	–		
Laumontite		– ———		
Illite	————————————————————————			
Smectite	—————————————			
Chlorite			————————————————	
Prehnite			———————————	
Pumpellyite		– – ———————————————————		
Epidote		– – ———————————————————		
Lawsonite				– – –
Sodic Amphibole				– – – – – –
Actinolite				————————
Albite	————————————————————————			
Calcite	– – – – – – – – – – – – – – – – – – –			
Quartz	————————————————————————			

Figure 3.21   Parageneses of secondary minerals in the Wakatipu metamorphic belt of New Zealand (data from Coombs *et al.*, 1959; Kawachi, 1975).

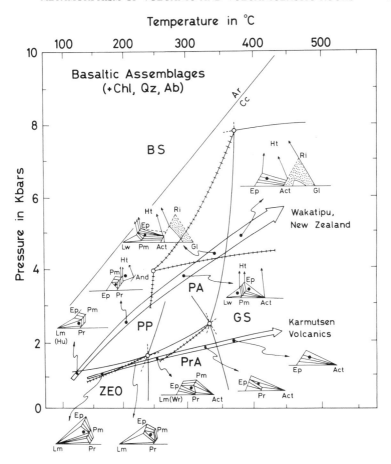

Figure 3.22 *P–T* diagram showing parageneses of minerals along estimated metamorphic gradients for (1) the Wakatipu belt of New Zealand; and (2) the Karmutsen metavolcanics, B.C.. The observed assemblages are plotted on to the chlorite-projection diagrams of Figure 3.2.

Coombs (1977) found complex, irregularly zoned distributions of laumontite, heulandite, albite, prehnite, pumpellyite, and clay minerals. They attributed the complexity to the interplay of many factors including differences in parent materials, in P_{CO_2}, and in P_{fluid}/P_{total} ratio, incomplete metamorphic reactions, variable permeability and ionic activity ratios in the fluid phase, and the effects of rising temperature following deep burial.

In the Upper Wakatipu district, four metamorphic zones are differentiated: I, PP; II, lawsonite–pumpellyite; III, PA; and IV, GS (Kawachi, 1975). However, distribution of these zones is not systematic, and these zones may not have formed in a single burial metamorphic event. To the east of the greenstone-ultramafic belt, the mafic volcanics and volcaniclastics exhibit prograde burial metamorphism from the PP through PA to GS facies, similar to those of other Torlesse greywacke basement rocks in South Island. However, to the west of the greenstone-ultramafic belt, minor lawsonite-albite facies rocks prograde through the sodic amphibole-bearing PA to GS facies, representing a slightly lower metamorphic gradient than that in the Sanbagawa belt of Japan.

The Brook Street Volcanics at D'Urville Island contain both PP and GS facies assemblages. The absence of PA facies rocks in this area is attributed to a local higher metamorphic gradient compared to most other parts of South Island (Sivell, 1984). Prehnite is generally absent in the lower-grade rocks due to higher CO_2 activity (Figure 3.14). Similarly, a 14 km-thick sequence of the Takitimu volcanics (Houghton, 1982) was subjected to metamorphism transitional between normal burial metamorphism of the 'Taringatura type' (Coombs, 1960) and thermal metamorphism of the Tanzawa Mountain type (Seki et al., 1969; see below for discussion). Two broad metamorphic zones of ZEO and PP facies are separated by a laumontite-out isograd. The sequence as shown in Figure 3.23 is characterized by gross overlap and sporadic distributions of all Ca–Al silicates. Several features need to be emphasized. (i) Two metadomains are differentiated based on contrasting assemblages contained in void spaces and shear fractures. The void spaces contain laumontite in the upper zone and prehnite in the lower zone, whereas the shear fractures have prehnite in the upper zone and epidote in the lower zone.

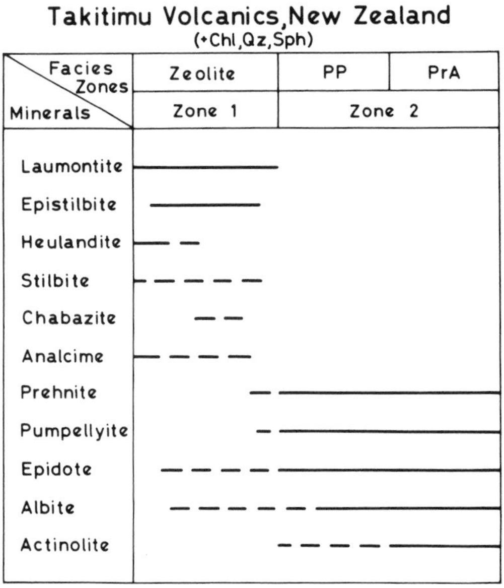

Takitimu Volcanics, New Zealand
(+Chl,Qz,Sph)

Minerals \ Facies Zones	Zeolite	PP	PrA
	Zone 1	Zone 2	
Laumontite	———————		
Epistilbite	—————		
Heulandite	——— —		
Stilbite	— — — — —		
Chabazite	— —		
Analcime	— — — —		
Prehnite	— ———————————		
Pumpellyite	— ———————————		
Epidote	— — — — ———————————		
Albite	— — — — — ———————		
Actinolite	— — — — ———————		

Figure 3.23 Distribution of some metamorphic minerals in the Takitimu volcanics of New Zealand (after Houghton, 1982, Figure 3).

Differences in water/rock and P_{fluid}/P_{total} ratios can account for such contrasts in mineral occurrence (Houghton, 1982). (ii) Both zones contain hematite, suggesting high f_{O_2} during metamorphism, which is consistent with the occurrence of andradite and widespread distribution of Fe^{3+}-rich epidote and prehnite. (iii) Prehnite is abundant in the lower zone volcanics, whereas elsewhere in New Zealand, including in the Brook Street Volcanics, prehnite has limited distribution. (iv) Occurrence of PrA and PP facies assemblages in porphyritic intrusive rocks suggests a higher metamorphic gradient than for a normal burial sequence and low CO_2 content of the metamorphic fluid. In fact, the assemblage Pr–Act–Ep–Pm (+ Ab + Qz + Chl) defines the discontinuous reaction bordering the PP and PrA facies for the model system (see Figure 3.8). The suggestion of higher geothermal gradient for the Takitimu volcanics is consistent with occurrence of yugawaralite and absence of PA facies rocks.

3.11.3.2 *Greentuff Formation of Japan.* The Miocene–Pliocene Greentuff Formation of Japan consists of a more than 3 km thick sequence of mafic to felsic lava, tuff, volcanic breccia and intercalated volcaniclastic sediments. Burial metamorphism of the Greentuff Formation was characterized by high activity of CO_2 in the ZEO to PP facies (Seki, 1973). Utada (1965) described three zeolite zones (clinoptilolite–mordenite, Am–Hu, Lm–Ep) at depths transitional from diagenetic to propylitic assemblages; the lack of PP facies assemblages is characteristic. Because these zeolites differ in composition, their occurrences are highly dependent on rock composition. Umegaki and Ogawa (1965) reported the formation of laumontite in tuffaceous sandstone and basaltic-andesitic tuff, and of both mordenite and clinoptilolite in interbedded rhyolitic tuff and tuff breccia. The zeolite depth zonation from mordenite (stilbite) through heulandite to laumontite and the absence of yugawaralite and wairakite is similar to that in the Taringatura sequence of New Zealand. Clay parageneses also vary with increasing depth from smectite through corrensite to chlorite + illite. The apparent lack of prehnite, pumpellyite, and actinolite in the assemblage may reflect high CO_2 content of metamorphic fluids.

3.11.4 *Thermal metamorphism*

Thermal metamorphism occurs whenever hot magmatic bodies are emplaced into cooler country rocks. Extent of thermal recrystallization depends on the nature and depth of the intrusive and the composition of country rocks. Because of thermal and chemical perturbations of the country rocks, well-zoned thermal aureoles ranging from very low- to high-grade are developed. Most of the previous petrological studies of thermal aureoles have emphasized higher-grade rocks (e.g. Kuniyoshi and Liou, 1976*b*). The two examples described below are specifically for very low-grade metamorphism around granitoid plutons.

3.11.4.1 *Tanzawa Mountain, Japan.* The Greentuff Formation at Tanzawa Mountain, Central Japan, has been thermally metamorphosed by the intrusion of quartz diorite (Seki *et al.*, 1969). Prograde metamorphic zones, developed concentrically to the pluton and listed in order outward from the pluton, were AM, GS, PP, laumontite–corrensite, and stilbite (clinoptilolite)–vermiculite zones. Subsequent retrograde metamorphism to ZEO and PP facies has affected all rocks, including the quartz diorite. The mineral assemblages of the Tanzawa Mountain significantly differ from the assemblages of burial metamorphism. Stilbite is of prime importance, in contrast to the common occurrence of heulandite in Taringatura, New Zealand. The occurrences of yugawaralite and wairakite in ZEO and PP facies rocks suggest that the Tanzawa metamorphism involved a higher metamorphic gradient than those in New Zealand (Figure 3.18). Moreover, the recently discovered occurrence of PrA facies assemblages in the transitional zones between the PP and GS zones and the lack of the PA facies further support this suggestion.

3.11.4.2 *Karmutsen Volcanics, Vancouver Island, B.C.* A thick (6000 m) pile of Upper Triassic Karmutsen metabasites from NE Vancouver Island, B.C. has been considered to be a typical example for 'burial metamorphism' with pronounced vertical zonal distributions of secondary minerals (e.g. Kuniyoshi and Liou, 1976a). However, recent detailed study of parageneses and compatibilities of secondary minerals and their zonal distribution with respect to the Coast Range Batholith suggests that these metabasites suffered thermal metamorphism by the intrusives. The progressive mineral assemblages range from the ZEO, PP, and PrA facies through GS and amphibolite to pyroxene hornfels facies toward the intrusive contact. Parageneses of very low-grade Karmutsen volcanics are shown in Figure 3.24 (Cho *et al.*, 1986; Cho and Liou, 1987). Compositions and parageneses of minerals are used to delineate the transition equilibria for ZEO to PP facies and for PP to GS facies. They are summarized along a metamorphic gradient in Figure 3.22 for comparison with that in New Zealand.

Characteristic mineral assemblages (Chl + Qz + Ab) include: (i) Lm + Pm + Ep for the ZEO facies; (ii) Pr + Pm + Ep, Pr + Pm + Cc, and Pm + Ep + Cc for the PP facies; (iii) Pr + Ep + Act + Cc for the PrA facies; and (iv) Act + Ep + Cc for the GS facies. Na-rich wairakite coexists with epidote or Al-rich pumpellyite in some PP facies samples. On the basis of the P–T grid for the model basaltic system described in the previous section, the assemblages and compositions of Ca–Al silicates are chemographically and theoretically interpreted. The results indicate that variations in the assemblages and mineral compositions are controlled by a sequence of continuous and discontinuous reactions. The transition from the ZEO to PP facies is defined by a discontinuous reaction (1): Lm + Pm + Qz = Pr + Ep + Chl + H_2O occurring at $P = 1.1 \pm 0.5$ kbar and $T = 190 \pm 20\,°C$; from the PP to PrA facies by a discontinuous reaction (2): Pm + Qz + CO_2 = Pr + Ep + Cc + Chl + H_2O; and from the PrA to GS by a continuous

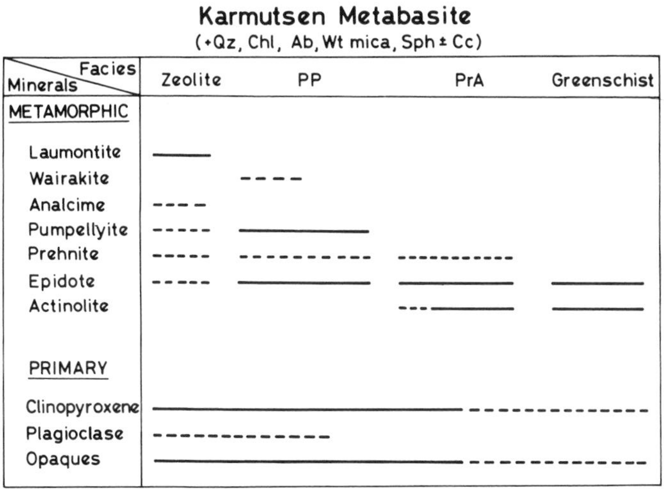

Karmutsen Metabasite
(+Qz, Chl, Ab, Wt mica, Sph ± Cc)

Minerals \ Facies	Zeolite	PP	PrA	Greenschist
METAMORPHIC				
Laumontite	——			
Wairakite		– – – –		
Analcime	– – – –			
Pumpellyite	– – – – –	——————		
Prehnite	– – – – –	– – – – – – – – –	– – – – – – – –	
Epidote	– – – – –	——————	——————	——————
Actinolite			– – – ——————	——————
PRIMARY				
Clinopyroxene	————————————		– – – – – – – –	– – – –
Plagioclase	– – – – – – – – – – – –			
Opaques	———————————		– – – – – – –	– – – –

Figure 3.24 Parageneses of secondary minerals in the Karmutsen metavolcanics, Vancouver Island, B.C. (data from Kuniyoshi and Liou, 1976*b*; Cho *et al.*, 1986; Cho and Liou, 1987).

reaction (3): $Cc + Chl + Qz = Ep + Act + H_2O + CO_2$. Thus, the first appearance of the GS facies assemblages governed by reaction (3) is highly dependent on both X_{Fe} and X_{H_2O}, compared to the laumontite- and pumpellyite-out reactions (1) and (2). The calculated metamorphic gradient in Figure 3.22 is higher than that for burial metamorphism in New Zealand. This together with the distribution of metamorphic zones suggests thermal metamorphism.

3.11.5 *Subduction zone metamorphism*

Several lines of evidence suggest that blueschists and related rocks are generated in high P/T tectonic environments and that blueschist terranes are associated exclusively with subducted lithologic slabs or related tectonic loading. Blueschists and related metamorphics occur in elongate, narrow zones where they are associated with rocks of oceanic affinities such as ophiolites and abyssal sea sediments. These belts are characteristically developed around the margins of the Pacific Ocean and also typify early stages in the Alpine (Tethyan) recrystallization. Many workers regard such metamorphic suites as subduction zone complexes that mark former lithospheric plate junctions. Early studies on BS facies metamorphism were confined mainly to Circum-Pacific and Himalayan-Alpine belts (see Ernst, 1975 for summary), and the Ural Mts., Russia (see Dobretsov and Sobolev, 1984, and Sobolev *et al.*, 1986, for summary). However, many new blueschist localities have recently been discovered; these include the Appalachian of the eastern USA (e.g. Laird and Albee, 1981), Seward Peninsula, Alaska (e.g. Forbes *et al.*, 1984), China (e.g. Zhang *et al.*, 1984), Oman (Lippard, 1985), Anglesey Island,

UK (Gibbons and Gyopari, 1986), Spitzbergen (Ohta *et al.*, 1986) and others. In spite of their ubiquitous occurrence in major suture zones, tectonic models for the genesis and emplacement of blueschists remain topics of controversy. Mineral parageneses and *P–T* constraints of blueschists and related rocks in two classic regions—the Franciscan Complex of California and the San-bagawa belt of Japan—are briefly described below to exemplify subduction-zone metamorphism.

3.11.5.1 *Franciscan Complex, California.* The Franciscan Complex of California has long been considered to be the type example of an accretionary complex formed by simple subduction, which includes coherent units of greywacke and metagreywacke that are separated by zones of melange (e.g. Hamilton, 1978). However, recent detailed lithological, palaeomagnetic, and other studies suggest that the Franciscan may be best explained by juxtaposition of many far-travelled terranes from different source areas (e.g. Blake and Jones, 1981). The Franciscan terranes include subduction complexes, frag-

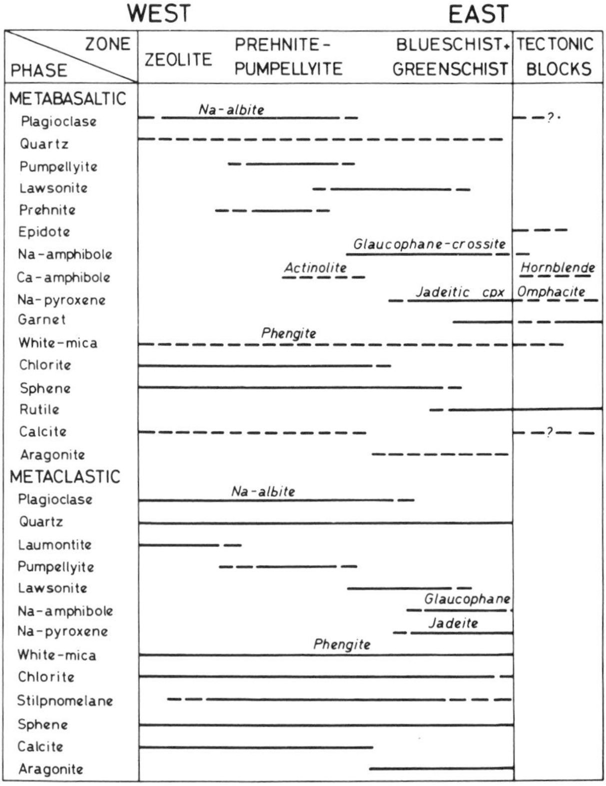

Figure 3.25 Parageneses of secondary minerals for Franciscan metabasaltic and metaclastic rocks in northern California (after Ernst, 1984).

ments of ocean floor or ocean islands, and greywacke-terranes of uncertain origin. A summary of metamorphism of two major lithologic units, grey-wackes and mafic volcanics, are shown in Figure 3.25.

Metamorphism of greywackes varies from west to east but not in an overall gradational fashion. Rocks of the westernmost Coastal belt contain patches and veins of laumontite. The chaotic greywacke + shale units of the Central belt are characterized by the occurrence of pumpellyite, but some melange blocks are of higher grade. The eastern Franciscan belt of coherent clastic units carries lawsonite + aragonite, and in the South Fork Mountain Schist, epidote appears at the expense of lawsonite. However, a systematic study of the most feebly recrystallized lawsonite-free meta-greywackes and meta-volcanics of the Type I rocks of Coleman and Lee (1963) in the Franciscan Complex has not been done.

In this section, three blueschist metamorphic zones in the Cazadero area are described (Maruyama and Liou, 1987). With increasing grade, they are lawsonite, pumpellyite, and epidote zones; their mineralogical variations are shown in Figure 3.26. Both the lawsonite and pumpellyite zones are equivalent to the Type II non-foliated metabasites of Coleman and Lee (1963), and the epidote zone includes Type III glaucophane schists. All metabasites contain minor amounts of quartz, albite, chlorite, aragonite, and sphene. The lawsonite zone metabasites are non-foliated, exhibit well-developed pillow structures, and contain relict pyroxenes. Both pillow cores and rims contain Fe-rich lawsonite

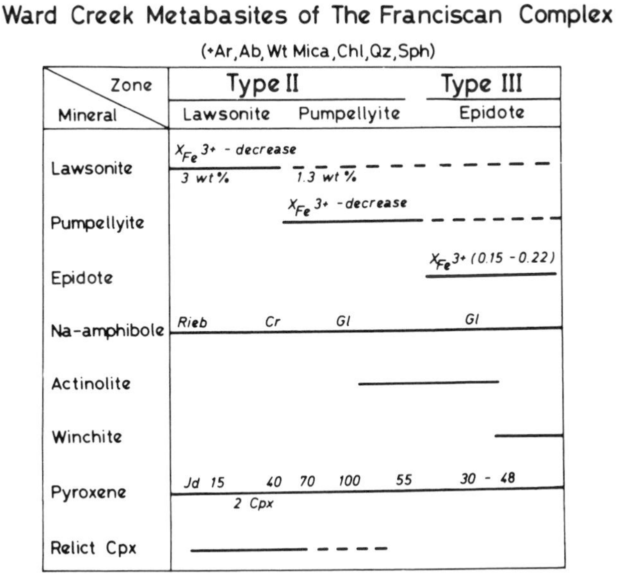

Ward Creek Metabasites of The Franciscan Complex

(+Ar,Ab,Wt Mica,Chl,Qz,Sph)

Zone / Mineral	Type II		Type III
	Lawsonite	Pumpellyite	Epidote
Lawsonite	X_{Fe}^{3+} - decrease 3 wt%	1.3 wt%	
Pumpellyite		X_{Fe}^{3+} - decrease	
Epidote			X_{Fe}^{3+} (0.15 - 0.22)
Na-amphibole	Rieb	Cr Gl	Gl
Actinolite			
Winchite			
Pyroxene	Jd 15 40 70 100 55		30 - 48
	2 Cpx		
Relict Cpx			

Figure 3.26 Parageneses of secondary minerals for the Ward Creek metabasites of the Franciscan Complex (after Maruyama and Liou, in prep.).

(up to 3 wt% Fe_2O_3), crossitic to riebeckitic amphibole, and two metamorphic clinopyroxenes. The pumpellyite zone metabasites are characterized by the appearance of pumpellyite at the expense of lawsonite and clinopyroxene, and by the assemblage Pm + Lw + crossite + jadeitic pyroxene (+ Ab + Qz + Chl + Sph). In the higher pumpellyite zone, minor actinolite appears. The strongly foliated epidote zone metabasites are characterized by the occurrence of (i) epidote together with pumpellyite and lawsonite, (ii) two amphiboles together with omphacite, and (iii) minor garnet. In the highest epidote zone rocks, winchite occurs.

Three discontinuous reactions were delineated for (i) pumpellyite-in, (ii) actinolite-in, and (iii) epidote-in, reactions, and their $P-T$ conditions were

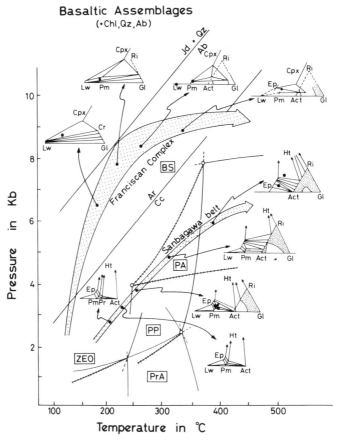

Figure 3.27 $P-T$ diagram showing parageneses of minerals along estimated metamorphic gradients for (1) the Sanbagawa belt, Japan; and (2) the Cazadero metabasites of the Franciscan Complex in California. The observed assemblages are plotted on to the chlorite-projection diagrams of Figures 3.2 and 3.3.

estimated using the isotopic temperatures of Taylor and Coleman (1968), and the Ca–Na pyroxene and glaucophane geobarometers described in the previous sections. The fossil geotherm recorded in these metabasites shown in Figure 3.27, indicates that an inflection point occurs in the P–T path for metamorphic recrystallization. Below the P and T of the inflection point, metamorphism proceeded with predominantly a pressure-increase from 4 to 7 kbar at nearly constant T of about 150–200 °C. For higher grade recrystallization above the inflection point, metamorphic temperature ranged from 200–350 °C at nearly constant P (7–8 kbar). Metamorphic gradient with similar inflection point also has been suggested for blueschists in New Caledonia (Yokoyama *et al.*, 1986), the Tauern Window (Selverstone and Spear, 1985), and the Sanbagawa belt described below.

3.11.5.2 *Sanbagawa Belt, Japan.* The Sanbagawa metamorphic belt consists of a thick sequence of schists, phyllites and associated meta-mafics and meta-ultramafics that extends over 1000 km in southwestern Japan. It has long been considered a once-subducted accretionary package which was uplifted by ridge-subduction in the Late Cretaceous (Uyeda and Miyashiro, 1974). A comparative study of the Sanbagawa belt and the Franciscan Complex by Ernst *et al.* (1970) concluded that the Sanbagawa belt was metamorphosed at relatively lower pressure and at temperatures equivalent to or slightly higher than those for the Franciscan Complex. Subsequent work supports their conclusion (e.g. Banno, 1986).

Progressive metamorphism of Sanbagawa basaltic rocks is characterized by a facies series ranging from the PP, through PA, GS to epidote amphibolite facies. Lawsonite sporadically occurs in low-grade metapelites and sodic amphiboles appear in Fe^{3+}-rich metabasites of both the PA and GS facies. Aragonite has not been found. The parageneses are not identical throughout the belt, but a general progression from fine-grained greenstones through interlayered greenschists and blueschists toward more coarsely crystalline epidote + garnet-amphibolites and clinopyroxene-bearing amphibolites is apparent. The metamorphic and structural discontinuities so typical of the Franciscan terrane are uncommon in the Sanbagawa belt. Most of the feebly recrystallized units are massive rather than schistose, so they are conveniently referred to as green-stones.

Zeolite facies metabasites in Central Kii Peninsula (Seki *et al.*, 1971) consist of laumontite with smectite and corrensite. The PP facies assemblages are widespread in areas south of the boundary between the Sanbagawa and Chichibu belts. Maruyama and Liou (1985) described the occurrence of the assemblage Ca–Na pyroxene ($X_{Jd} = 0.20$) + Pm + Chl at the transition between the PP and PA facies.

The PA facies metabasites of the Sanbagawa belt have received the most detailed study (e.g. Nakajima *et al.*, 1977; Nakajima, 1983). Common metabasites contain assemblages of Pm + Act + Ep (+ Ab + Chl + Qz + Sph), whereas some Fe^{3+}-rich metabasites have additional crossite, riebeckite, or riebeckitic actinolite together with haematite. Three subzones are delineated and separated by two discontinuous reactions, (1) Ht + Act + Chl = Ri + Ep + F at higher T and (2) Pm + Ht = Ep + Chl + Act + F at lower T. Using the epidote geothermometer and the sodic amphibole geobarometer described in the previous sections, the PA facies metabasites of the Sanbagawa belt are estimated to have recrystallized at 250–330 °C and pressures of 3 to 5.2 kbar. Their parageneses are summarized along a metamorphic gradient in Figure 3.27 for comparison with those of the Franciscan Complex.

3.12 Conclusions

Very low-grade metamorphism is a ubiquitous process in the upper portion of the Earth's crust. Depending on the tectonic setting and physico-chemical conditions, various types of metamorphism are hence produced. Incipiently recrystallized metabasites possess similar characteristics in textures and in the extent of recrystallization. They contain mineral assemblages consisting of combinations of Ca–Al silicates including various Ca–zeolites, prehnite, pumpellyite, epidote, lawsonite, sodic amphibole, and pyroxene, and actinolite together with chlorite, albite, and quartz. Repetition of depth zonations of mineral assemblages in time and space strongly suggests that the changes in composition and parageneses of minerals in various tectonic settings are fairly systematic, and a number of metamorphic facies can be defined in terms of their $P-T$ conditions. Each index assemblage has a definite position in the $P-T$ field; and accordingly, any change in the basaltic assemblage indicates a change in P, T, and/or chemical potential of the mobile components in the fluid phase. Such successive and systematic changes in mineral assemblages along a certain metamorphic gradient can be expressed by a sequence of continuous and/or discontinuous reactions. Experimental studies of these reactions combined with chemographically deduced relations provide a petrogenetic grid applicable to natural parageneses and delineate the physical conditions for numerous very low-grade mineral facies in low μCO_2 environments. Figures 3.18, 3.22 and 3.27 summarize the parageneses and compositions of secondary minerals along metamorphic gradients suggested for ocean-floor, hydrothermal, thermal, burial, and subduction-zone metamorphism. Observed assemblages in ZEO, PP, PrA, PP, BS and GS facies metabasites from several classic terranes are presented.

Sliding equilibria involving phases with continuous changes in composition are common for very low-grade metamorphic reactions. Many facies boundaries are defined by such continuous reactions in which both compositions and proportions of the reacting phases vary in $P-T$ space. To apply the proposed petrogenetic grid such as Figures 3.22 and 3.27 and to discuss compositional trends with increasing metamorphic grade, the following steps are necessary:

(i) to identify minerals of low-variance assemblages in a certain metamorphic domain; (ii) to analyse compositions of these phases and determine their equilibrium compositions; (iii) to present both compositions and parageneses of Ca–Al silicates in ACF or its modified projections; (iv) to delineate both continuous and discontinuous reactions for progressive metamorphic sequences; (v) to apply experimentally determined or thermodynamically calculated reactions to determine the $P-T-X$ relations; and (vi) to use at least two independent thermometry and barometry methods for $P-T$ estimates.

It is obviously one of the challenging tasks of petrologists to delineate the metamorphic facies boundaries and to correlate these data with natural parageneses. Many data have been accumulated and integrated interrelating the experimental and natural stability and compositional relations of the index mineral assemblages. A quantitative petrogenetic grid is suggested. Further study of sliding equilibria involving carbonates in very low-grade metamorphosed rocks is needed.

Acknowledgements

The research described in this paper has been supported by National Science Foundation EAR 82-04298 and 85-07988. This paper has had critical reviews by Drs W.G. Ernst, M. Frey, Mary Keskinen, Peter Schifflman, and Heather Ponader, whom we thank for their help.

4 Organic material and very low-grade metamorphism

MARLIES TEICHMÜLLER

4.1 Introduction

Organic material in sediments reacts very sensitively to rise of rock temperature and pressure. Since the degree of coalification (maturation), called *rank*, is an irreversible process, organic matter plays an important role in estimations of the degree of rock diagenesis and metamorphism.

This chapter describes the different types of organic matter occurring in rocks and their different coalification tracks. Besides the various causes of coalification (temperature, time and pressure) the coalification process in general is treated with its different changes at different rank stages. Modern rank parameters, especially optical (microscopic) methods which can be applied to finely-dispersed organic material in rocks other than coal, are described. Anthracitization and graphitization with their different stages, rank parameters and geological causes are treated separately because they correspond to very low- and low-grade metamorphism of clastic rocks.

4.2 Types of organic matter in sedimentary rocks

Organic matter is understood here as substances which contain organic carbon. A rock consisting almost entirely of organic matter is coal occurring in more or less thick layers or seams. Clastic rocks, mainly those of clayey and silty composition, often contain coaly inclusions, ranging in size from megascopically visible bands or lenses to finely-dispersed particles visible only under the microscope.

Coal is not homogeneous. After Stopes (1935), the various microscopic constituents of coal are called *macerals*. Macerals are comparable with minerals in inorganic rocks. A maceral is defined as 'a microscopically recognizable individual constituent which does not contain any mineral substances resolvable with the microscope' (*International Handbook of Coal Petrography*, 1971, 1985). Macerals are distinguishable in polished sections under incident light and oil immersion, mainly on the basis of their reflectance, fluorescence and morphography. An internationally acknowledged classification of macerals, the *Stopes–Heerlen System* has been produced by the International Committee of Coal Petrology, which has edited the *International Handbook of Coal Petrography* in three editions with supplements

(1963, 1971 and 1975) in English, French, German and Russian. New macerals have since been detected, mainly by means of the fluorescence mode. *Textbook of Coal Petrology* (Stach *et al.*, 1982) gives the most recent and comprehensive information about the macerals and their origin.

Table 4.1 shows the maceral groups and the macerals which are identifiable in polished sections of low-rank coals under incident white and short-wavelength light. The huminite/vitrinite group[†] is characterized by its origin, mainly from lignin and cellulose, its relatively high oxygen content and an intermediate reflectance. Liptinites derive from hydrogen-rich plant materials such as sporopollenin, cutin, suberin, resin, waxes and plant oils. They are distinguished by the lowest reflectance and by a relatively strong fluorescence when irradiated with ultraviolet, violet or blue light. Inertinites are characterized by a high reflectance and a high carbon content, caused, *inter alia*, by charring, mouldering and fungal attack before or during the peat stage.

Macerals change with increasing degree of coalification (rank), becoming more and more alike, mainly due to a different rate of reflectance increase

Table 4.1 Maceral groups and macerals

Group	Maceral	Characteristics
Vitrinite	Telinite	Cell walls
	Collinite {telo- / detro-	Amorphous (gel or gelified tissue/detritus)
	Corpocollinite	Cell fillings
	Vitrodetrinite	Detritus
Liptinite	Sporinite	Spores, pollen
	Cutinite	Cuticles
	Suberinite	Suberinized cell walls (cork)
	Fluorinite	Plant oils
	Resinite	Resins, waxes, latex
	Alginite	Algae
	Bituminite	Amorphous (bacterial, algal, faunal)
	Chlorophyllinite	Chlorophyll
	Liptodetrinite	Detritus
	Exudatinite	Secondary exudates
Inertinite*	Fusinite	Cell walls (charred, oxidized)
	Semifusinite	Cell walls (partly charred)
	Sclerotinite	Fungal cell walls
	Macrinite	Amorphous gel (oxidized, metabolic)
	Inertodetrinite	Detritus
	Micrinite	Secondary relics of oil generation (mainly)

*A small part of inertinite derives from melanin-rich plant and animal material ('primary inertinite'). A greater part attains its inertinitic properties only during the early coalification process ('rank inertinite').

[†]Huminites are the precursors of vitrinites in the peat and lignite stages.

during coalification (cf. Figures 4.11, 4.12). Figure 4.1 shows photomicrographs of three coals at different stages of coalification. In the stage of meta-anthracites (corresponding to anchimetamorphism of rocks) the microscopic picture appears to be almost homogeneous (see section 4.4.1.2).

In clastic rocks, macerals are sometimes called *phytoclasts* (Bostick, 1979) or *organoclasts* (Alpern, 1980).

Oil source rocks usually contain the same or similar macerals as coal seams, but with a different proportion of the individual macerals. Whereas in most coal layers vitrinite is the dominant maceral group, the liptinite group usually predominates in oil source rocks in which the macerals alginite and bituminite are the most important, because they are oil-prone.

At low rank stages, in normal reflected light, the macerals of the vitrinite and inertinite groups are easily recognizable in polished sections of rocks other than coal, whereas most macerals of the liptinite group are hard to distinguish from the mineral matrix because both are highly translucent and thus do not reflect the light. Only bituminites may be easily distinguishable, especially under crossed nicols, when they appear black in contrast to the lighter mineral matrix. The other liptinites become easily visible under irradiation with short-wavelength light, due to their fluorescence. Therefore, fluorescence microscopy has gained a great importance for the microscopic study of oil source rocks. Figure 4.2 shows an example of the appearance of low-rank macerals in clastic rocks under white and blue incident light. At higher rank stages ($> 0.5\%$ R_m vitrinite reflectance) liptinites become better distinguishable from the translucent mineral matrix, due to increase in reflectance. Some liptinites

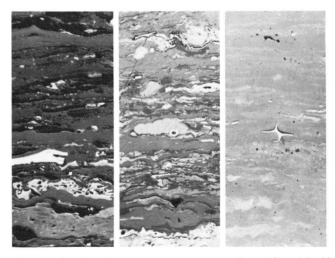

Figure 4.1   Increase of rank as shown under the microscope. From left to right: high volatile bituminous coal (38% volatile matter of vitrite), medium volatile bituminous coal (28% volatile matter of vitrite), anthracite (7% volatile matter of vitrite). Vitrinite grey, inertinites white, liptinites dark-grey. Polished sections, oil immersion, 200 × .

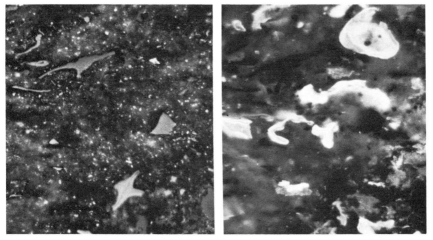

Figure 4.2 Carboniferous claystone with finely dispersed organic matter; (left) in normal white light with easily visible particles of vitrinite and inertinite; (right) the same field under blue light irradiation with easily visible liptinites (sporinite). Polished section, oil immersion, 500 ×.

disappear in overmature (see section 4.3.1) source rocks as a result of complete volatilization, others change to vitrinite and/or to inertinite (especially granular micrinite) as the solid, carbon-rich relics of oil generation and due to a disproportional reaction.

The term *kerogen* is often used for organic substances in hydrocarbon source rocks and it is sometimes mistaken for the term 'maceral'. But it must be borne in mind that 'kerogen' is a chemical term, introduced by organic geochemists, for organic matter which is not soluble in organic solvents. Of the macerals, only some bituminites and some exsudatinites are highly soluble. It is only for this reason that most macerals belong to the 'kerogen' of chemists. On the other hand, some kerogen in oil source rocks is of submicroscopic size (p. 134) and, therefore, does not belong to the macerals which are defined as microscopically recognizable constituents (see above). This submicroscopic organic matter is very finely distributed between the minerals, mainly adsorbed on or incorporated with clay minerals from which it can be separated by treatment of the rock with hydrochloric and hydrofluoric acids. The part of the submicroscopic organic matter in source rocks which is non-soluble in organic solvents appears in strew slides of isolated kerogen as part of the *amorphous fraction*.

For practical purposes most oil geologists study strew slides of the isolated total organic matter, either after concentration by froth flotation (Robert, 1974) or after treatment with HCl and HF. Others study the isolated kerogen after treatment of the rock not only with acids, but also with organic solvents. The following constituents are distinguished in these various concentrates in transmitted light (Hunt, 1979):

Woody   Corresponding to vitrinite
Coaly   Corresponding to inertinite
Herbaceous   Corresponding to liptinite, except alginite and bituminite
Algal   Corresponding to alginite
Amorphous   Corresponding to bituminite and the part of organic matter
             which remains submicroscopic in the untreated rock.

It should be noted here that, unlike the organic petrologist, the organic geochemist distinguishes in source rocks four types of organic matter which display different *evolution paths* in a diagram plotting the H/C against the O/C atomic ratios (Durand, 1980; cf. Figure 5.15, this volume). Type 1 is characterized by the highest hydrogen content due to a high proportion of algal, zooplanktonic and bacterial matter. Prototype is the Tertiary Green River Shale of the Uinta Basin, USA. Type 2 with a medium H-content represents a mixture of type 1 and type 3 organic matter. Prototypes are the Jurassic source rock of the Posidonia Shale in Germany and the Toarcien of France, respectively. Type 3, with a relatively low hydrogen content, is distinguished by a high amount of organic matter from terrestrial plants, similar to the parent matter of coals. The prototype is the Cretaceous source rock of the Douala Basin, Cameroun. Type 4 contains C-rich material as inertinites and reworked vitrinites. The oil yield decreases from type 1 to type 4.

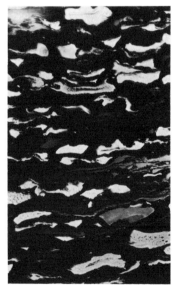

Figure 4.3 Photomicrographs of high-volatile bituminous coal in a polished thin section; (left) in reflected light (sporinite dark-grey, inertinites white); (right) in transmitted light (sporinite white, inertinites and part of vitrinite black, and—in contrast to the reflected light picture—not distinguishable from each other). Polished thin section, oil immersion, 500 × .

In former times, up to the fifties of this century, coal was often studied in thin sections, especially by North American coal petrologists. In low-rank bituminous coals, vitrinites are characterized by a red colour in transmitted light, whereas liptinites appear yellow and all inertinites are black (opaque). Therefore, the latter are no longer distinguishable from each other in transmitted light (Figure 4.3). Low-rank coals, up to the stage of high volatile bituminous A coals and the boundary to the medium-volatile bituminous coal stage ($1.2\% R_m$ vitrinite reflectance) become translucent at a thickness of about 15 μm. Higher-rank coals need lower thicknesses, down to 5 μm. As the absorption index increases with increasing rank, translucency finally becomes impossible, at least in practice. Therefore, organic matter in anchimetamorphic rocks can be studied only in reflected light.

It is important to note here, that, as far as coaly inclusions in clastic rocks are concerned, the reflected light method must be used also at lower rank stages because the normal thickness of a rock thin section (20 μm) is too high to reach translucency of most coaly matter. Thus, vitrinites remain opaque and are no longer distinguishable from inertinites and opaque minerals such as pyrite. Only high translucent liptinites like algae may become recognizable in rock thin sections, if they are not underlain by opaque organic or inorganic constituents. In thin sections of oil shales, the submicroscopic organic material usually causes a brownish colour of the clay matrix.

The feature of superimposition of macerals even in thin sections of pure coal has led to different results for micropetrographic analyses and to different nomenclatures for coal constituents in thin sections (Thiessen–Bureau of Mines System) as compared with macerals in polished sections (Stopes–Heerlen System). A correlation of both nomenclatures became possible through a comparative study of polished thin sections in transmitted and in reflected light (Teichmüller, 1954). Figure 4.3 shows photomicrographs of a polished thin section from a high volatile bituminous coal in transmitted and reflected light.

4.3 Coalification

4.3.1 *The process in general*

The term 'coalification' designates the development of organic matter from the peat stage through the stages of lignite, sub-bituminous coal, bituminous coal, anthracite and meta-anthracite to the stage of semi-graphite (Table 4.2). This development is characterized, *inter alia*, by an increase of the C-content and of vitrinite reflectance as well as by a decrease of volatile matter released during pyrolysis.

From the beginning of the sub-bituminous coal stage the alteration of organic matter is so severe that it can be compared with the metamorphism of inorganic rocks, although the temperature–pressure conditions at this rank

E

Table 4.2 Coalification stages according to the German (DIN) and North American (ASTM) classifications, their distinction by different physical and chemical rank parameters and the applicability of the parameters at the different rank stages.

Rank (German)	Rank (USA)	Refl. Rm$_{oil}$	Vol. M. d.a.f. % Vitrite	Carbon d.a.f.	Bed Moisture	Cal. Value Btu/lb (kcal/kg)	Microscopic Characteristics	Applicability of Different Rank Parameters
Torf	Peat	0.2	68 / 64	ca. 60	ca. 75		free cellulose, details of initial plant material often recognizable, large pores	
Weich-	Lignite	0.3	60 / 56				no free cellulose, plant structures still recognisable, cell cavities frequently empty formation of rank inertinite	
Matt-	Sub-Bit. C / B	0.4	52 / 48	ca. 71	ca. 35 / ca. 25	7200 (4000) / 9900 (5500)	geochemical gelification and compaction takes place, vitrinite is formed, formation of exudatinite 1st coalification jump of liptinites	
Glanz-	C / A	0.5 / 0.6	44	ca. 77	ca. 8–10	12600 (7000)	formation of micrinite	
Flamm-	B	0.7 / 0.8	40					
Gasflamm-	A		36				2nd coalification jump of liptinites rapid rise of red/green quotient of sporinite fluorescence	
Gas-		1.0	32					
	Medium Volatile Bituminous	1.2 / 1.4	28 / 24	ca. 87		15500 (8650)	beginning of 3rd coalification jump, rapid rise of liptinite reflectance	
Fett-	Low Volatile Bituminous	1.6	20				Rm sporinite = Rm vitrinite	
Ess-		1.8	16					
Mager-	Semi-Anthracite	2.0	12				Rmax liptinite > Rmax vitrinite	
Anthrazit	Anthracite	3.0	8	ca. 91		15500 (8650)	Rmax liptinite > Rmax inertinite	
Meta-Anthr.	Meta-A.	4.0	4				Rmax vitrinite > Rmax inertinite	

stage produce only weak diagenetic changes in the associated rocks. The reason is that organic matter reacts much more sensitively to rise of temperature and pressure than do minerals. Therefore, the term 'metamorphism of coal' which is often used, particularly in the English language, does not even correspond with very low-grade metamorphism of rocks.

The rock petrologist should note that many organic geochemists use the following terms to designate the main stages of hydrocarbon maturity (Durand, 1980):

Term	Maturity	Vitrinite reflectance
Diagenesis	Immature	< 0.5–0.6%
Catagenesis	Mature	0.5–0.6% to 2.0–2.3%
Metagenesis	Overmature	> 2.0–2.3%

In contrast to mineral transformations during diagenesis and metamorphism, coalification is irreversible and not dependent on factors such as ion concentrations, pH, Eh and partial pressure of water.

Although many authors have shown relationships between degree of coalification and rock diagenesis and metamorphism (e.g. Kisch, 1974, 1983; Héroux et al., 1979; Teichmüller et al., 1979; Frey et al., 1980), many exceptions to the rules exist. For instance, in areas of strong but short heating, as in the roofs of igneous intrusive bodies, coalification may proceed in advance of mineral transformations (Stadler and Teichmüller, 1971; Wolf, 1975; Teichmüller et al., 1979; see also Figure 7.5, this volume).

Coalification comprises chemical and physical changes. The chemical alterations are caused mainly by the increase of rock temperature and—to a minor degree—by the duration of heating. Physical changes are caused mainly by lithostatic and tectonic pressure (see section 4.3.3).

The most striking petrographic alteration during the coalification process is the diagenetic gelification which takes place at the stages of sub-bituminous coals: the coal loses its former brown colour and its soft and dull appearance and is converted to a black, hard, and lustrous gel-like product. The porous huminites seen under the microscope are converted to dense vitrinites. This process seems to be mainly a colloidal one, initiated by the formation of bitumen which serves as a solvator (see below).

Products of the coalification process are not only the solid coals at their different rank stages but also fluids and gases. It is well known that during the early stages of coalification mainly water and CO_2 are given off (according to van Heek et al., 1971, at coalification temperatures between 30–100 °C), whereas at later stages (corresponding to temperatures up to 200 °C) varying quantities of methane are released. Recent geochemical and fluorescence microscopical studies have shown that this heavy loss of methane is preceded by the generation of $\pm$ fluid bitumen (Brooks and Smith, 1967; Leythaeuser and Welte, 1969; Teichmüller, 1974a, b). This process is called *bituminization*, and is part of the coalification process in the range of sub-bituminous and high-volatile bituminous coals. Bituminization obviously causes not only the diagenetic gelification but is also responsible for the coking properties of bituminous coals (Teichmüller, 1974a, b, 1982/84; Ottenjann et al., 1982; Lin et al., 1986). As a matter of fact, Rouzaud (1984) observed liquid-like droplets of bitumen in coking coals under the high resolution transmission electron microscope. This method (TEM), combined with electron diffraction measurements, appears to become very important for the further clarification of the coalification process: according to Rouzaud and Oberlin (1983) the basic structural units of coal are aromatic layers of less than 10 Å size. Their local orientation to parallel oriented layers and their aggregation to *clusters* varying in size from 50 Å to 100 Å occur just before the release of liquid hydrocarbons ends.

4.3.2   Rank stages and rank parameters

The chemical as well as the physico-structural changes during coalification are not at all uniform but vary in the different rank stages. This is the reason why different rank parameters have to be used to determine the different rank stages. Moreover, it must be considered that the various macerals behave differently during the coalification process. Figures 4.11 and 4.12 demonstrate the different increase of optical reflectance for the three maceral groups with increasing rank (decreasing volatile matter), and Figure 4.4 shows the different coalification tracks of the macerals sporinite (liptinite group), vitrinite, and macrinite/fusinite (inertinite group) in terms of the atomic ratios H/C:O/C. It was for these reasons that vitrinite analyses have been introduced instead of coal bulk analyses for comparative coalification studies (Teichmüller and Teichmüller, 1949), vitrite being the maceral association (microlithotype) which consists of at least 95% of the maceral vitrinite. Vitrinite is the predominant maceral in most coals, and changes relatively uniformly during coalification. Moreover, it is relatively easily isolated from bright coal layers.

4.3.2.1   *Chemical rank changes and parameters.* Rank changes of isolated huminites/vitrinites were studied on a rank series of some hundred coal

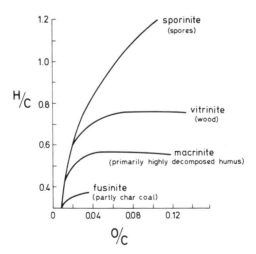

Figure 4.4   Chemical coalification paths (H/C:O/C) of various macerals (after van Krevelen, 1961).

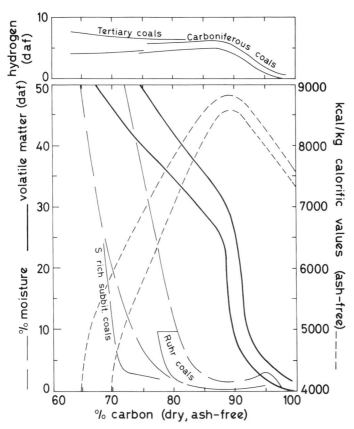

Figure 4.5 Increase in coalification of vitrite, based on various rank parameters, plotted against carbon (after Patteisky and Teichmüller, 1960).

samples ranging in rank from the lignite (*Braunkohle*) stage to the stage of semi-graphite on the basis of various chemical and physical rank parameters, such as carbon, oxygen and hydrogen content, volatile matter, moisture and calorific value (Patteisky and Teichmüller, 1960). The main results are shown in Figure 4.5. They indicate that moisture of vitrites (as mined) and— depending on the moisture content—the calorific value (moist, ash-free) are the best rank parameters for low-rank coals (60–87% C); i.e. for lignites, sub-bituminous and high-volatile bituminous C and B coals (*Weichbraunkohle, Mattbraunkohle, Glanzbraunkohle, Flammkohle, Gasflammkohle*, and *Gaskohle* according to the German coal classification, cf. Table 4.2). The strong decrease of moisture reflects the strong decrease of porosity at these low-rank stages.

Beginning at the stage of medium volatile bituminous coal with about 28% volatile matter (87% C) up to the anthracite stage with 8% volatile matter (91% C), the amount of volatile matter which escapes from the vitrite during pyrolysis up to 900 °C (proximate analysis) is a good rank parameter. The volatile matter decreases with increasing rank. It belongs to the aliphatic fraction of the vitrinite 'molecule' which consists of an aromatic nucleus surrounded by aliphatic molecular groups (Figures 4.6, 4.34). Thus, the decreasing volatile matter is an indirect measure for the increasing aromatization of vitrinite during coalification.

The anthracite stages are characterized by a rapid decrease of hydrogen from 4 to 0.8% H (dry, ash-free).

Figure 4.6 shows the changes in the chemical constitution of vitrinite. The hexagons in the uppermost row represent the aromatic rings and the lines are the aliphatic groups which, at higher rank stages, are gradually removed while the aromatic rings coalesce into larger clusters. Thus, coalification represents a rise of aromatization and of condensation. The rising orientation of the elementary molecular units parallel to the bedding plane is shown in the second row of Figure 4.6 (see also Figure 4.34).

4.3.2.2 *Optical rank changes and parameters.* The most important petrographic rank parameter is (i) the *optical reflectance* of vitrinite, which increases with rank (Figure 4.7). This method has the great advantage that particles of some microns in size can be measured. Therefore, rank determination became possible on coal particles from borehole cuttings and—most important to the rock petrologist—also on finely dispersed vitrinite inclusions in rocks other than coal.

Vitrinite reflectance depends on the refractive index and the absorption index of vitrinite and on the refractive index of the immersion oil according to the Beer equation:

$$R = \frac{(n - n_0)^2 = n^2 K^2}{(n + n_0) + n^2 K^2}$$

<table>
<tr><td>high volatile bituminous coal
(~35% vol.m., 82% C)</td><td>low volatile bituminous coal
(~20% vol.m., 89% C)</td><td>meta-anthracite
(~2% vol.m., 95% C)</td></tr>
</table>

Figure 4.6 Increase of aromatization and condensation, and ordering of the aromatic units parallel to the bedding plane during coalification of vitrinite (after Teichmüller and Teichmüller, 1984).

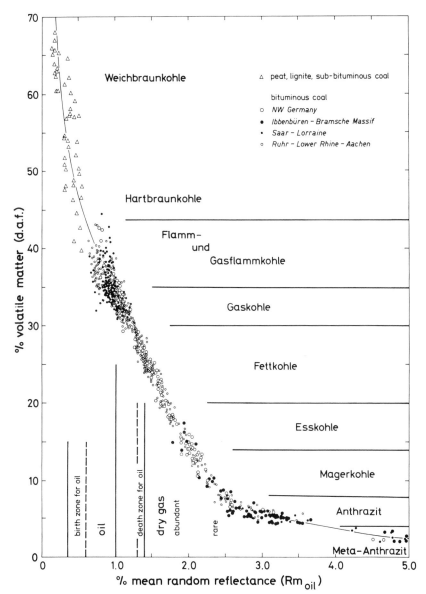

Figure 4.7 Relationship between volatile matter of vitrite concentrates and vitrinite reflectance during coalification from peat to anthracite stage, with relation to hydrocarbon occurrences.

where n and K are the refractive index and the absorption index respectively of vitrinite, and n_0 is the refractive index of the immersion oil.

The refractive index n is a function of atomic density and increases with degree of aromatization. The absorption index K depends on the number of delocalized electrons, since their mobility facilitates absorption of electro-

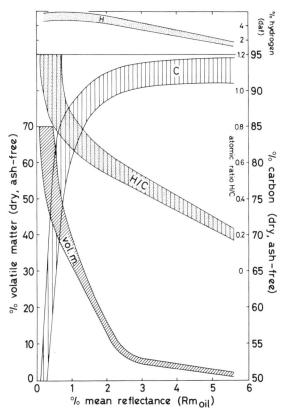

Figure 4.8 Relations between vitrinite reflectance and various chemical rank parameters (after Teichmüller and Teichmüller, 1979a).

magnetic waves. Electron mobility increases with the size of the aromatic lamellae and complexes, i.e. with the degree of condensation. It is for these reasons that a close relationship exists between increasing vitrinite reflectance and decreasing volatile matter in the stages of bituminous coals and anthracites (Figure 4.7). Figure 4.8 shows the relationships between the various chemical rank parameters and vitrinite reflectance.

Optically, vitrinite reflectance is the percentage of incident light that is reflected from vitrinite on a polished surface. It is commonly measured on spots of 2–3 μm diameter in monochromatic green light (546 nm) with oil immersion objectives. A photomultiplier is used to measure the reflectance intensity which is determined through comparison with standards of known reflectance (e.g. sapphire, diamond, glasses). The light source must be stabilized. The method is described in detail, *inter alia*, by Davis (1978) and by Stach *et al.* (1982). Its usage has been much facilitated by computer techniques.

The so-called mean or random reflectance (R_m) is measured in non-polarized light. Due to load pressure vitrinite commonly displays bireflectance, the degree of which rises with rank. Therefore, maximum (R_{max}) and minimum (R_{min}) reflectance are measured in polarized light (the polarizer in the incident beam set at 45° position) by rotating the microscope stage. If possible, 50–100 single measurements should be made to calculate an average value of R_m, R_{max} and R_{min}.

According to Ting (1978) the following relation exists:

$$\text{maximum reflectance} = 1.066 \times \text{random reflectance.}$$

According to Davis (1978) measurements of random (or mean) reflectance have the advantage that they are less time-consuming and allow very small particles ($< 5\,\mu$m) in rocks to be measured more exactly and more easily than determining the maximum reflectance. Moreover, the latter become very difficult and often worthless in case of optically biaxial vitrinites (cf. p. 128). After a thorough comparative study, including dispersed vitrinite in clastic rocks and biaxial optical properties, Hevia Rodriguez (1977) concluded that the mean random reflectance (and not the mean maximum reflectance) of vitrinite is the most appropriate optical rank parameter.

As shown in Figures 4.7, 4.10 and 4.29, the rate of increase of vitrinite reflectance depends on the rank range. It is low at low rank stages and increases in the stages of bituminous coals and anthracites, for which vitrinite reflectance is a reliable rank parameter.

(ii) *Optical anisotropy* of vitrinite is caused by the progressive adjustment of the aromatic nuclei into the bedding plane, due to load pressure (cf. Figures 4.6 and 4.34). In thin sections, this pressure-induced anisotropy is already clearly developed at the stage of sub-bituminous coals. Reflectance anisotropy (or bireflectance) becomes measurable at the stage of high volatile bituminous A coal (at about 1.0% R_{max}). It becomes noticeable at the stage of medium volatile bituminous coal (28% volatile matter, 1.3% R_m) and is more pronounced in the anthracite stages (Figure 4.9). It is defined as the difference (or as the ratio) between the maximum reflectance (R_{max}) and the minimum reflectance (R_{min}). According to Murchison (1978) the degree of bireflectance permits to estimate the degree of ordering of the molecular structure, especially of the alignment of the aromatic layers. This alignment is favoured through the bituminization process which corresponds to the *liquid phase* of bituminous coals according to Hirsch (1954) (cf. section 4.4.1.3). The liquid phase is comparable with the plastic phase during the coking process which can be regarded as an artificial coalification. In both cases, viscosity is the prerequisite for a good ordering of the aromatic lamellae.

According to Rouzaud and Oberlin (1983, p. 129) reflectance anisotropy is caused by a statistical molecular orientation which, for its part, is 'due to either flow stresses applied to a visco-elastic material or to shear stresses (due to lithostatic and tectonic pressure) applied to a porous brittle solid' in the anthracite stages (cf. Figure 4.32).

In tectonically undisturbed coal beds, vitrinite usually has optical uniaxial

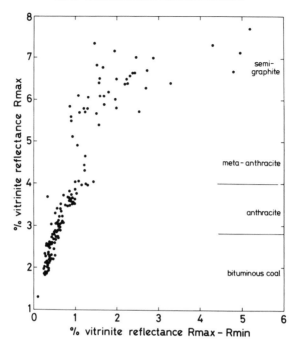

Figure 4.9 Relationship between maximum reflectance and bireflectance $(R_{max} - R_{min})$ of vitrinites from clastic rocks and coal seams in northwestern West Germany (after Teichmüller *et al.*, 1979).

negative properties (cf. Figure 4.24), and, therefore, R_{max} can be measured in all directions, whereas the real R_{min} value is measurable only in sections parallel to the optical axis, that is, in sections perpendicular to the bedding plane. Exceptions occur around hard inclusions in coal (e.g. pyrite globules, inertinites), where the normal pattern is disturbed, and biaxial anisotropy of vitrinite has developed due to locally transverse directed pressure. The same effect is very common in clastic rocks where the molecular structure of vitrinitic inclusions has been disturbed by local pressure from the relatively hard adjacent minerals which are pressed into the relatively soft vitrinite during rock compaction. In these cases the true R_{max} value of vitrinite can be measured only in one particular direction which is seldom encountered in a polished rock section. Therefore, R_{max} values measured in rocks other than coal commonly scatter widely. It is for these reasons that, especially at high rank stages (meta-anthracite, semi-graphite), one should consider only the 10–20% highest R_{max} values to calculate an average (Cook *et al.*, 1972). Clearly, the same holds for R_{min} determinations for which the 10–20% lowest recorded values should be averaged.

British and North American coal petrologists usually measure R_{max} at all rank stages, whereas in continental Europe R_m is usually measured in

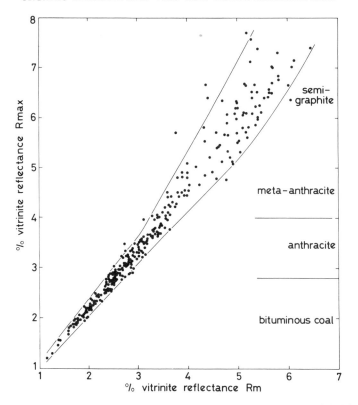

Figure 4.10   Relationship between maximum and mean (random) reflectance of vitrinites from West German coals (after Teichmüller and Teichmüller, 1981).

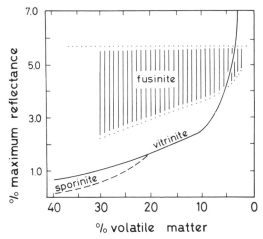

Figure 4.11   Development of maximum reflectance of the three maceral groups during coalification (decreasing volatile matter) for bituminous coals and anthracites (modified after Alpern and Lemos de Sousa, 1970; and Alpern, 1980).

bituminous coals and even in anthracites, because, at these stages, bi-reflectance is still relatively low (Figures 4.9 and 4.29) and the R_m measurements are less time-consuming. According to Ragot (1977) the following relation exists for vitrinites sectioned arbitrarily:

$$R_m = \frac{2R_{max} + R_{min}}{3}$$

Figure 4.10 shows the relationship between R_m and R_{max} values measured in coals and other rocks of NW Germany: up to about 3.5% R_m and 4.0% R_{max} (boundary between anthracite and meta-anthracite of the German scientific classification, cf. Table 4.2) the relationship is more or less linear and relatively close. Bustin (1984, p. 27) found good relationships even up to 6% R_{max}.

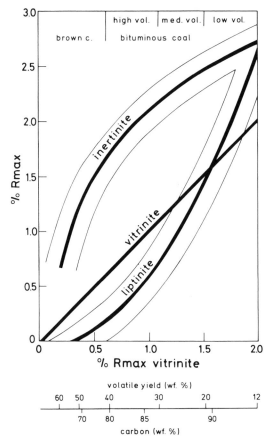

Figure 4.12 Coalification tracks in terms of maximum reflectance for the three maceral groups from the brown coal to the low volatile bituminous coal stage (modified after Smith and Cook, 1980).

(iii) The increase in *reflectance of liptinites and inertinites* with increasing rank differs considerably from the increase in vitrinite reflectance as demonstrated in Figures 4.11 and 4.12. Figure 4.11 shows that increase of liptinite reflectance is especially strong in the bituminous coal range between 1.0 and 1.6% vitrinite reflectance. At about 1.6% R_m the reflectance of sporinite becomes the same as the reflectance of vitrinite. In meta-anthracites, vitrinite reflectance surpasses the reflectance of inertinites (fusinites) at about 4–5.5% R_{max} (Alpern and Lemos de Sousa, 1970). According to Smith and Cook (1980) many inertinites increase their reflectance drastically in the range between brown coal and high volatile bituminous coal (0.2–0.9% R_{max} vitrinite), as is shown in Figure 4.12. This result implies the generation of liquids and/or gases already at these low rank stages and, therefore, is of great importance for hydrocarbon exploration (Smith *et al.*, 1984).

(iv) In sedimentary rocks devoid of vitrinite, such as most carbonate rocks and oil reservoirs, reflectance measurements are sometimes made on *exsudatinites* ('*migrabituminen*'). Exsudatinite is a secondary maceral deriving

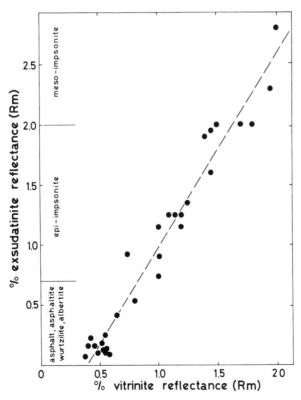

Figure 4.13 Relationship between mean (random) reflectance of exsudatinite (from '*migrabitumen*') and of vitrinite (after Jacob, 1985).

Figure 4.14   Photomicrograph of strongly coalified graptolite periderm ($10.5\%\ R_{max}$, $2.0\%\ R_{min}$) embedded in Ordovician schist of the Soest-Erwitte 1 borehole, West Germany (after Teichmüller, 1978). Polished section, oil immersion, 500 ×.

from petroleum-like exudates generated in coals and source rocks during bituminization. Jacob (1985) correlated vitrinite reflectance with reflectance of exsudatinites (measured on various *'migrabitumen'*), as is shown in Figure 4.13. A fairly good relation exists for the overmature stage in which the exsudatinites of impsonites display a higher reflectance than the corresponding vitrinites, whereas at low rank stages ($< 0.8\%\ R_m$ vitrinite) the reflectances of exsudatinites are commonly lower than the reflectances of vitrinites and, moreover, may vary considerably in the same rock.

(v) In Pre-Devonian rocks which do not contain classic vitrinite from higher plants, *coalified microfossils* may be used for comparative reflectance measurements. Thus, the reflectance of graptolite periderm (Figure 4.14) has been measured since 1976 by Watson (1976) in Scotland; Kurylowicz *et al.* (1976) and Burne and Kantsler (1977) in Australia; Teichmüller (1978), Clausen and Teichmüller (1982) and Teichmüller and Teichmüller (1982) in Germany and Sweden; Kemp *et al.* (1985) and Oliver (1986) in the British Isles; Goodarzi (1984) in Turkey; and Goodarzi and Norford (1985) in Canada. Kemp *et al.* (1985) and Oliver (1986) studied the relationship between graptolite reflectance, illite 'crystallinity' and index minerals in very low-grade metamorphic Ordovician and Silurian rocks of the British Isles. Reflectance values of more than $3\%\ R_{max}$ were measured on graptolites from anchimetamorphic rocks. Figure 4.15 shows the strong increase of graptolite reflectance from the Welsh Borderland in the southeast to the Welsh Basin in the northwest, together with the corresponding index minerals in rocks of Arenig and Wenlock ages after Oliver (1986). Reflectance measurements on chitinozoans (Goodarzi, 1985*b*) indicate another fossil group with potential for use in rank measurements. In some Pre-Devonian rocks vitrinite-like organolites occur which are supposed to be derived from large algae. Peat *et al.* (1978) describe such 'giant filaments' up to 100 μm long from the Proterozoic of the Northern Territory, Australia.

(vi) Other microscopic methods to estimate rank of organic matter in polished sections are measurements of *fluorescence intensity* (Jacob, 1964) and

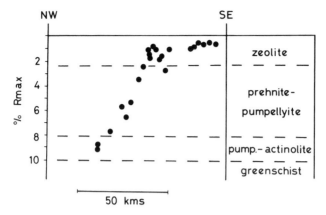

Figure 4.15 Increase of graptolite reflectance in Ordovician and Silurian rocks from the Welsh borderland in the southeast to the Welsh Basin in the northwest, in comparison with metamorphic facies (modified after Oliver, 1986, by kind permission of the author).

of spectral fluorescence properties (Ottenjann *et al.*, 1974, 1975; Ottenjann, 1982). Although these methods are applicable only at low rank stages (up to about 1.3% R_m vitrinite reflectance), they represent a fairly good complement to the reflectance method, since vitrinite reflectance increases relatively slowly and scatters widely at low rank grades (Figures 4.7 and 4.21). Figure 4.16 shows the commonly used fluorescence rank parameters. The classic excit-

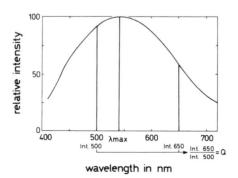

Figure 4.16 Microscopic fluorescence rank parameters measured on liptinites under incident light (excitation wavelength 365 ($\pm$ 30) nm (after Ottenjann *et al.*, 1974, 1975; and Ottenjann 1980).

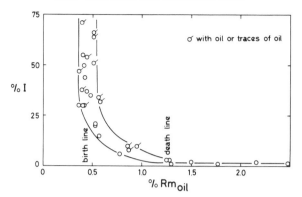

Figure 4.17 Loss of fluorescence intensity (%*I*) of the mineral-bituminous groundmass in the Posidonia Shale of Western Germany in the oil window (0.5–1.3% R_m) (after Teichmüller and Ottenjann, 1977).

ation wavelength used for fluorescence measurements is 365 nm (ultraviolet). Recently, blue and even green light excitations were also introduced because they cause higher fluorescence intensities at λ_{max}.

Fluorescence microscopy of organic matter in rocks has gained its greatest importance for oil source rocks, especially for the estimation of their oil proneness and their maturity (Alpern, 1970; Teichmüller and Ottenjann, 1977; Robert, 1979; Hutton *et al.*, 1980). In these rocks, but also in other liptinite-bearing rocks, fluorescence intensity of liptinites decreases with increasing rank and their fluorescence colour shifts to longer wavelengths, i.e. to the red (Teichmüller, 1982/84). These changes are strongest in the *oil window* when petroleum generates from the kerogen (cf. Figures 4.7 and 4.18). The same is valid for the *mineral-bituminous groundmass* of source rocks. Figure 4.17 shows an example from the Posidonia Shale, a Jurassic source rock of Germany and France (Toarcian of the Paris Basin). The reason is that in polished sections of immature and mature source rocks the submicroscopic part of organic matter causes a striking *organic fluorescence* of the mineral matrix (mineral-bituminous groundmass of Teichmüller and Ottenjann, 1977). The organic nature of this fluorescence is indicated by the property of *alteration*, that is by the change of fluorescence intensity and of fluorescence colour during irradiation, as well as by the loss of fluorescence intensity and the red shift of fluorescence colour with increasing maturity. The same features are observed in many faunal relics with an inorganic frame (carbonates, silica, apatite, phosphates) and with submicroscopic organic components in between, e.g. in conodonts (see below). This organic fluorescence extinguishes at the *oil death line*, corresponding to a vitrinite reflectance of 1.3–1.5% R_m (Figures 4.7, 4.17 and 4.18).

Fluorescence alteration is measured to distinguish between immature, mature and overmature source rocks (Teichmüller and Ottenjann, 1977; Leythaeuser *et al.*, 1980).

Hufnagel (1977) and Kantsler and Cook (1979) propose the use of fluorescence intensity of microfossils like dinoflagellate cysts and acritarchs to estimate rank. These microfossils have the advantage that they are most abundant in marine carbonate rocks where vitrinite is least abundant and commonly absent.

Measurements of vitrinite fluorescence on polished coal sections have recently gained great interest, mainly through problems of coking properties (Teichmüller, 1982/84; Ottenjann et al., 1982; Lin et al., 1986). The primary visible fluorescence of huminites/vitrinites extinguishes in the sub-bituminous coal stage (at about 0.5% R_m vitrinite reflectance). There a new, secondary fluorescence begins to develop due to newly generated bitumen (see section 4.3.1). The secondary vitrinite fluorescence reaches its maximum at about 1.0% R_m and extinguishes (is no longer measurable) at about 1.3% R_m (Ottenjann et al., 1982, who used an excitation wavelength of 365 nm). Impregnations with secondary bitumen are also known from the vitrinite of jets which represent driftwoods in oil source rocks.

(vii) As indicated in section 4.2, the translucency of vitrinite and liptinite macerals in thin sections depends on the rank of coal, but, as it depends also on the thickness of the thin section, translucency measurements on thin sections are theoretical. On the other hand, *translucency and colour of* isolated *spores and pollen* (palynomorphs) in strew slides permit the estimation of the rank of organic matter in rocks other than coal.

After Kuyl et al. (1955) had first drawn attention to the darkening and shrinkage of pollen exines with increasing depth in oil boreholes of Venezuela and had registered a so-called 'black zone' in which all pollen remained opaque and no oil was present, Gutjahr (1966) and Grayson (1975) developed quantitative measurements of absorption and translucency of isolated pollen. But their methods were not accepted in practice. Instead, a simple but rather subjective method is used by oil geologists for maturity studies, that is the determination of the *Thermal Alteration Index* (TAI) (Staplin, 1969). According to colour and translucency, five stages of TAI are distinguished. They are correlated with other microscopic rank parameters and with volatile matter in Figure 4.18.

(viii) Epstein et al. (1977) introduced a new optical method of determining rank of organic matter in rocks other than coal, the so-called *Conodont Colour Alteration Index* (CAI). Conodonts are marine microfossils (0.1–1.0 mm size) occurring from the Cambrian to the Triassic, mainly in calcareous rocks. The change of translucency and colour of isolated unweathered conodonts in transmitted light is based on traces of submicroscopic organic matter which is finely distributed within the inorganic carbonate–apatite skeleton of these microfossils. In the range of diagenesis and very low-grade metamorphism (50–300 °C) five stages of CAI are distinguished according to colour (Figure 4.18), in a similar way to the TAI-values for palynomorphs, although the colour changes of conodonts begin later and, in contrast to palynomorphs,

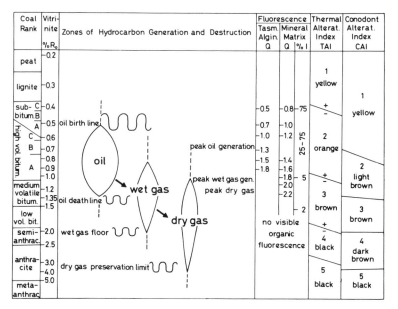

Figure 4.18 Correlation of various microscopic rank parameters (after Teichmüller, 1986) with relation to the zones of hydrocarbon generation and destruction (after Dow, 1974).

conodonts continue to change their colour at higher stages of metamorphism (300–550 °C). They first become grey (CAI = 6), then opaque (CAI = 7), then crystal clear (CAI = 8). These changes result from loss of organic carbon, release of water of crystallization, and crystallization of the inorganic substances. A great advantage of the conodont method is that it can be used in carbonate rocks in which coal macerals, especially vitrinites, are very rarely present. The conodonts can be extracted from these rocks by organic-acid treatment and are further concentrated by heavy liquid and/or magnetic separation.

Epstein *et al.* (1977) used an Arrhenius plot of experimental data (obtained from heated conodonts) and of field data to assess temperature ranges for the various CAI values. Colour alteration begins at 50 °C and continues until about 550 °C.

Based on the fundamental work of Epstein *et al.* (1977) the CAI method has been applied in many areas to clarify patterns of diagenesis and metamorphism and to draw isograd maps, often for hydrocarbon and mineral exploration. Such studies have been carried out in North America by Harris (1979), Harris *et al.* (1980), Wardlaw *et al.* (1984), Wardlaw and Harris (1984) and Patrick *et al.* (1985). In Scandinavia and the British Isles, Bergström (1980), Aldridge (1984, 1986) and Oliver *et al.* (1984) have used the method to clarify questions of very low-grade metamorphism of the Caledonides. Patrick *et al.* (1985) found that temperatures estimated from CAI values are about 100 °C lower (average 350 °C) than those assessed from calcite–dolomite geo-thermometry (average 460 °C) in Precambrian to mid-Palaeozoic rocks of Alaska, the rocks having been metamorphosed to the blueschist facies in Jurassic times.

4.3.3   Causes of coalification

4.3.3.1   *Temperature and time.* Today it is generally accepted that increase in rock temperature is most important for rise of rank, especially for the chemical changes during coalification. Rock temperature increases with depth of subsidence or burial, thus causing an increase of degree of coalification in tectonically undisturbed sequences. This relation is well known to coal geologists as Hilt's Law.

The rate of rank increase depends on (i) the geothermal gradient, (ii) the subsidence rate or heating rate, (iii) the heat conductivity of the underlying and accompanying rocks, (iv) the rank parameter and the rank range which the coal has reached.

(i)   The influence of the geothermal gradient is obvious in boreholes which encounter young sediments, e.g. in the Upper Rhine Graben with gradients varying between $42 °C\,km^{-1}$ and $77 °C\,km^{-1}$ (Figure 4.19).

(ii)   A rapid subsidence rate causes a lower rank at a given depth and temperature (Lopatin, 1976a, b). Rates need to be lower than $10\,m\,Ma^{-1}$ if reflectance equilibrium is to be maintained during continuous subsidence.

(iii)   Rocks with a high heat conductivity (especially salt, but also sandstones in contrast to claystones) cause a retardation of chemical rank increase with depth (Figure 4.20). On top of salt domes, rank increase is promoted by dam of heat in the less conductive overlying rocks (Robert, 1985).

(iv)   Using vitrinite reflectance as a rank parameter, the increase of rank with depth is much lower in the range of lignites to high volatile bituminous coals than at higher rank stages ($> 1\% R_m$, $< 34\%$ volatile matter). It is especially rapid in the stages of anthracites and meta-anthracites (German classification). Figures 4.21 and 4.22 show these relationships. On the other side, using moisture as rank parameter, its decrease is strongest during the very low rank stages (peat to sub-bituminous coals, cf. Figure 4.5).

Generally vitrinite reflectance depends on the maximum temperature which has affected the coaly matter for a certain time. The influence of heating time increases with temperature. For temperatures below 50 °C, the time factor can even be neglected. An example is the Moscow lignite of Mississippian age which never reached higher temperatures.

At higher temperatures a certain time span is necessary to reach rank

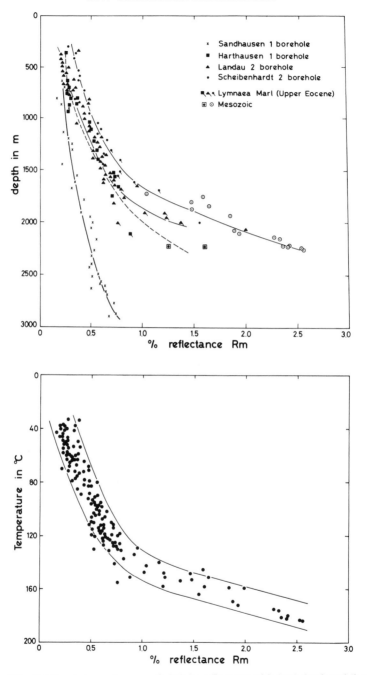

Figure 4.19 Relation between increase of vitrinite reflectance with (top) depth and (bottom) temperature in boreholes of the Upper Rhine Graben with various geothermal gradients. Sandhausen 1.42 °C km^{-1}, Harthausen 1.70 °C km^{-1}, Landau 2.77 °C km^{-1}, Scheibenhardt 2.77 °C km^{-1} (after Teichmüller and Teichmüller, 1981).

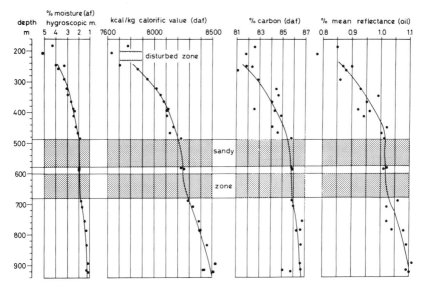

Figure 4.20   Retardation of rank increase (based on vitrite analyses) with depth, caused by a sandy zone in the Teufelspforte borehole (Saar District) (after Teichmüller and Teichmüller, 1968).

equilibrium. In the Tertiary of the Upper Rhine Graben, for example, heating of 2.3 Ma duration was insufficient to attain equilibrium (Espitalié, 1979). Moreover, in rapidly sinking basins the equilibrium temperature for a given depth and geothermal gradient may not yet have been reached. Therefore, high heating rates (caused by rapid sinking) retard the chemical coalification processes with respect to temperature. Kantsler *et al.* (1978) reported that in the offshore area of Gippsland Basin, Australia, which has undergone extremely rapid sedimentation, there is a lag of reflectance increase behind temperature.

Using reaction kinetics (Arrhenius equation), the relation between temperature, time and rank of coal was first demonstrated by Karweil (1956) on the basis of volatile matter, and later by many authors on the basis of vitrinite reflectance (Lopatin, 1971; Bostick, 1973; Hood *et al.*, 1975; Bostick *et al.*, 1979; Waples, 1980; and others). Figure 4.23 shows the relationships between vitrinite reflectance, maximum temperature and *effective heating time*, the latter defined as the time span during which the coal was within 15 °C of its maximum burial temperature (Hood *et al.*, 1975).

Because of the effect of duration of heating and of varying palaeogeothermal histories it is impossible to relate a particular degree of coalification to a precise temperature. Typical coalification temperatures are approximately 100–170 °C for bituminous coals and 170–250 °C for anthracites.

The relationships between rank of coal, rock temperature and heating time allow certain reconstructions of the geothermal history. Subsidence curves, in

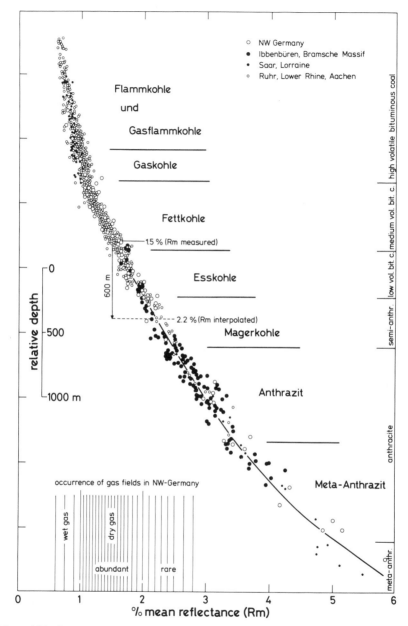

Figure 4.21 Increase in vitrinite reflectance with relative depth and with rank stages (German and US classifications) in 45 deep West German boreholes (after Teichmüller and Teichmüller, 1984).

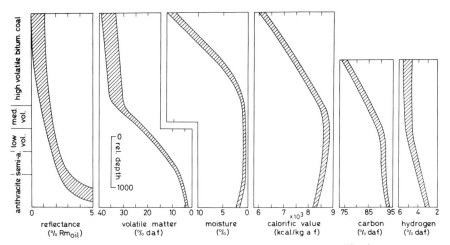

Figure 4.22 Increase in rank with depth on the basis of different coalification parameters (*af* = ash-free, *daf* = dry, ash-free) (after Teichmüller and Teichmüller, 1979*a*).

which burial depths *v.* absolute time intervals are plotted, are needed to estimate the effective coalification temperatures. Mathematical models, based on the Arrhenius equation (Karweil, 1956; Tissot and Espitalié, 1975), or on empirical data from deep boreholes which encountered young sediments (Buntebarth, 1978/79) permit reconstruction by computation of the geothermal history, if these subsidence curves and the rank gradients ($\% R_m \, \text{km}^{-1}$) are known (e.g. Buntebarth *et al.*, 1982). Two books dealing with reconstructions of palaeogeothermics on the basis of coalification studies have recently appeared (Robert, 1985; Buntebarth and Stegena, 1986).

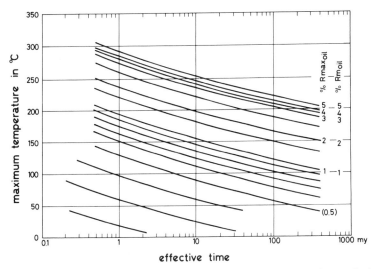

Figure 4.23 Relation between maximum temperature, effective heating time and vitrinite reflectance (R_{max}, R_m) (after Bostick *et al.*, 1979).

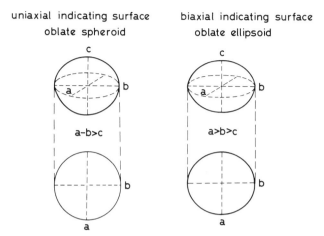

Figure 4.24 Reflectance indicating surfaces for optically uniaxial and biaxial vitrinites (after Stone and Cook, 1979).

4.3.3.2 *Pressure.* The influence of overburden pressure on coal rank is greatest in the very low rank stages (peat to sub-bituminous coals) in which physical rank, in terms of decreasing porosity (expressed as decreasing moisture of the as-mined coal) increases rapidly (Figure 4.5). Chemical reactions are retarded by hydrostatic and lithostatic pressure. The reason is that the removal of coalification gases is hindered, as has been proved by, for example, coalification experiments under different static pressures (Huck and Patteisky, 1964; Davis and Spackman, 1964; Horváth, 1983; Goodarzi, 1985a).

As mentioned earlier, anisotropy of vitrinite is caused by pressure, normally by load pressure. Tectonic pressure may disturb the ordering of the aromatic lamellae parallel to the bedding plane (Petrascheck, 1954), thus giving rise to biaxial anisotropy. Figure 4.24 shows the difference between reflectance indicating surfaces for optically uniaxial and biaxial vitrinites after Stone and Cook (1979), and Figure 4.32 shows the influence of tectonic pressure on the anisotropy of vitrinite layers in a phyllite.

Levine and Davis (1984) studied the relations between tectonics and reflectance anisotropy in Carboniferous bituminous coals ($1.5-1.9\%$ R_{max}) of the Broad Top Coal Field, situated in the Valley and Ridge province of the Appalachians in Pennsylvania. The R_{max} axes were oriented more or less parallel to the fold axes. The plunging of the R_{min} axes suggested the latest tectonic compression to be oriented toward the west. In strongly deformed areas the reflectance indicatrices had even changed from the usual uniaxial negative to a biaxial positive geometry. Figure 4.25 shows the mutually perpendicular planes used for the reflectance measurements. The tectonically caused anisotropy was superimposed on a pre-tectonic anisotropy pattern.

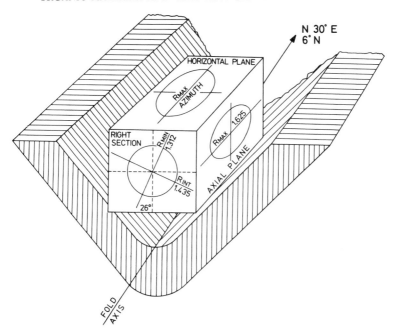

Figure 4.25   The presumed indicating surfaces of vitrinite reflectance in relation to the direction of a syncline axis. Example from the Pennsylvanian Appalachians (after Levine and Davis, 1984).

Shearing is especially effective to promote the increase of maximum reflectance at high rank stages and, thus, to enhance anthracitization and semi-graphitization (see section 4.4.1). The disappearance of oxygen and other sterically hindering groups in the vitrinite 'molecule' facilitates the growing and ordering of the aromatic units in the direction of graphite formation (cf. Figure 4.34).

Bustin (1983) measured anomalously high reflectance values which are restricted to very narrow films immediately adjacent to or within shearing zones of the Canadian Rocky Mountains; representative maximum values were 2.76–3.09% R_m against normal values of 0.72–0.97% R_m. These highly localized and sharp rises of rank up to the anthracite stage are explained by short-acting, very high temperatures (350–650 °C) as a consequence of frictional heating during stick–slip faulting.

4.4   Anthracitization and graphitization

4.4.1   *The process of anthracitization and semi-graphitization*

As will be discussed in Chapter 7, very low-grade metamorphism corresponds mainly to the meta-anthracite rank stage which is reached by a strong anthracitization and leads to the semi-graphite stage. Anthracitization and

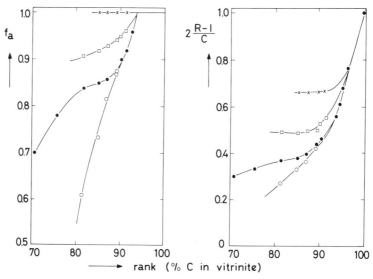

Figure 4.26 Aromaticity (f_a) and ring-condensation index $[2(R-1)/C]$ of sporinite (open circles), vitrinite (filled circles), macrinite (squares) and fusinite (crosses) (after van Krevelen, 1961).

semi-graphitization represent the end of the coalification process and vary in many respects from the earlier coalification processes.

4.4.1.1 *Chemical changes.* Chemically anthracitization and semi-graphitization are characterized by a strong decrease in hydrogen and in the atomic H/C ratio (Figure 4.8). The carbon content rises more rapidly with depth than in the range of high-rank bituminous coals (Patteisky and Teichmüller, 1960). Likewise, aromaticity and condensation of the aromatic units increase much more rapidly than before. Figure 4.26 shows the increase of the aromatization factor $(fa$ = aromatic carbon/total carbon) and of the ring condensation index $(2(R-1)/C$, where R/C = number of rings per C atom), plotted against carbon content for various macerals according to van Krevelen (1961).

The strong decrease of volatile matter with depth which took place in the bituminous-coal range slows down (Figure 4.22) because functional groups, especially those containing oxygen, have practically disappeared. The moisture content *increases* up to about 4% H_2O in the meta-anthracite stage, then decreases rapidly. The calorific value decreases considerably, due first to increase of moisture, then to failing ignition (cf. Figure 4.5).

4.4.1.2 *Optical changes.* Petrographically, anthracites are distinguished by a great hardness and a bright lustre, the latter attaining a metallic yellowish appearance in meta-anthracites. The megascopically visible stratification is much less pronounced than in bituminous coals. In many cases meta-anthracites and semi-graphites appear dull and dusty as a consequence of intensive mylonitization (Raben and Gray, 1979a, b).

Under the microscope, in non-polarized light, anthracites and meta-anthracites appear much more homogeneous than do bituminous coals (cf. Figure 4.1). The liptinite macerals have disappeared, partly due to volatiliz-ation, partly due to approximation of their optical properties to those of vitrinite and inertinite. In polarized light, especially under partly crossed nicols, the former micropetrographic heterogeneity becomes visible (Figure 4.27). Cutinite and sporinite attain a higher maximum reflectance and a stronger bireflectance than the surrounding vitrinite (Figure 4.28). The maximum reflectance of inertinites may be equal to or even lower than the reflectance of vitrinite, the latter rising particularly strongly in the anthracite stages. Often it is only the typical structure of fusinite, semi-fusinite and sclerotinite (open cell cavities) which reveals the inertinitic character. The maximum reflectance of vitrinite first strongly approximates the maximum reflectance of inertinites, and then even surpasses it at a rank stage

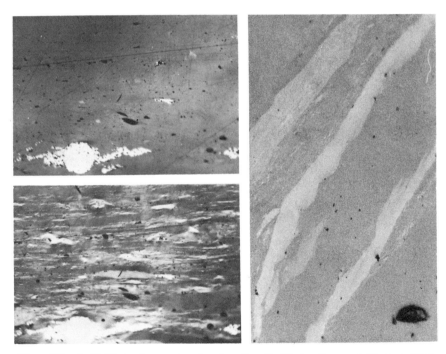

Figure 4.27 (Left) Photomicrographs of meta-anthracite (Carboniferous of Piesberg, West Germany) with 2.5% volatile matter. Top, in non-polarized light; bottom, under crossed nicols. White inclusions are pyrite. Polished section, oil immersion, 250 ×. **Figure 4.28** (Right) cutinite (white) and vitrinite (grey) in Carboniferous anthracite from Ibbenbüren, West Germany (6.6% volatile matter, 2.8% R_{max}, vitrinite). The picture demonstrates the inversion of maximal reflectance from vitrinite to cutinite. Polished section, oil immersion, 1 polar, 500 ×.

corresponding to about 4.5% R_{max} (Alpern and Lemos de Sousa, 1970). This development is shown in Figure 4.11 and may be explained, together with the much higher anisotropy of vitrinite, through the originally higher hydrogen content of vitrinite compared with inertinite (see below).

The behaviour of macerals during meta-anthracitization and semi-graphitization is comparable with the coking properties of the different macerals in coking coals: vitrinites and liptinites (the latter if not volatilized completely) are transformed to highly anisotropic graphitoid coke lamellae, whereas true inertinites, as the name indicates, show inert behaviour.

In natural coke, relatively large graphitoid crystallites were observed, formed from hydrogen-rich vitrinite (desmocollinite) at lower temperatures than small crystallites, formed from (more strongly reflecting) hydrogen-poor vitrinite (telocollinite) at higher temperatures (Teichmüller, 1973). This example was quoted by Diessel *et al.* (1978) to explain the different degrees of meta-anthracitization and semi-graphitization recognized in the same polished section of a high-pressure schist (pp. 149–151).

A rapid increase of vitrinite reflectance and bireflectance with depth is characteristic of the anthracite stage, at least up to meta-anthracites with about 6% R_{max} (Figures 4.21 and 4.29). This strong rise of reflectance is also striking in binary diagrams plotting vitrinite reflectance *v.* decreasing volatile

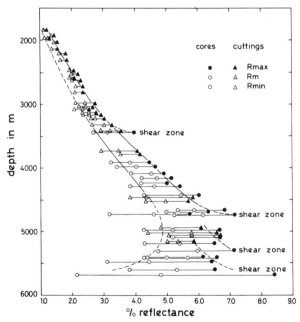

Figure 4.29   Increase in vitrinite reflectance (R_{max}, R_{min}, R_m) and of vitrinite anisotropy (R_{max} − R_{min}) in the Münsterland 1 borehole, West Germany. Note the increase in R_{max} and in anisotropy near tectonic shear zones (after Teichmüller and Teichmüller, 1979a).

matter or v. increasing carbon content (Figures 4.7, 4.8). It reflects the rapid rise of the absorption index, and partly also of the refractive index, the latter running through a maximum at the stage of meta-anthracite.

Figure 4.29 shows a broad scatter of reflectance values (not only R_m but also R_{max}), beginning at the meta-anthracite stage with about 6% R_{max}; in the Münsterland 1 borehole which allowed reflectance measurements over a depth difference of more than 4000 m, R_{max}, R_m and R_{min} increase more or less continuously with depth until a value of 6% R_{max} is reached. Thereafter the R_{max} values begin to scatter between 6–7% without increasing further with depth. At the same stage R_{min}-values begin to decrease, resulting in a strong increase of bireflectance. Since graphite is distinguished by a very low minimum reflectance (0.5% R_{min} under oil) and an extremely high bireflectance (17.8% R_{max}, according to Kwiecinska, 1980), the point at which R_{min} begins to decrease may be explained as the beginning of pre-graphitization. The reversal of R_{min} values was first described by Ragot (1977). Figure 4.30 shows a modification of Ragot's diagram, in which the reflectance values R_{max}, R_m and

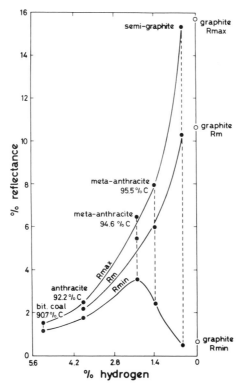

Figure 4.30   Relation between hydrogen content and vitrinite reflectance (R_{max}, R_{min}, R_m) at high rank stages. Note the start of decrease in R_{min} in the meta-anthracite stage (after Ragot, 1977; modified by Teichmüller and Teichmüller, 1981).

R_{min} are plotted against hydrogen content as the most important chemical rank parameter for anthracitization. According to Figure 4.30, the reversal of R_{min} takes place at the stage of meta-anthracite with about 95% C and 1.5% H. These results require a revision of a still-common opinion that vitrinite reflectance and bireflectance increase steadily during anthracitization and meta-anthracitization.

The broad scatter of reflectance values at the anthracite and meta-anthracite stage, has also been described by other authors (Dahme and Mackowsky, 1951; McCartney and Ergun, 1967; Alpern and Lemos de Sousa, 1970; Diessel et al., 1978; Raben and Gray, 1979a, b). It may be partly due to the optically biaxial nature of vitrinite, especially for vitrinite particles in clastic rocks (see section 4.3.2.2). Moreover, as shown in Figure 4.29, shearing movements cause a rise in R_{max} and of anisotropy, and this may take place very locally, even at a microscopic scale. Raben and Gray (1979b) explain the wide scatter of R_{max} values in highly deformed anthracites and meta-anthracites of the Narragansett Basin (New England, USA) by various proximities to planes of shear.

Another reason for the widely-scattered R_{max} values measured on very high-rank organic matter in rocks is the fact that the various macerals, formerly easily distinguishable by their various reflectances (see section 4.3.2.2) are now more or less similar in optical appearance (cf. Figure 4.1). In fact, liptinites and hydrogen-rich vitrinites, formerly distinguished by relatively low reflectance

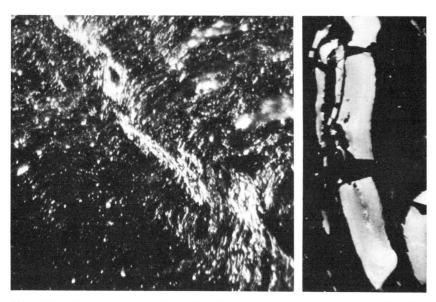

Figure 4.31 Photomicrographs of Ordovician/Silurian anchimetamorphic rocks from Almaden, Spain, with strongly reflecting, finely dispersed meta-bituminite, enriched in a shear zone (left) and graptolite periderm (right). Polished sections, oil immersion, 1 polar, 500 ×.

values, now attain even higher maximum reflectance and stronger bireflectance than the formerly highly-reflecting inertinites and hydrogen-poor vitrinites (Figures 4.11, 4.27 and 4.28). As it is difficult and sometimes impossible to recognize the former under the microscope, all particles tend to be measured, especially by inexperienced persons. From this point of view it is interesting that, according to Saupé *et al.* (1977), anchimetamorphic rocks of Ordovician–Silurian age in Almaden, Spain, bear meta-bituminite in the semi-graphite stage (10–12% R_{max}), whereas graptolite periderm is still in the meta-anthracite stage (about 5% R_{max}) (Figure 4.31). Similarly, Teichmüller and Teichmüller (1979*b*, p. 328) found particles in the semi-graphite as well as in the meta-anthracite stage in the same anchimetamorphic Palaeozoic rocks of the Variscan Rhenish Mountains west of the Rhine (Figure 4.32).

Diessel *et al.* (1978) distinguished optically as many as three stages of rank which may occur side by side in the same Tertiary metamorphic rock from New Caledonia (cf. Figure 4.33):

  (i) 'Coal' (mainly vitrodetrinite) with reflectance values between 1.4–6.5% R_{max} and 1.2–3.15% R_{min}

 (ii) 'Transitional matter' with reflectance values varying between 7.2–9.8% R_{max} and 1.3–2.0% R_{min}

Figure 4.32 Microfolded and sheared vitrinite layers in the stage of semi-graphite with strong bireflectance (6.8% R_{max}, 1.5% R_{min}) from Lower Devonian phyllite of the Our Valley, Luxembourg. Photomicrograph, polished section, oil immersion, 1 polar, 270 ×.

(iii) 'Graphite' with 13.0–15.0% R_{max} and 0.55–0.41% R_{min}: this probably formed from marine algae.

Attention must be drawn to the definition for 'graphite' given by Diessel *et al.* (1978, p. 63): 'the term "graphite" is used in this paper to describe carbonaceous material of crystalline morphology, with the optical anisotropy and reflectance properties of graphite as summarized by Kwiecinska *et al.* (1977) and with layered two-dimensional (para-crystalline or turbostratic) and three-dimensional (graphite sensu stricto) structure'. This means that the 'graphite' of Diessel *et al.* (1978) may belong, at least partly, to what other authors call semi-graphite. This has been confirmed by X-ray diffraction diagrams, in which these organic-matter concentrates display irregular peaks, whereas true dimensional graphite with $d_{002} = 3.354 \text{Å}$ and sharp $d_{110,112}$, and d_{114} lines (cf. Figure 4.35) first appear at the epidote isograd. These authors conclude, from a comparison between optical and X-ray studies, that during pre-graphitization 'graphitized' vitrinite (= 'graphite' in Figure 4.33) may attain the same high values of reflectance and anisotropy as true, three-dimensional graphite. Nevertheless, these vitrinites do not have the same high

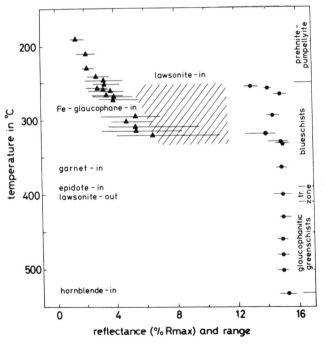

Figure 4.33 Relationship between maximum reflectance of finely-divided organic matter, maximum rock temperature and index minerals in Tertiary high-pressure schists of New Caledonia. Coal stages are indicated by triangles, graphite by circles. The transition zone is shaded (modified after Diessel *et al.*, 1978).

density nor the same X-ray diffraction pattern. The X-ray studies revealed only incipient ordering of C-atoms, and the density is lower than for real graphite ($1.91–2.22 \, g/cm^3$ against $2.25 \, g/cm^3$). The development of a three-dimensional graphite crystal structure, although affording more energy, 'is not accompanied by a further increase of reflectance' (Diessel et al., 1978, p. 73).

Vitrinite anisotropy of meta-anthracites and semi-graphites may be influenced by the direction of tangential pressure, thus allowing conclusions about the stress regime in folded areas. Hower and Davis (1980) found the maximum-intermediate reflectance plane of a meta-anthracite to be essentially parallel to the axial plane of the synclinorium in the intensely folded southern anthracite field of Pennsylvania.

4.4.1.3 *Changes of the ultra-fine structure.* There is no doubt that the coal remains in a non-crystalline phase up to the stage of semi-graphite. This is the result of X-ray, electron and laser diffraction studies. One of the earliest X-ray investigations (Hirsch, 1954) showed that low-rank coals (up to about 85% C) are characterized by an *open structure* in which the aromatic lamellae (size 7–8 Å) are connected by cross-links and are randomly arranged in all directions, resulting in a high porosity (Figure 4.34, above). Higher-rank bituminous coals and semi-anthracites (in the range between about 85–91% C) have a *liquid structure* in which the former cross-links have decreased considerably and the lamellae (size 9 Å) are better oriented. So-called crystallites (X-ray units) comprise two or more lamellae. Pores are practically absent. The *anthracitic structure* of Hirsch is present in coals with more than 91% C (the boundary between bituminous coals and anthracite, according to Patteisky and Teichmüller, 1960). All bridges have disappeared and the orientation and size of the lamellae have strongly increased. The porous system is oriented in the same direction as the lamellae (cf. Figure 4.34, below).

The *crystallites* of X-ray and electron diffraction are stacks of aromatic lamellae which may vary in height and horizontal size and in the space between the single lamellae. These stacks still display a *turbostratic texture* in the semi-graphite stage, in other words, although the lamellae are ordered more or less parallel to each other, the position of C-atoms of adjacent layers is still arbitrary. A complete three-dimensional orientation is only reached through true graphitization (formation of graphite crystals).

According to McCartney and Ergun (1965) and Ragot (1977), small graphite crystals may already occur in meta-anthracites and semi-graphites. Their abundance increases with degree of semi-graphitization. Figure 4.35 shows a series of X-ray diffraction curves from vitrinites of different rank ranging from the lignite to the semi-graphite stage, in comparison with the graphite diagram. The most striking feature is the successive approach to the high graphite peak for (002) (the latter indicating a dense packing of the lamellae 001 with a space difference of only 3.354 Å for real graphite). This approach is particularly strong in the stages of anthracite and meta-anthracite. The

F

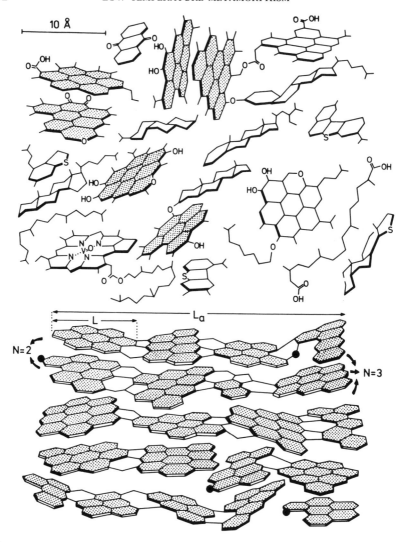

Figure 4.34 Structural molecular models of low-rank coal (above), and of high-rank coal (below), the latter with typical turbostratic arrangement of the aromatic units (shaded). N = number of layers, L = diameter of aromatic units, La = diameter of a wrinkled layer of aromatic units. Black circles indicate defects (e.g. hetero-atoms) at the margin of the aromatic layers, causing the zigzag texture (after Oberlin et al., 1980).

diameter of the layers and the number of C-atoms per layer also increase considerably during anthracitization (van Krevelen, 1961, pp. 332–333).

    According to van Krevelen (1961, p. 340) 'crystallite growth and graphitiz-ation can proceed only if two essential conditions are fulfilled: the system of cross-linking uniting the crystallites should not be too strong, and neighbour-

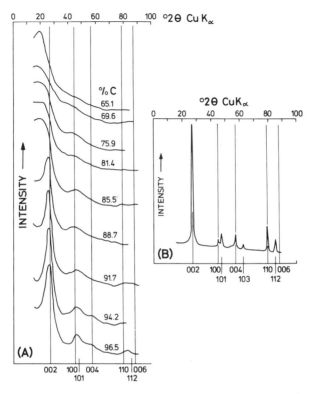

Figure 4.35   Change of X-ray diffraction curves obtained from vitrinites (A) with increasing rank (C-content, cf. Figure 4.5), in comparison with the X-ray diffraction curve of graphite (B) (modified after van Krevelen, 1961).

ing crystallites should lie in near-parallel orientation'. This is normally the case with formerly hydrogen-rich macerals (liptinites, H-rich vitrinites) in anthracites.

High-resolution transmission electron microscopy has thrown new light on the ultrafine structure of coal, especially of high-rank coals and coaly inclusions in other rocks (Oberlin *et al.*, 1980; Rouzaud and Oberlin, 1983). The method has been combined with X-ray, electron diffraction, and, most recently, with optical diffraction using a laser beam (Oberlin *et al.*, 1984). The high-resolution electron microscope permits direct visualization of the aromatic units within the coal 'molecule'. Figure 4.36 (top) shows an anthracite in bright field where individual aromatic layers (black) become visible at a magnification of 1 500 000 ×. The layers are arranged in clusters of 50–100 Å size around large pores (300–600 Å) which are flattened parallel to the bedding. For comparison, a true natural graphite at the same magnification is shown beneath.

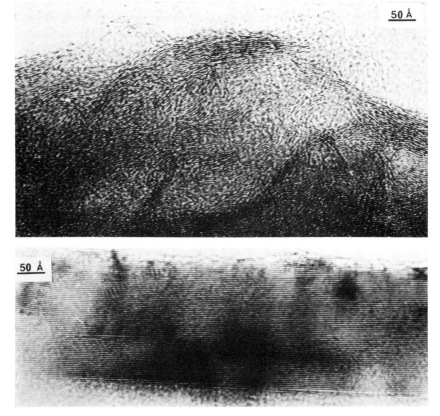

Figure 4.36   Permian meta-anthracite (7.4% R_{max}, 3.4% R_{min}) from Korea (top), and graphite from Korea (bottom) under the high resolution transmission electron microscope, bright field. At the high magnification of 1 500 000 × the black lines represent the aromatic layers and carbon layers, respectively. By courtesy of Dr J.N. Rouzaud, Orléans.

Although in anthracites the interlayer spacing has become more or less constant, the structure is still turbostratic. The layer diameter grows to 50–100 Å in meta-anthracites. It grows suddenly to $L_a = 1000$ Å at the semi-graphite stage. At the same time the number of layers in one individual stack increases from 2–3 to more than 40, and the single layer becomes more planar. Finally, the turbostratic structure disappears suddenly and real graphite with a three-dimensional arrangement of the layers is formed, combined with a drastic increase in layer diameter. Both processes, semi-graphitization and graphitiz-ation, take place suddenly (as 'jumps') and require high temperatures and pressures in combination with shearing and/or stretching, according to Bonijoly Roussel (1980) and Bonijoly *et al.* (1982). These authors emphasize the role of anthracite porosity* for graphitization. The pores in anthracites are first surrounded by more or less 'crumbled' aromatic layers which become flattened as the pores coalesce and grow to larger and thinner (perpendicular to the bedding plane) units. At the stage of semi-graphite, partial graphitization takes place when the formerly crumbled aromatic layers become larger, straight and stiff. Graphitic shells are formed around hard inclusions.

To sum up, the ultrafine structure of coal changes especially strongly during meta-anthracitization and semi-graphitization, during which the height of the aromatic crystallites increases, their layer diameter increases and the interlayer spacing decreases considerably.

4.4.2 Rank stages and rank parameters

Different classifications of anthracites and meta-anthracites exist. The *economic classifications* vary in the different countries and are based on the volatile matter of the bulk coal, either dry, ash-free (daf) or dry, mineral matter-free (dmmf). In Germany no economic distinction is made between various anthracites: '*Anthrazitkohlen*' are defined as coals with less than 10% volatile matter. However, in the USA (ASTM classification, cf. Table 4.2) a distinction is made between semi-anthracite (14–8% volatile matter), anthracite (8–2% volatile matter) and meta-anthracite (< 2% volatile matter). According to the 'scientific rank classification' based on vitrite analyses (Patteisky and Teichmüller, 1960; *International Handbook of Coal Petrography*, 1963) the boundary between bituminous coal (*Steinkohle*) and anthracite (*Anthrazit*) is characterized by the beginning of a very strong decrease in hydrogen and a stronger increase in carbon in comparison with the C-increase of high-rank bituminous coals. According to the scientific classification, the limit between very low-volatile bituminous coal (*Magerkohle*) and anthracite (*Anthrazit*) is characterized by the values 91% C, 4% H, 2.5% O, and 8% volatile matter. These values correspond to vitrinite reflectances of 2.8% R_{max} and 2.5% R_m.

From about 93.5% C, the C content (plotted against volatile matter) begins to rise especially strongly. Therefore, the boundary between anthracite (*Anthrazit*) and meta-anthracite (*Meta-Anthrazit*) was defined by the values 93.5% C, 2.5% H, and 4.0% volatile matter. These values correspond to a vitrinite reflectance of 4% R_{max} and 3.5% R_m.

Another natural boundary for rank stages lies where the formerly strong hydrogen decrease slows down markedly. This limit is defined as the boundary between meta-anthracite (*Meta-Anthrazit*) and semi-graphite (*Semi-Graphit*) with the values 96.5% C, and 0.8% H.

At this high coalification stage, vitrinite reflectance may vary considerably

---

*An inherent surface area of 300 m²g⁻¹ coal (!) has been measured in anthracites *v.* 25–40 m²g⁻¹ coal in bituminous coals.

between 5 and 10% R_{max}. Here, the R_{min} values become decisive. They should be less than 2% R_{min} (Teichmüller *et al.*, 1979).

According to Ragot (1977), in the anthracite stage, R_{min} first increases with increasing R_{max} (as in bituminous coals). At a stage which is characterized by a hydrogen content of about 1.5% H (corresponding to about 3.5% R_{min}, 6% R_{max} and 5% R_m) R_{min} begins to decrease strongly until it reaches a value of less than 0.5% in graphite (cf. Figure 4.30). The reversal of the R_{min} curve corresponds approximately to the boundary between anthracite and meta-anthracite of the American ASTM classification, whereas, according to the scientific classification, it occurs in the meta-anthracite stage.

There is no doubt that at the highest stages of coalification the most common chemical and optical rank parameters, such as volatile matter and vitrinite reflectance, lose much of their former applicability. Instead, hydrogen, carbon, and, in the meta-anthracite stage, minimum reflectance, become the better indicators of rank (cf. Table 4.2).

The transmitted-light methods can no longer be applied, except for the Conodont Alteration Index (CAI) (p. 135) which, according to Epstein *et al.* (1977), distinguishes the beginning of very low-grade metamorphism by the CAI value 5, whereas values of CAI of 8 were determined in greenschists and marbles interbedded with garnet–mica schists (Schönlaub and Zezula, 1975).

The increasing degree of crystallite ordering (pre-graphitization and graphitization) can best be measured with X-ray or electron diffraction. Therefore these methods should be preferred for organic matter in very low-rank metamorphic rocks, although they are time-consuming, the more so since the organic matter has first to be separated from the mineral matrix. Acknowledged quantitative boundaries based on these methods are still lacking. Ragot (1977) proposed a measure of the degree of semi-graphitization as follows:

$$g = \frac{3.44 - \bar{d}_{002}}{3.44 - 3.354}$$

where 3.44 Å is the minimum interplanar spacing for (d_{002}) in cokes, formed at a temperature of 1600 °C, 3.354 Å is the interplanar spacing (d_{002}) for true graphite, and $\bar{d}_{002}$ is the average interplanar spacing between two layers for the various degrees of semi-graphitization.

4.4.3   Causes of meta-anthracitization and semi-graphitization

Causes of meta-anthracitization and semi-graphitization are high temperatures and pressures, often combined with shear stress. As for the earlier coalification in the range of bituminous coals, rock temperature is the decisive factor, but pressure plays a much greater role than before.

According to Figure 4.23, depending on the effective heating time, the maximum temperature for the boundary between anthracite and meta-anthracite (3.5% R_m and 4.0% R_{max}) can vary between about 190 °C (for a theoretical time span of 400 million years) and about 300 °C (for a time of 500 000 years). It is 250 °C for 10 Ma and 280 °C for 1 Ma, which might be the most reasonable coalification times for most meta-anthracites, the formation of which has not been influenced by shearing stress. In flat-lying rock series, meta-anthracites normally occur only where additional heat supply was possible, as in the roof of intrusive bodies or in zones of an anomalously thin crust. The reason is that the high temperatures which are necessary for meta-anthracitization are not commonly reached through subsidence in regions of normal geothermal gradients (3–4 °C 100 m^{-1}).

In NW Europe, meta-anthracites and semi-graphites are found above large intrusive bodies. Well-known examples are the Vlotho and Bramsche Massifs, where reflectance values of more than 4% R_m were measured for the Mesozoic (Deutloff et al., 1980), and coal layers of Westphalian-D age have been mined with values of 6–7% R_{max} and 2% R_{min}, respectively. These meta-anthracites (with less than 3% volatile matter) are associated with pyrophyllite-bearing quartzitic dykes (Stadler and R. Teichmüller, 1971).

Jordan and Koch (1975, 1979) could trace a pluton in the subsurface which is connected with the Brocken Massif in the Harz Mts. The shape of this pluton is approximately defined by the 3% R_{max} iso-reflectance line in the Mississippian. The 7% R_{max} isoline measured in the Devonian follows the margin of the contact zone.

In the Sesia-zone of the Western Alps (Falletti), plant remains in Tertiary phyllites are in the meta-anthracite and semi-graphite stages, probably due to the intrusion of tonalites. Comparative studies of organic and inorganic metamorphism lead to the conclusion that the boundary between anchi- and epizone has been reached (Stadler et al., 1976).

Even above very young intrusions, formation of meta-anthracite takes place, as in the Cerro Prieto geothermal field of Mexican California (maximum geothermal gradient 160 °C km^{-1}!) where coalification is still continuing and seems to be at its maximum above an intrusion which is less than 600 000 years old. The vitrinite reflectance increases from 0.12% (peat stage) at 240 m (60 °C) to 4.1% R_m (meta-anthracite) at 1700 m (350 °C) (Barker, 1979). The biotite zone is reached at 2000 m (!) with reflectance values of 4–6% R_m. No semi-graphitization ($R_{min} < 2\%$) has taken place, probably due to lack of sufficient pressure and of shearing.

Shearing stress is undoubtedly very important for the formation of meta-anthracites and semi-graphites. Under tectonic pressures, shearing and stretching facilitate the ordering of the aromatic lamellae in the direction of graphite formation. Bonijoly Roussel (1980) and Bonijoly et al. (1982) have studied the influence of pressure on graphitization in experiments, finding that under a pressure of 5 kbar, semi-graphitization of meta-anthracite took place at 1000 °C, compared to 2200 °C without pressure. Temperatures which are suggested for the formation of true graphite in nature (see section 4.4.4) imply natural temperatures for semi-graphitization of more than 300 °C. Such temperatures correspond to the boundary from very low-grade to low-grade metamorphism according to Winkler (1979).

As already mentioned an excellent and basic paper dealing with causes of pre-graphitization (although not so called) in metamorphic rocks has been published by Diessel et al. (1978) (cf. section 4.4.2.1). The authors studied the coalification of finely-dispersed organic matter in Tertiary high-pressure schists of New Caledonia in relation to metamorphic index minerals, maximum temperature and pressure, with the main finding that relatively low temperatures but high pressures caused by rapid tectonic thickening retarded coalification but accelerated graphitization. A temperature of 335 °C was necessary to form optical graphite at pressures ranging between 3–5.5 kbar (Figure 4.33). Graphitization was complete in the epidote zone at temperatures of 390 °C and pressures of 6.3 kbar. The authors conclude that the high pressures favoured the formation of graphite at relatively low temperatures.

4.4.4   Graphitization

Although graphitization does not belong to coalification sensu stricto, it has already often been mentioned in this chapter as the final process of organic-matter metamorphism. According to Schüller (1961), graphite is indicative of low-grade metamorphism (see Chapter 7). Its formation affords high temperatures and pressures as well as tectonic shear.

Graphite is formed through the three-dimensional arrangement of the C-atoms into a true hexagonal crystal structure. The diameter of the layers increases drastically and their vertical distance (d_{002}) shrinks to 3.354 Å.

According to Kwiecinska (1980), the rhombohedral form of graphite is found in microcrystalline graphites and in semi-graphites (up to 35%). The rhombohedral modification is unstable and transforms to hexagonal graphite at higher temperatures.

Rare types of graphite may be formed by thermal cracking of tars and gases, leading to deposition of pure carbon in voids and cracks (Raben and Gray, 1979a). This type is often associated with natural cokes which have been formed through contact metamorphism of coal seams by magmatic dykes or sills (Teichmüller, 1973).

For the microscopist it is very difficult to distinguish between semi-graphite and true graphite in polished sections of metamorphic rocks. Graphite is difficult to polish, because translations of the basal layers (001) take place very easily. This leads to cryptocrystalline and microcrystalline structures (Ragot, 1977, Plate VII) on which reflectance measurements become impossible. According to Kwiecinska (1980) R_{max} values measured on the polished basal layer surface of a large graphite crystal are about 20% lower than measured on the untreated surface. The true R_{max} is 17.8% (Kwiecinska, 1980) facing a true R_{min} value of 0.2%. Other authors give values of 15.6% R_{max} and 0.8 R_{min} (Ragot, 1977, p. 77), or 16% R_{max} and 0.8% R_{min} (Ergun, 1967). Figure 4.37

Figure 4.37   Photomicrograph of graphite crystals occurring in a paragneiss of Proterozoic age at Kropfmühl, near Passau, Bavaria. Polished section, oil immersion, 1 polar, 110 ×.

shows a photomicrograph of well-developed, large crystals of graphite from a graphite mine near Kropfmühl, Bavaria, and in Figure 4.36 (bottom) natural graphite is depicted under the high-resolution transmission electron micro-scope. The perfect ordering of the carbon atoms into parallel layers and the distance between single layers are clearly visible at a magnification of 1 500 000 ×.

In experiments, temperatures of about 3000 °C are necessary to form real graphite from coal. But at high hydrostatic pressures combined with gliding, experimental graphitization is already possible at temperatures of 1600 °C (Bonijoly Roussel, 1980). In nature, progressive graphitization caused only by rising temperatures is thermodynamically impossible, even at very long coalification times (Bonijoly et al., 1982). There, real graphite can be formed only under high pressure and stress in combination with high temperatures. Hamilton et al. (1970) described graphite which formed from *Glossopteris* leaves as a result of contact metamorphism of rocks of hornblende–hornfels facies (550–700 °C). On the basis of X-ray diffraction studies and by comparison with metamorphic mineral facies, Landis (1971) states that true graphite is formed at 450 °C and pressures of 2–6 kbar. Diessel et al. (1978) found true graphite in the lawsonite–epidote transition zone, corresponding to 390 °C and 6.3 kbar (cf. Figure 4.33), and Taylor (1971) even quotes temperatures of about 300 °C for graphite formation in rocks which were under shear stress. In many metamorphic rocks graphite has been destroyed due to oxidation with free hot water (Taylor, 1971).

In contrast to the electron microscopic observations of Bonijoly et al. (1982), and also in contrast to Landis (1971) who concluded from studies of metamorphic rocks in New Zealand that graphitization occurs in stages ('jumps'), Itaya (1981) suggests a continuous change of the 'crystal' structure from the bituminous coal stage to the graphite stage. Nevertheless, Figure 9 of his paper (plot of H/C atomic ratio $v. d_{002}$) shows two breaks in the curve, the first corresponding to a drastic decrease in d_{002} at the beginning of the semi-graphite stage, the second break representing the attainment of the graphite structure with $d_{002} = 3.35$ Å at an atomic ratio H/C of 0.05. According to Rouzaud (pers. comm.), X-ray diffraction diagrams represent mean statistical data (from many particles whose crystalline structure may vary) and, therefore, cannot reveal the sudden changes of single particles which become recognizable under the high resolution electron microscope.

4.5   Conclusions

During very low-grade metamorphism of rocks, organic matter reaches the coalification stage (rank) of meta-anthracite, according to the scientific coal classification. This stage corresponds to the stages of high-rank anthracite + low-rank meta-anthracite of the commercial coal classification used in North America. According to the commonly-used coalification parameters, meta-anthracites are distinguished by the following values:

    4.0–1.25% volatile matter
  93.5–96.5% carbon
    2.5–0.8% hydrogen
    4–> 5(− < 10%) maximum vitrinite reflectance (R_{max})
    3–> 2% minimum vitrinite reflectance (R_{min})
  > 3.5% mean (random) vitrinite reflectance (R_m).

However, these classical rank parameters must be considered to lose some of their former significance when the meta-anthracite stage is reached, because first indications of pre-graphitization become more important. A remarkable change in the ultra-fine structure has been revealed by X-ray and electron diffraction diagrams as well as by direct observations under the high resolution transmission electron microscope. Due to a resolution power of 3 Å it is possible to see single aromatic layers in meta-anthracites.

Meta-anthracites have a turbostratic texture, i.e. the aromatic lamellae lie more or less parallel to each other, but the C atoms of adjacent layers still have an arbitrary position. With increasing meta-anthracitization and pre-graphitization, the space between single lamellae decreases and the diameter of the layers as well as the number of C atoms per layer increase.

The most common method for rank measurements in relation to rock metamorphism is reflectance measurement under the microscope. Using this method it must be considered that organic constituents (macerals) follow different coalification tracks (Figures 4.11, 4.12). In the meta-anthracite stage, liptinites (like alginite, sporinite and cutinite) have reached higher reflectance values and a stronger anisotropy than the vitrinites, and maximum vitrinite reflectance surpasses the reflectance of inertinites. These reversals of former behaviour are due to the easier graphitizability of the hydrogen-richer constituents. This is the reason why the coal petrologist may sometimes find organic inclusions which, according to optical parameters, are in the stages of meta-anthracite, semi-graphite and even graphite side by side in the same sample of anchimetamorphic rocks.

Irrespective of these primarily-induced causes, values of vitrinite reflectance measured in metamorphic rocks tend to vary considerably due to biaxial optical properties which are caused by tectonic stress and also by very local deformations of the ultra-fine structure due to pressure of surrounding minerals during rock compaction. Consequently, it is practically impossible to measure true R_{max} values. Approaches are possible if the 10–20% highest values measured are averaged, as are the 10–20% lowest values to approach the real R_{min}. In the meta-anthracite stage, R_{min} begins to decrease instead of increasing as before. A value of $< 2\%$ R_{min} is proposed to designate the boundary between meta-anthracite and semi-graphite.

Instead of vitrinite reflectance, graptolite reflectance determined on polished rock sections and/or the colour of isolated conodonts in transmitted light may be applied as microscopic rank parameters for very low-grade metamorphic rocks.

Rock temperatures leading to the formation of meta-anthracites are estimated to lie between 200–350 °C. Since pre-graphitization and graphitization are strongly promoted by high lithostatic pressure, especially if combined with shearing stress and gliding, natural graphitization temperatures may vary between 300–700 °C. Modern electron microscopic and diffraction studies revealed that both semi-graphitization and graphitization occur suddenly as 'jumps' and, initially, also very locally within a single particle.

The correlation between rank of organic matter and mineralogical changes during very low-grade metamorphism will be treated in Chapter 7 of this volume.

5 Fluid inclusion studies during very low-grade metamorphism

JOSEF MULLIS

5.1 Introduction

Fluid inclusions in minerals from very low-grade metamorphic rocks, particularly quartz, are assumed to have retained a constant composition and volume since the time of trapping. Thus, every inclusion represents a small amount of the palaeofluids present in the rock at some specific time during metamorphism. Several fluid inclusion generations within a crystal record the evolution of fluid composition which occurred in the rock during a part of the metamorphic history.

Detailed fluid inclusion studies of very low-grade metamorphic rocks started some fifteen years ago, but to date, relatively little work has been done on this topic (see reviews in Crawford, 1981a; Roedder, 1984; Crawford and Hollister, 1986).

The dominant fluid species encountered in most low-grade metamorphic rocks are HHC (higher hydrocarbons than methane), CH_4, H_2O and CO_2, although in some rocks N_2 and H_2S may also become significant components. Minor species, such as H_2, CO or SO_2, are generally rare and, if present, are only recorded in small concentrations. Aqueous solutions usually contain chlorides and are occasionally associated with very small amounts of dissolved sulphates, carbonates or other salts. Very low-grade metamorphic fluids, therefore, generally contain species in the C–O–H–N–S–salt system, although in most cases fluid composition can be meaningfully represented by the system C–O–H. On a regional scale, very low-grade metamorphic fluids can be grouped into three fluid zones (Figure 5.1); (i) a *higher hydrocarbon zone*, characterized by fluids with 1 to > 80 mole % HHC; (ii) a *methane zone* comprising CH_4–H_2O mixtures and containing 1 to > 90 mole % methane but < 1 mole % HHC; and (iii) a *water zone* containing H_2O–CO_2–CH_4–N_2 fluid mixtures with 80 to > 99 mole % H_2O, < 10 mole % CO_2 and < 1 mole % CH_4 (Mullis, 1979). Exceptions are observed in fine-grained sedimentary rocks, for example which may contain significant amounts of organic matter (such as black shales) and may liberate copious quantities of dissolved HHC, CH_4 or N_2 in the water zone (Price, 1981, 1982; Price et al., 1981).

To achieve an understanding of the chemical equilibria and phase behaviour of fluid inclusions during trapping, the constraining $P-\bar{V}-T-X$ data and

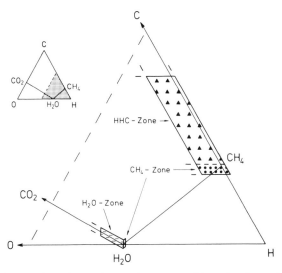

Figure 5.1 Fluid zonation of very low-grade metamorphic rocks shown in the C-O-H projection.

kinetics of chemical equilibria of the systems involved should be well known. However, chemical equilibria in the C–O–H–N–S–salt systems for $P–T$ conditions of very low-grade metamorphism are only poorly known through few experimental and theoretical studies. Most investigations have been undertaken at temperatures of more than 300 °C and pressures of more than 1 kbar where C–O–H fluids were considered to be in equilibrium with oxygen buffers (Eugster and Wones, 1962), graphite buffer (French and Eugster, 1965) or both together (Eugster and Skippen, 1967). Additionally, experimental or theoretical studies on equilibria of C–O–H fluids with N, S and minerals containing iron have been made (Holloway and Reese, 1974; Ohmoto and Kerrick, 1977; Frost, 1979; Holloway, 1981).

This contribution reviews the principles of fluid inclusion studies in quartz of fissures and rocks of very low-grade metamorphism. Host mineral and fluid inclusion characterization are initially presented. After briefly describing some current analytical methods applied in fluid inclusion studies, the thermo-optical techniques are emphasized for estimation of fluid density and composition. Based on such a methodical approach, the use of fluid inclusions for geothermometry and geobarometry in very low-grade metamorphism is then evaluated. Finally, several examples of fluid inclusion studies in very low-grade metamorphic terrains are reported and discussed in terms of fluid immiscibility, quartz growth, thermobarometry and fluid formation mechanisms.

Abbreviations used in this chapter are given in Table 5.1.

Table 5.1 Abbreviations

T	Temperature
P	Pressure
V	Volume
Th, Ph	Temperature and pressure of homogenization
Tt, Pt	Temperature and pressure of trapping
Tm	Temperature of melting
Te	Temperature of eutectic
Tp	Triple point
Cp	Critical point
n	Number of moles
M	Molecular weight
X	Mole fraction
$\bar{V}$	Molar volume
wt %	Weight %
hydr	Hydrate
clathr	Clathrate
vol	Volatile part
aq	Aqueous part
equiv	Equivalent
s, l, v	Solid, liquid, vapour

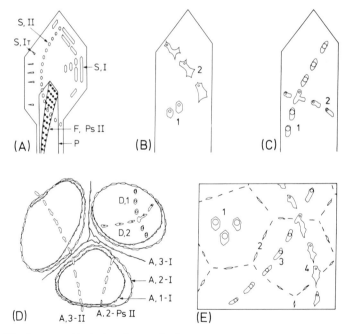

Figure 5.2 Quartz and inclusion characterization.

Quartz typology

A: Fibre quartz (F), overgrown by prismatic quartz (P) and skeletal quartz (S). B, C: Prismatic quartz. D: Detrital quartz (D), overgrown by two authigenic quartz generations (A). E: Recrystallized quartz in massive rocks.

5.2   Characterization of host minerals and their fluid inclusions

Fluid inclusion studies should rely on optical examination of the host mineral as well as on type and texture of inclusions within the host. For confident use in petrology, the fluids must have been preserved without chemical or physical change since trapping. Quartz seems to best fulfil these conditions. Thus criteria regarding quartz morphology, inclusion typology, possible alteration of fluid inclusions as well as relative trapping chronology are discussed in this section.

5.2.1   *Quartz morphology*

Several quartz types may be distinguished in very low-grade metamorphic rocks, including detrital quartz, authigenic quartz, recrystallized quartz and fissure quartz (Figure 5.2). Detrital quartz is often overgrown by authigenic quartz rims and these may contain several distinct fluid inclusion generations (Figure 5.2*D*). Fissure quartz often displays different generations of growth and may occur as several varieties, including milky quartz, simple prismatic

---

Inclusion typology

Primary inclusions   (I), trapped during quartz growth:

(i) in skeletal quartz along faces, edges and corners (A: S, I) and also in 'tubes' behind solid inclusions (A: S, I_T).

(ii) In authigenic quartz overgrowths, along the boundary between detrital and authigenic quartz (D: A, 1–I), along the boundary between the first and second authigenic quartz generation (D: A, 2–I) and also along crystal boundaries of authigenic quartz of the same generation (D: A, 3–I).

Pseudosecondary inclusions   (Ps II), trapped in already crystallized quartz, but partially or totally overgrown by prismatic or skeletal quartz (A: F, Ps II), or overgrown by a second authigenic quartz rim (D: A, 2–Ps II).

Secondary inclusions   (II), trapped after quartz growth. All inclusions in B, C and E, showing differences in shape, textural position and crosscutting relations. Inclusions that appear to be primary are mostly very early pseudosecondary inclusions (B: 1) or 'relics' after quartz recrystallization (E: 1).

Relative chronology of inclusions

Primary inclusions are dated with respect to their host mineral, pseudosecondary with respect to their overgrowth. Some of the most common criteria for dating secondary inclusions are:

(i) Textural position of inclusions: 'Isolated' inclusions are generally older than inclusions arranged along well defined trails (B: 1/2; E: 1/2).

(ii) Shape of inclusions: Aqueous inclusions with a negative quartz shape are usually formed earlier at higher temperature than large, poorly recrystallized inclusions (B: 1/2). However, gas-rich inclusions may be contemporaneous or younger.

(iii) Crosscutting relations of inclusion trails: The inclusion content from an earlier trail has been replaced by the fluid of a younger trail with conservation of the cavity at the intersection (C: 1/2; D: D, 1/2; E: 3/4).

(iv) Concentration of inclusions along the grain boundary of the recrystallized quartz. Such inclusions are earlier than inclusion trails cross-cutting grain boundaries (D: A, 1–I/A, 2 − Ps II; D:A, 3–I/A, 3–II; E:2/3).

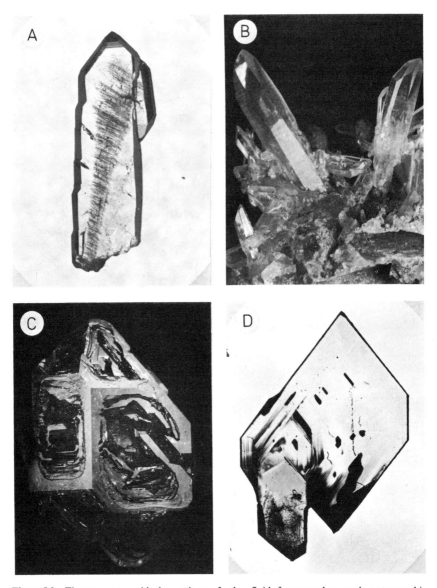

Figure 5.3 Fissure quartz as ideal containers of palaeofluids from very low-grade metamorphic rocks of the external parts of the Central Alps. *A:* Fibre quartz, syncinematically grown during opening of Alpine fissure. Planes inside the fibre quartz consist of pseudosecondary fluid inclusions (Val d'Illiez, methane zone; length of crystal is 7 mm). *B:* Prismatic quartz containing secondary inclusions. (Vättis, water zone; size of crystal group is 7 cm). *C:* Window-shaped skeletal quartz (Val d'Illiez methane zone; length of crystal is 32 mm). *D:* Thick section of a prismatic quartz overgrown by sceptre shaped skeletal quartz. Several growth zones become visible after X-ray irradiation. The prismatic quartz contains secondary inclusions, the skeletal quartz predominantly primary inclusions showing negative crystal shape (black) (Val d'Illiez, methane zone; length of crystal is 11 mm).

quartz, fibre quartz or skeletal quartz (Figure 5.3). Faces of earlier growth stages may sometimes be covered with solid inclusions and, as a result, multiple growth stages become discernible. Where such criteria are lacking, different growth stages are readily distinguishable by hot cathodolumines- cence (Matter and Ramseyer, 1985) or X-ray irradiation (Figure 5.3D). Fibre- and skeletal quartz varieties have particular significance for fluid inclusion studies and are described here in more detail.

Fibre quartz (Figures 5.2A, 5.3A, 5.10A) typically crystallizes during the opening and enlargement of fissures. Growing quartz crystals, fixed between the two fissure walls, are repeatedly cracked by fissure extension and resealed, forming quartz fibres (Lemmlein, 1946; Stalder and Touray, 1970; Mullis, 1976b). This corresponds to the 'stretched crystals' of Durney and Ramsay (1973). During this 'crack-seal' mechanism (Ramsay, 1980), fluids are systema- tically trapped in the 'growing' crystal and record the physico-chemical evolution of the fluid from which the fibre is crystallized. Fibre quartz may be overgrown by both the prismatic and skeletal quartz varieties which are widespread in very low-grade metamorphic terrains.

Skeletal quartz (Figures 5.2A and 5.3C, D) characteristically displays crystal growth on edges and corners. The most common forms are 'sceptre'- and 'window-shaped' quartz crystals (Gambari, 1868; Bombicci, 1898; Stalder and Touray, 1970; Mullis, 1976a). The internal morphology of these two varieties is characterized by Brazil twinning, complex structural defects and multiple alternations of thin lamellae oriented parallel to the terminal r, z and m faces. All these criteria suggest discontinuous but rapid crystal growth from a gas- enriched, emulsion-like fluid.

5.2.2 Types of fluid inclusions

The most common classification scheme for fluid inclusions is based on their origin and recognizes primary, secondary and pseudosecondary inclusions. Detailed descriptions are given in Yermakov (1965) and Roedder (1981, 1984).

Primary inclusions are usually arranged within crystals along growth faces, edges and corners (Figures 5.2A, 5.3D, 5.4A–C) or upon abraded surfaces of detrital grains (Figure 5.2D). These inclusions often form during rapid crystal growth. If such inclusions are enriched in gas, then they often show negative crystal shapes displaying extreme variation in size from < 10 μm to in excess of 1000 μm. Solid inclusions sometimes disturb the growth of more or less perfect crystal faces causing formation of small tubes, which are commonly aligned along the growth direction of quartz and filled with fluid.

By contrast, secondary inclusions (Figures 5.2, 5.4D–F) generally occur in trails along healed fractures (caused by brittle deformation) which may cross- cut grain and crystal boundaries. Fluid inclusions in different trails are often different in composition, shape and size, and are, therefore, indicative of relative trapping chronology.

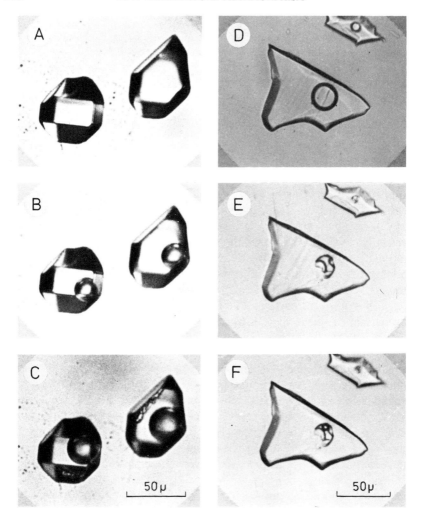

Figure 5.4 Methane-bearing fluid inclusions at temperatures between -150 and $+25\,°C$ (Val d'Illiez, Switzerland).

A–C Primary methane-rich inclusions:

A $+25\,°C$ Supercritical methane.

B $-100\,°C$ Methane vapour bubble in liquid methane. $Th_{CH_4} = -92\,°C$. $\rho_{CH_4} = 0.267\,g/cm^3$.

C $-150\,°C$ Solid carbon dioxide in liquid methane. Sublimation temperature of CO_2 solid $= -108\,°C$.

D–F Secondary water-rich inclusions:

D $+25\,°C$ Bubble of supercritical methane in aqueous chloride solution.

E $-50\,°C$ Supercritical methane (outlined by a dark rim) surrounded by ice and methane clathrate. $Tm_{ice} = -1.5\,°C$. Salinity $= 2.6$ wt% equiv. NaCl.

F $-100\,°C$ Methane vapour bubble within liquid methane, both surrounded by ice and methane clathrate. $Th_{CH_4} = -93\,°C$. $\rho_{CH_4} = \sim 0.272\,g/cm^3$.

Inclusion trails terminated at growth boundaries of a host crystal are named pseudosecondary (Yermakov, 1949). The cracks along which these occur were healed during further quartz growth so that they are secondary with respect to the host crystal, but primary with respect to the enclosing overgrowth (Figures 5.2A, D and 5.3A).

5.2.3 *Stretching, leakage and necking-down*

Despite the fact that quartz undergoes effectively no chemical interaction with the trapped fluid (except during the annealing processes within fluid inclusions), changes in fluid composition and density may occur either after trapping in nature or during sample preparation and microthermometry measurements in the laboratory.

Stretching. Stretching is a permanent deformation of the host crystal caused by internal overpressure and is commonly observed. During heating in the laboratory the effects of stretching become evident by a rise of the homogenization temperature during successive heating/cooling runs. Such behaviour is often observed in methane-bearing aqueous inclusions whose fluid pressure may increase considerably with temperature. In large, aligned or poorly sealed aqueous inclusions stretching may also occur through expansion of ice upon freezing of the aqueous fluid. In both these cases of stretching, the bulk density decreases and, therefore, the vapour bubble in the inclusion increases at room temperature after each run. In order to avoid potentially inaccurate measurements, the consistency of the bubble volume should be checked both before and after heating and freezing runs. Additionally, a check on the consistency of the homogenization temperature between different runs should give an indication of stretching if present. Criteria useful for the recognition of stretched inclusions are given by Bodnar and Bethke (1980, 1984).

Leakage. When the internal fluid pressure exceeds the strength of the host mineral, inclusions may leak by partial or total decrepitation. Partially decrepitated inclusions are usually surrounded by a halo of newly formed 'satellite' inclusions, whilst totally decrepitated fluid inclusions are often refilled by a younger fluid of a different chemical composition, and are widespread in the methane zone. These phenomena are often associated with fissures which have undergone rapid enlargement, causing considerable pressure decrease in the fissure system (Mullis, 1975, 1976a). A pressure decrease during retrograde crystal growth of > 1 kbar with respect to earlier-formed high-density methane-rich fluid inclusions is common. According to the experimental investigations of Leroy (1979) and Pécher and Bouiller (1984), fluid inclusions greater than 35 μm can be expected to decrepitate with an internal overpressure of 800–900 bar. Similar decrepitation phenomena in medium- and high-grade metamorphic rocks are discussed by Poty (1969),

Rich (1975), Hollister and Burruss (1976), Burruss (1977), Touret (1977), Kreulen (1977), Crawford et al. (1979), Hollister et al. (1979) and Swanenberg (1980).

Comparable partial and total decrepitation can be produced in the laboratory as first described by Lemmlein (1956) in synthetic crystals of NaNO$_3$. During heating as internal fluid pressure increases with temperature, the confining pressure remains at the ambient room condition resulting in leakage. Therefore, aqueous and gaseous inclusions containing high density methane will tend to decrepitate before bulk homogenization of the inclusions.

The importance of recognizing the potential effects of leakage is well demonstrated by the common misinterpretation of fluid inclusions in zones of dynamic (ductile deformation) and static recrystallization fabrics from high grade anchimetamorphic and epimetamorphic rocks. During these processes it is well known (Voll, 1976; Wilkins and Barkas, 1978) that fluids are eliminated from within quartz grains and are expelled towards grain boundaries. Those fluid inclusions that occasionally remain within the recrystallized quartz characteristically display well-shaped negative quartz forms and are often misinterpreted as primary inclusions. However, the composition of these fluid inclusions can be shown to have been inherited from the pre-recrystallization inclusions, even though there is a marked lowering of density with respect to the earlier inclusions. These inclusions are, therefore, newly formed inclusions but contain a relic of a former fluid. The interpretation of such phenomena is confirmed in heating experiments over several weeks carried out by Gratier (1982, 1984) and Pécher and Bouiller (1984).

Loss of matter from fluid inclusions can also occur by diffusion. Molecular hydrogen, because of its small size, has frequently been considered to have diffused outside of fluid inclusions (e.g. Hollister and Burruss, 1976, Roedder and Skinner, 1968). However, recent Raman microprobe studies (Dubessy et al., 1980) and thermodynamic calculations (Dubessy, 1984) indicate that hydrogen loss from fluids entrapped in quartz is negligible at the P–T conditions of very low-grade metamorphism.

Necking-down. During annealing and recrystallization of relatively large inclusions and fractures within quartz, progressively smaller inclusions form in an attempt to reduce the amount of surface energy. This process is called necking-down (Roedder, 1962a). As long as the trapped fluid remains homogeneous, necking-down does not influence the composition or density of the system. However, if phase changes occurred prior to completion of necking-down, the new inclusions may show important differences in composition and density from the precursor inclusions. Fluid inclusion trails displaying variations in gas–liquid ratios can sometimes be interpreted in this manner.

Detailed optical study of thick sections either irradiated with X-rays or bombarded with electrons for cathodoluminescence, may reveal complex colour zonation in fissure, authigenic and recrystallized quartz. This colour banding usually represents multiple stages of growth. Careful observation of the textural relationships which the primary, pseudosecondary and secondary inclusions display with these growth zones enables a relative chronology or fluid inclusion sequence to be established (Figures 5.2 and 5.3A, D).

5.2.4   Relation between fluid inclusion trapping and metamorphism

During prograde metamorphism fluid inclusions are gradually expelled from the host minerals as recrystallization proceeds. Moreover, fissure quartz precipitation generally takes place during or after the peak temperature is reached. Therefore, to obtain any information regarding the peak temperature event during metamorphism it is important to use the earliest fluid inclusion generation for measurement. For a given locality the earliest inclusions are found within syncinematically-grown fibre quartz.

Alternatively, the relative timing of fluid entrapment with respect to metamorphic events can be achieved by comparing fissure minerals with metamorphic rock-forming minerals of the same paragenesis. For example, Poty (1969) and Poty and Stalder (1970) reported greenschist facies minerals from Alpine fissures that occur in both greenschist facies host rocks of the Mont Blanc Massif and also in amphibolite facies host rocks of the Penninic area. This indicates that fluids in fissure quartz of the Mont Blanc Massif must have been trapped in the vicinity of the temperature maximum, whereas fluid inclusions from the Penninic area record retrograde conditions.

Daughter minerals may also be found within fluid inclusions. If present, the type and sequence of those minerals precipitated parallels the paragenesis of metamorphic minerals in both the fissure and the host-rock. Such daughter minerals are, therefore, also suited for estimating relative trapping time with respect to metamorphism.

5.3   Analytical methods

The small size and different varieties of positions that fluid inclusions may have inside a given host makes it difficult to characterize their contents. Several methods are thus usually used to analyse fluid inclusions. The choice of the analytical method depends on the particular problem to be solved and on the quality and quantity of the trapped fluid. There are two fundamentally different approaches to the analysis of fluid inclusions: microanalytical and bulk analytical techniques. Microanalytical techniques have the great advantage that *individual* fluid inclusions can be analysed, thus giving information on specific generations of inclusions. Bulk analytical techniques, although providing additional different information, suffer from the problem of only giving average values of several contained inclusion populations.

5.3.1 *Microanalytical non-destructive methods*

A given sample usually displays several fluid populations of different compositions that may have been trapped at different times. To maximize the retrieval of geochemical information from individual fluid inclusions, microanalytical non-destructive methods must first be employed before studying samples by destructive methods. Well-established microanalytical non-destructive methods include microthermometry and Raman spectroscopy.

Equipment for microthermometry consists of an optical microscope combined with a heating and freezing stage. Different stages have been designed by Poty *et al.* (1976), Werre *et al.* (1979) and Shepherd (1981). Full details of the technique are given by Yermakov (1965), Roedder (1962*b*, 1984), Hollister and Crawford (1981) and Shepherd *et al.* (1985). In essence, the procedure involves measuring the temperature of phase transitions during heating (see section 5.4). Accurate calibration and careful measurements provide a good estimation of fluid composition and density if experimental data of the relevant unary, binary or multicomponent system are known. Unfortunately, small amounts of gas species are often difficult or impossible to detect by microthermometry. Raman spectroscopy can be used to supplement information provided by microthermometry.

In micro-Raman spectroscopy the sample is illuminated by a visible monochromatic laser radiation which is focused by a high-magnification objective of a standard optical microscope. The same objective collects all the scattered radiation which is analysed by a monochromator (Rosasco *et al.*, 1975; Delhay and Dhamelincourt, 1975). This technique enables crystals and single molecules of the C–O–H–N–S system (Dhamelincourt *et al.*, 1979) free or enclosed in non-fluorescent transparent minerals to be identified (Touray *et al.*, 1985). Mole fractions of the volatiles can be derived from intensity measurements. The sensitivity and accuracy of the analyses depend on fluid density, size and depth of the inclusions, refractive index of minerals (Dubessy *et al.*, 1987) and the relative Raman scattering cross-section of the different compounds (Cheilletz *et al.*, 1984; Wopenka and Pasterias, 1986).

An additional, convenient technique for identifying the presence of hydrocarbons within fluid inclusions and of different populations of them is UV-excited fluorescence microscopy as described by Kvenwolden and Roedder (1971) and Burruss (1981*a*).

Despite the advantage of individual fluid inclusion study these non-destructive methods do have shortcomings. Micro-Raman spectroscopy cannot be used to analyse non-transparent or fluorescent minerals. Moreover, the lower detection limit for trace elements in fluid inclusions using the micro-Raman spectroscopy technique is relatively high, as the volume of individual fluid inclusions is small.

To circumvent some of those shortcomings, destructive methods may be used for the investigation of fluid inclusions.

5.3.2   *Micro- and bulk-analytical destructive methods*

Several destructive methods suitable for the analysis of individual fluid inclusions have been developed during the last 15 years. The ion microprobe is suitable for ion-detection in complex brines (Nambu *et al.*, 1977). A wide range of metals and some non-metals can be detected by emission spectroscopy after liberating and then 'vaporizing' the inclusion contents by laser-beam ablation (Tsui and Holland, 1979; Bennett and Grant, 1980). Ionization of the liberated matter by laser probe mass spectrometry (LPMS) allows the semi-quantitative and quantitative analyses of solids and liquids (Deloule and Eloy, 1982). Simple laser decrepitation can also be employed for determination of the released H_2O and CO_2 by a capacitance manometer (Sommer *et al.*, 1985).

The contents of fluid inclusions in bulk samples can be liberated for analysis by crushing or heating. A simple method is to use a crushing stage for detecting the presence and identifying the types of compressed volatiles (Deicha, 1950; Roedder, 1970). This stage can be used in the laboratory as well as the field. In crush–leach methods the sample is crushed in the presence of water. The resulting dilute solution can be analysed by atomic absorption spectrometry, emission flame photometry, colorimetry or other methods to determine the relative amounts of dissolved anions and cations (Roedder *et al.*, 1963; Poty *et al.*, 1974). Neutron activation analysis prior to crushing (reviewed in Touray, 1976) prevents possible contamination of the inclusion composition. Using gas chromatography, various components, such as H_2O, CO_2, CH_4, N_2, H_2S, HHC, etc., of any gas mixture in fluid inclusions can be detected and quantified (Barker, 1966; Touray and Lantelme, 1966; Zimmermann, 1966). Alternatively, mass spectrometry is also well suited for determining some gaseous species. If sufficient material is available, D/H, $^{13}C/^{12}C$, $^{18}O/^{16}O$ or $^{15}N/^{14}N$ ratios can be determined as a step towards elucidating the source of the fluids trapped within the inclusions (Hoefs and Stalder, 1977; Hoefs and Morteani, 1979).

However, as with the microanalytical techniques the use of bulk destructive methods is also subject to several shortcomings (Cheilletz *et al.*, 1984):

(i)   The analysed fluids often represent mixtures of different fluid populations
(ii)   Fluids analysed may be contaminated by the host or by some daughter minerals
(iii) Complete extraction of all the trapped fluids is difficult.

Nevertheless, with continual technical improvements and with careful choice of suitable material, reliable results can be obtained. Cheilletz et al. (1984), for example, were able to demonstrate that the results of measurements from four analytical techniques (microthermometry, Raman spectroscopy, mass spectrometry and gas chromatography) of water-enriched fluid inclusions scatter within 5%. Additionally, a new approach for analysing individual inclusions or their populations by computerized mass spectrometry is being developed by Barker and Smith (1986). This method is based on inclusion decrepitation under vacuum during slow heating. In this technique the mass spectra, and hence composition of the released fluids, is recorded in minimum time steps of 25 μs.

5.4   Estimation of fluid composition and density by microthermometry

A first, very informative approach for simultaneous optical examination and non-destructive analytical investigation of individual fluid inclusions is microthermometry. As this method enables the composition and density of the main types and generations of fluid inclusions in a given sample to be estimated, this method is discussed here in some detail.

5.4.1   *The principle of microthermometry*

The principle of microthermometry is illustrated for many unary systems in Figure 5.5, and some properties of selected species are given in Table 5.2. Assume, for example, a liquid-like dense fluid A that has been trapped at temperature Tt_A and pressure Pt_A. During cooling, the inclusion will move

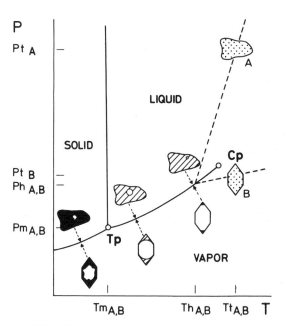

Figure 5.5   Phase transitions during heating of fluid inclusions with high (*A*) and low (*B*) density unary fluids. Tp = triple point; Cp = critical point. Inclusions: black = solid phase; hatched = liquid phase; white = vapour phase; stippled = fluid phase.
Phase transitions: Tm, Pm = temperature and pressure of melting; Th, Ph = temperature and pressure of homogenization; Tt and Pt = temperature and pressure of trapping.

Table 5.2 Triple points and critical points of some selected one-component systems. Data from Allamagny (1976)

Species	Triple point $T(°C)$	P(bar)	Critical point $T(°C)$	P(bar)
N_2	-210.00	0.13	-146.95	34.00
CH_4	-182.47	0.12	-82.62	45.96
CO_2	-56.57	5.18	31.06	73.82
C_2H_6	-183.27	0.11×10^{-4}	32.27	48.84
C_3H_8	-187.68	3.3×10^{-9}	96.67	42.50
H_2S	-87.5	0.23	100.05	89.37
C_4H_{10}	-138.29	4×10^{-6}	152.03	37.96
H_2O	$+0.015$	6.11×10^{-3}	374.12	221.2

along an isochore (line of constant density) until it reaches the liquid–vapour curve, where a small vapour bubble appears. For a vapour-like dense fluid, such as trapped in inclusion B at temperature Tt_B and pressure Pt_B, a film of liquid will form and wet the inclusion walls as the inclusion cools below the intersection on the liquid–vapour curve. Both these bubbles will change in size as further cooling takes place, until freezing occurs well below the triple point Tp. The sequence of phase transitions must be measured while the temperature is slowly increased because nucleation of new phases during cooling is usually very sluggish, reflecting metastability (Roedder, 1971). During heating the species melts at temperature $Tm_{A,B}$. The liquid and vapour phase of the inclusion A will homogenize into the liquid phase at temperature Th_A and pressure Ph_A. Similarly, inclusion B will homogenize into the vapour phase at temperature Th_B and pressure Ph_B. The same principle can be applied, with the exception of H_2O, to phase transitions in any unary fluid system.

If, however, one volatile component inside a given fluid mixture is visible, e.g. unmixed liquid and gaseous CO_2 inside a water-rich solution, temperature and pressure of solid–vapour or liquid–vapour phase transitions may vary but their sequence (as shown in Figure 5.5) remains similar to that shown for unary systems.

5.4.2 Salt determination

As the melting point of ice in saline solutions is related to its salt content, freezing studies are used to estimate the salinity of aqueous solutions. For pure aqueous NaCl solutions the measured Tm can be converted directly to weight% (wt%) using melting data established by Potter et al. (1978). However, in natural settings, fluid–rock equilibria involving complex multicomponent systems are more likely to be present. The most abundant species normally encountered are NaCl, KCl, $CaCl_2$ and $MgCl_2$. In most such complex salt systems accurate salt determination is not possible for three main reasons.

(i) The eutectic or the first melting temperature (Te) of aqueous chloride systems must be measured to obtain diagnostic information on the salt composition. However, this is often difficult to observe due to the small size of the fluid inclusions. In some instances this problem can be overcome by sequential freezing (Mullis and Stalder, 1987);

(ii) Complex salt solutions cannot always be separately resolved, because of the similarity of Te of some different solutions; and

(iii) There is a lack of comprehensive experimental data directly applicable to many of these complex salt solutions.

Nevertheless, dissolved salts may be qualitatively characterized and semi-quantitatively estimated. The exact eutectic temperature of a given salt system depends upon the type of dissolved salts present (Table 5.3). Addition of any other salt to an existing water-salt system lowers Te to a given value. Therefore, the eutectic behaviour can be used to provide useful constraints on the dominant type of dissolved ionic species present. If, for example, the measured Te in a fluid inclusion lies in the range between -20 and $-25\,°C$ then the dissolved salts are dominated by the $H_2O-NaCl-KCl$ ternary system (see Table 5.3). Alternatively, a measured Te value of around or slightly below $-50\,°C$ indicates the $H_2O-NaCl-CaCl_2$ ternary salt system is dominant

Table 5.3 Eutectic temperatures and compositions of some selected chloride and non-chloride aqueous solutions

Dissolved species	Eutectic temperature (°C)	Eutectic composition (wt%)	Reference
KCl	-10.7	19.5%	1
NaCl	-20.8	32.2%	2
$MgCl_2$	-33.6	20.6%	1
$CaCl_2$	-49.8	30.2%	1
LiCl	-74.8	?	3
NaCl-KCl	-22.9	20.2% NaCl 5.8% KCl	1
NaCl-$MgCl_2$	-35	1.6% NaCl 22.7% $MgCl_2$	1
NaCl-$CaCl_2$	-52.0	1.8% NaCl 29.9% $CaCl_2$	1
NaCl-$CaCl_2$-$MgCl_2$	-57	?	4
Na_2SO_4	-1.2	4.0%	1
Na_2CO_3	-2.1	5.9%	1
$NaHCO_3$	-2.3	6.3%	1
NaF	-3.5	4.1%	1
NaBr	-28.0	40.3%	1

Data from: (1) Schäfer and Lax (1962)
            (2) Potter et al. (1987)
            (3) Borisenko (1977)
            (4) Luzhnaya and Vereshtchetina (1946)

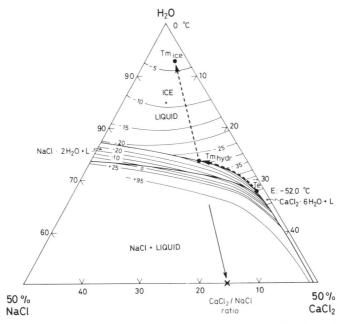

Figure 5.6 NaCl–CaCl$_2$–H$_2$O system. Compositions in wt %. Dashed line shows the melting path of a given aqueous chloride solution as described in section 5.4.2. Te = temperature of eutectic; Tm_{hydr} = melting temperature of hydrate; Tm_{ice} = melting temperature of ice. Figure slightly modified after Crawford (1981b); data from Yanatieva (1946).

since calcium is by far the most common bivalent cation in natural aqueous chloride solutions.

Now, consider in detail an example of how we can determine the salt composition of a given ternary aqueous chloride brine in a fluid inclusion. During microthermometric heating of a low to moderate saline solution a complex melting sequence can be observed (Figure 5.6). At the eutectic temperature of $-52\,°C$, CaCl$_2$·6H$_2$O melts. As heating continues, the composition of this initial fluid will move along the cotectic curve (separating the NaCl·2H$_2$O + liquid field from the ice + liquid field) until hydrohalite (NaCl·2H$_2$O) melts at the point Tm_{hydr}. Further heating results in the evolution of the fluid along the tie line connecting the melting point of the hydrate with the water pole of the ternary system until ice melts at Tm_{ice} and defines the final fluid composition. The CaCl$_2$/NaCl ratio can now be estimated by simple extrapolation of the H$_2$O–Tm_{hydr} tieline on the NaCl–CaCl$_2$ join. Unfortunately, hydrate melting is optically difficult to identify but sometimes it can be fixed by sequential freezing and heating (Haynes, 1985).

Chemical analyses of fluid inclusions from Alpine fissure quartz of the Western and Central Alps which may be typical of very low-grade metamorphic fluids (Poty et al., 1974; Mullis, 1976a) have shown that the most common anion present is chloride. These predominantly chloride solutions have the following average cation ratios: K^+/Na^+, 0.04–0.2; Ca^{2+}/Na^+, 0.001–01; Mg^{2+}/Na^+, 0.0001–0.01. This indicates that the most abundant monovalent cation is sodium, whilst the dominant divalent cation is calcium. Salt systems in the methane and water zone may, therefore, be characterized either by the binary system H_2O–NaCl or by the ternary system H_2O–NaCl–$CaCl_2$, providing salinities are expressed in NaCl or in NaCl and $CaCl_2$ equivalents respectively.

Eutectic temperatures are also useful for distinguishing other salts from chlorides. For aqueous chloride solutions, $Te < -10\,°C$ whilst for aqueous sulphate, hydrocarbonate and carbonate solutions $Te > -10\,°C$. The only eutectic temperatures that are similar to those of aqueous chloride brines are those of K_2CO_3 and some bromide, iodide and fluoride solutions, but the amount of such salt species in natural systems is usually negligible. Thus, melting temperatures of aqueous solutions displaying $Te < -10\,°C$ can be considered to be indicative of aqueous chloride brines. Experimental data for three- or four-component aqueous chloride brines probably quite typical of natural systems (Luzhnaya and Vereshtchetina, 1946), suggest that the amount of dissolved chlorides may more or less correspond to wt % NaCl equivalents derived from the melting temperature of ice. The error in estimating the salt content of naturally occurring brines in terms of NaCl equivalents is usually $< 5\%$ of the amount present. This is true for most aqueous chloride mixtures, but not for end-member fluid compositions other than H_2O–NaCl.

In addition, salt estimation by freezing point depression is affected by the type and amount of dissolved gas in the aqueous solution. During freezing, gas hydrates (clathrates) remove water from liquid solution. As a result the brine becomes increasingly saline and shows a lower Tm_{ice} than would be normally expected from the H_2O–salt system. If all CO_2 dissolved in aqueous solution at room temperature (< 2–3 mole %), reacts to form the CO_2 clathrate, the increase in salinity of the residual aqueous solutions is about 15–25% (Collins, 1979). If CH_4 is the volatile species, and if all the dissolved CH_4 (< 0.3 mole% at room temperature), and also some part of the methane bubble, reacts to form clathrate, the overestimation of the salt content will be $< 5\%$ (Mullis, 1976a). Bozzo et al. (1973) and Collins (1979) demonstrated that increasing salt content lowers Tm_{hydr}. Thus, in presence of CO_2 liquid and CO_2 gas, Tm_{hydr} may be used for salinity determination. This method only works when no CH_4 or other gases are present.

5.4.3 Gas estimation

Volatiles occur both in the non-aqueous and aqueous part of fluid inclusions. They can be characterized by their melting temperature (Tm_{vol}), by melting behaviour of the clathrate (Tm_{clath}) and by liquid–vapour homogenization (Th), (see Figure 5.5 and Table 5.2).

Pure single-phase volatile systems are recognized by melting of solid N_2, CH_4, H_2S or CO_2 at exactly their triple point. Liquid–vapour homogenization temperatures then range between their critical and triple point values depending upon volatile density (Figures 5.5, 5.11A).

Volatiles, however, rarely occur as pure species in natural systems, and mixtures of N_2, CH_4, CO_2 and of small amounts of H_2S are much more typical, although difficult to recognize. Nevertheless, thermo-optical investigations can often provide valid estimations of fluid speciation and composition. If the melting (sublimation) of a given volatile species shifts towards lower temperatures than its solid–liquid–vapour invariant point, then additional gases are present. Figure 5.7 shows that the effect of the critical behaviour of increasing methane content in CO_2–CH_4 mixtures is to lower the critical temperature of this mixture below $31.4\,°C$ (critical point of CO_2). At the same time the melting (sublimation) temperature of CO_2 falls below its triple point of $-56.6\,°C$. Using both Th_{CO_2} and Tm_{CO_2} of the CO_2–CH_4 mixtures, fluid composition can be determined (Arai $et\ al.$, 1974, in Burruss, 1981b; Heyen $et\ al.$, 1982). Impurities in a CH_4-rich fluid may also be detected. In the ternary CH_4–CO_2–N_2 system small amounts of CO_2 raise the critical temperature of the mixture to values possibly above $-82.4\,°C$ (critical temperature of CH_4), whereas dissolved N_2 in the mixture lowers it. If both dissolved CO_2 and N_2 are present, however, each may counteract the temperature effect of the other. Thus, the presence of CH_4–CO_2, CH_4–N_2 or CH_4–CO_2–N_2 mixtures could easily lead to an incorrect estimation of the density of assumed CH_4- or CO_2-

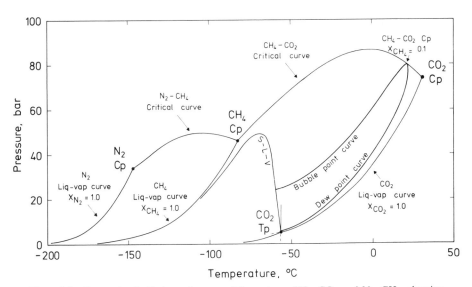

Figure 5.7 Composite P–T phase diagram of the systems CH_4–CO_2 and N_2–CH_4, showing bubble point and dew point curve of the system CH_4–CO_2 at $X_{CH_4} = 0.1$. Constructed after Swanenberg (1979, 1980) and Burruss (1981b).

bearing inclusions respectively. Chemical analysis by either Raman spectroscopy or by gas chromatography is, therefore, recommended in cases where dissolved CO_2 or N_2 are suspected. Small amounts of CO_2 dissolved in CH_4-rich or N_2-rich inclusions may be detected by melting of CO_2 solid at temperatures of at least $< -100\,°C$ (Figure 5.4C). Knowing the approximate volume occupied by CO_2 solid at $-150\,°C$ (CO_2 density of $1.56\,g/cm^3$) the CO_2 content can be estimated (Guilhaumou et al., 1981). Even small amounts of H_2S dissolved in CO_2, CH_4 or N_2 are detectable by H_2S melting (sublimation) below its triple point of $-85.5\,°C$ (Guilhaumou et al., 1984; Dubessy et al., 1984). The presence of CH_4 in the $CO_2–CH_4–H_2O$ system can easily be recognized by clathrate melting at $> 10\,°C$, the CO_2 liquid–CO_2 vapour–CO_2 clathrate–H_2O liquid invariant point (Collins, 1979).

Natural fluids of very low-grade metamorphic rocks are either enriched in methane (methane zone) or in aqueous chloride solutions (water zone). Methane-rich inclusions of the methane zone contain generally some dissolved HHC, CO_2, N_2 and H_2S (Stalder and Touray, 1970; Mullis, 1976a; Dhamelincourt et al., 1979; Dubessy et al., 1984; Guilhaumou and Beny, 1985). As all these species together rarely exceed 10 mole % of the fluid, the trapped volatiles may be considered to approximate to the pure methane end-member for pressure estimation. Inclusions typical of the water zone are usually very poor in volatiles. Exceptions to this generalization include low-grade metamorphic carbonates or black shales that may provide fluids containing larger amounts of either CO_2 (Soom, 1986), or mixtures of CO_2–CH_4–N_2.

5.4.4   Estimation of fluid density and composition

The approximate *bulk density* of the inclusions ρ_l is obtained by the following equation:

$$\rho_l = V_{aq}\cdot\rho_{aq} + V_{vol}\cdot\rho_{vol}$$

where V_{aq} and $V_{vol}(= 1 - V_{aq})$ are the optically estimated volumetric fraction of the aqueous and volatile parts of the inclusions.

The volume fraction of volatiles in the $H_2O–CO_2–NaCl$ system may also be calculated by iteration procedures described in Parry (1986). The density of the aqueous chloride solution (ρ_{aq}) is calculated by the regression equation of Potter and Brown (1977):

$$\rho_{aq} = \frac{1000\cdot\rho_0 + M_{NaCl}\cdot m_{NaCl}\cdot\rho_0}{1000 + A_0\cdot m_{NaCl}\cdot\rho_0 + B_0\cdot m_{NaCl}^{3/2}\cdot\rho_0 + C_0\cdot m_{NaCl}^2\cdot\rho_0}$$

where ρ_0 is the density of water at $40\,°C$ in $g/cm^3 = 0.99164$, M_{NaCl} the molecular wt of NaCl, m_{NaCl} the molality of NaCl, $A_0 = 17.45$, $B_0 = 1.71$ and $C_0 = 0.040$. The density of volatiles (ρ_{vol}) is obtained by the liquid–vapour homogenization temperature as shown for the system CH_4 in Figure 5.11A.

The *bulk molar volume* of the inclusion $(\bar{V}_1)$ is estimated by the equation

$$\bar{V}_1 = \frac{\Sigma M_i \cdot X_i}{\rho_1}$$

where M_i and X_i are the molecular weight and mole fraction of the species i, respectively.

The approximate *bulk fluid composition* is obtained by calculating the mole fractions X_i of the species i.

$$X_i = \frac{n_i}{n_{vol} + n_{H_2O} + n_{NaCl}}$$

$$n_{vol} = \frac{V_{vol} \cdot \rho_{vol}}{M_{vol}} + n'_{vol}$$

$$n_{H_2O} = \frac{V_{aq} \cdot \rho_{aq}(100 - wt\% NaCl)}{M_{H_2O} \cdot 100}$$

$$n_{NaCl} = \frac{V_{aq} \cdot \rho_{aq} \cdot wt\% NaCl}{M_{NaCl} \cdot 100}$$

n_i is the number of moles of species i. M_{vol}, M_{H_2O} and M_{NaCl} are the molecular weight of the main volatile species, water and NaCl, respectively. Wt % NaCl and n'_{vol} are the weight percent NaCl and the number of moles of volatile species dissolved in the aqueous part at room temperature. The CO_2 content dissolved in aqueous solution coexisting with liquid CO_2 at 25 °C and 63 bar is around 5.5 wt % (Wiebe and Gady, 1940). This value decreases with increasing salinity and decreasing pressure. The content of dissolved CH_4 and N_2 at room temperature is < 0.3 wt % and can be estimated using the experimental data of Czolbe (1975), Haas (1978) and Blount *et al.* (1980). The quantity of H_2O vapour dissolved in volatiles at room temperature is neglected for the purpose of this calculation.

Useful Fortran programs for calculation of fluid properties from microthermometric data on fluid inclusions have been published by Nicholls and Crawford (1985).

5.5 Geothermometry and geobarometry

Fluid inclusions in quartz can be considered as closed systems of more or less constant composition (isoplethic) and constant volume (isochoric) since entrapment. Such systems can be completely characterized by their $\bar{V} - X$ properties (Burruss, 1981b). Geothermometric and geobarometric information can be derived if experimental $P - \bar{V} - T - X$ data of the fluids involved are available for the geological $P - T$ conditions of interest. As unmixing phenomena are common in the methane zone, thermobarometry will be discussed for inclusion systems trapped in both the one-phase and the two-phase region.

5.5.1 Fluid inclusions trapped in the one-phase region

Fluid inclusions trapped at equilibrium conditions in the one-phase region display, at room temperature, one, two or more phases with constant volumetric ratios. If composition and density (or molar volume) of a given inclusion fluid are known and if the experimental $P-\bar{V}-T-X$ properties of the involved fluid are available the isochores of that fluid may be calculated (see Roedder, 1984, Table 8-1). Since entrapment must lie somewhere along the evaluated isochore at some temperature greater than the homogenization temperature, the actual trapping temperature must be evaluated along the isochore by an independent geobarometer. This method of pressure correction is demonstrated in Figure 5.10B. An independently determined pressure of 1.5 kbar, for example, yields a trapping temperature at 375 °C. This temperature lies 126 °C above the homogenization temperature of 249 °C, where the isochore originates from the liquid–vapour curve of the system H_2O–NaCl. In the same manner the trapping pressure of the fluid can be estimated by an independent geothermometer.

5.5.2 Fluids trapped in the two-phase region

The solubility of CH_4 in aqueous chloride solutions is very low and, therefore, if dissolved CH_4 is present in any significant quantities, water–methane immiscibility can be expected below temperatures of 300 °C (Welsch, 1973; Price, 1979; Figure 5.8). From this, it follows why the common occurrence of methane-bearing aqueous solutions together with methane-rich inclusions within a given host is widespread in the methane zone.

General rules of immiscibility applied to fluid inclusions have been detailed by Pichavant et al. (1982) and Ramboz et al. (1982). The constraints for real immiscibility are:

(i) The two immiscible phases are trapped at the same time, pressure and temperature
(ii) The molar volumes of the immiscible phases are different, but related to each other
(iii) The fugacities of a given species in the liquid phase are equal to those in the vapour phase:

$$f\ CH_4^{L,P,T} = f\ CH_4^{V,P,T} \text{ and } f\ H_2O^{L,P,T} = f\ H_2O^{V,P,T}$$

The isopleth of each fluid delineates the $P-T$ plane into two areas. The low-temperature side corresponds to the two-phase field for the composition of the isopleth. The homogenization temperature of a fluid with a known composition occurs along this isopleth (Figure 5.9). Applying the phase rule to the CH_4–H_2O system, the degree of freedom (F) of the liquid–vapour equilibrium (coexistence of the two immiscible fluid phases) is $F = C - P + 2 = 2$, where C and P refer to the number of components and phases, respectively. Conse-

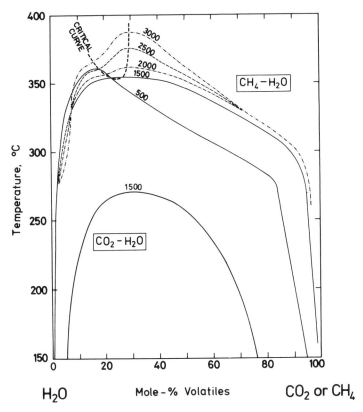

Figure 5.8 Temperature-composition sections for the CH_4–H_2O system at 500, 1500, 2000, 2500 and 3000 bar and for the CO_2–H_2O system at 1500 bar. CH_4–H_2O data for the water-rich part from Sultanov *et al.* (1972), Welsch (1973) and Price (1979), and for the methane-rich part from Welsch (1973). CO_2–H_2O data from Takenouchi and Kennedy (1964). Note that the two-phase fields are enlarged with increasing salinity.

quently, if the composition of the two immiscible fluid end-members is known, the *P–T* conditions of the system are defined. The isopleths of each fluid end-member intersect in one point representing the *P–T* conditions of immiscibility. As a result, the *homogenization temperature* of water-rich and methane-rich inclusions reflects saturation conditions and corresponds to the temperature at which unmixing and trapping took place (Figures 5.9, 5.10*B*) (Mullis, 1976*a*, 1979; Roedder and Bodnar, 1980; Pagel *et al.*, 1986).

Simultaneous trapping of the two unmixed homogeneous fluids in a given sample is rarely observed and in any case is often difficult to distinguish from necking down in the two-phase region or for heterogeneous trapping phenomena. Nevertheless, some characteristic relationships between inclusion texture, host morphology and fluid composition provide constraints for immiscibility or essentially simultaneous trapping of water-rich and methane-

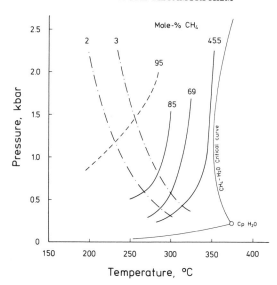

Figure 5.9 Liquid-vapour isopleths of the system CH_4-H_2O, showing solubility of methane in salt-free water and of salt-free water in methane. Methane concentrations are given in mole %. Full line isopleths after Welsch (1973); dashed isopleth interpolated after Welsch (1973); dash-dot isopleths calculated after Blount et al. (1980). With increasing salinity and at constant pressure, isopleths are displaced towards higher temperatures.

rich fluids. In quartz crystals from Val d'Illiez (Mullis, 1975, 1976a; Mullis et al., 1973) at least four episodes of immiscibility events are detected, each documented by the presence of a water-rich and a methane-rich inclusion generation. The recognition of such genetically related fluid inclusion populations is fundamental for the correct application of geothermometry and geobarometry to fluid inclusion studies.

5.5.3 Methods of geothermometry

Three methodological approaches for estimating the trapping temperature are proposed.

(i) $Tt = Th$ of the immiscible fluids. If simultaneous trapping can be demonstrated, the homogenization temperature of the two immiscible fluids is equal to the trapping temperature as shown above. Unfortunately, Th of the methane-rich inclusions cannot usually be observed because the inclusions are almost totally filled with methane.

(ii) $Tt < Th$ of water-rich inclusions. In most cases, simultaneous trapping of the two immiscible fluids cannot be readily demonstrated. However, as

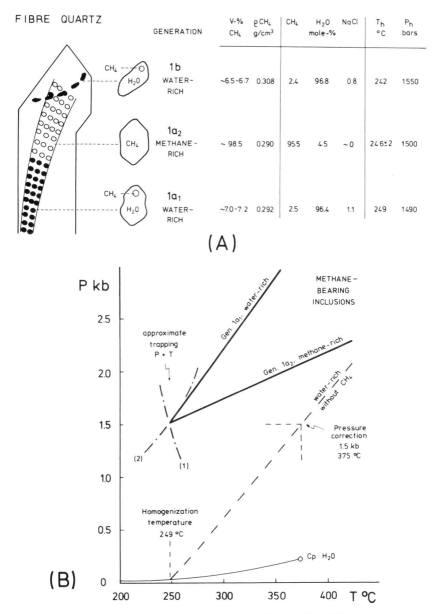

FIBRE QUARTZ

	GENERATION	V-% CH$_4$	ρCH$_4$ g/cm^3	CH$_4$	H$_2$O mole-%	NaCl	T$_h$ °C	P$_h$ bars
CH$_4$ / H$_2$O	1b WATER-RICH	~6.5-6.7	0.308	2.4	96.8	0.8	242	1550
CH$_4$	1a$_2$ METHANE-RICH	~98.5	0.290	95.5	4.5	~0	246±2	1500
CH$_4$ / H$_2$O	1a$_1$ WATER-RICH	~7.0-7.2	0.292	2.5	96.4	1.1	249	1490

(A)

(B)

Figure 5.10 Geobarometry in fibre quartz from Alpine fissures (Val d'Il!iez, Switzerland). *A:* Homogenization temperatures (*Th*) and derived homogenization pressures (*Ph*) as approximate trapping conditions of water-rich and methane-rich fluid inclusions in fibre quartz. *B:* Isochores of the two immiscible fluid generations 1a$_1$ and 1a$_2$ in fibre quartz belonging to the system CH$_4$–H$_2$O–NaCl (dark solid lines). The two isochores originate from the intersection of the two liquid-vapour isopleths or the bubble (1) and dew point (2) curves of the corresponding compositions at *Th* of 249 °C, indicating fluid immiscibility and approximate to real conditions of trapping. Long dashed line: Isochore originating from the liquid-vapour curve of the system H$_2$O–NaCl at *Th* of 249 °C and containing 1.1 mole% NaCl but no methane. Applying a conventional pressure correction at 1.5 kbar, the trapping temperature is 375 °C.

water-rich and methane-rich inclusions are successively trapped in a sequence of annealed fractures, as shown in fibre quartz (Figure 5.10A), the water-rich inclusions can be expected to be saturated with methane immediately before the transition to the methane-rich inclusions. Thus, *Th* of water-rich inclusions $(1a_1)$ can be interpreted as the approximate trapping temperature of both the water-rich and the methane-rich inclusion generations.

(iii)   *Interpolation of Th from adjacent CH_4-bearing water-rich inclusions.* If the methane-rich inclusions $(1a_2)$ shown in Figure 5.10A, are directly preceded and followed by methane-bearing water-rich inclusions (generations $1a_1$ and $1b$), then the approximate trapping temperature of the methane-rich fluid inclusion generation may be derived by interpolation of the homogenization temperatures from the adjacent methane-bearing water-rich fluid generations $(1a_1$ and $1b)$.

5.5.4   *Methods of geobarometry*

Three methodological approaches are proposed for estimating the trapping pressure.

(i)   *Isochore intersection method.* When water- and methane-rich fluids are genetically related but separately trapped, then their isochores (Figure 5.10B) originate from the intersection point on the corresponding isopleths (Figure 5.9), thereby providing the temperature and pressure of entrapment.

(ii)   *Th-isochore 'intersection' method.* If the isochore of the water-rich inclusion population is difficult to define, the pressure may be evaluated along the methane-rich isochore (Figure 5.11B) at the homogenization temperature of the water-rich inclusions, assuming it to be the approximate trapping temperature. The latter is deduced either from the homogenization temperature of the water-rich inclusions formed directly before or from interpolated homogenization values of the earlier and later adjacent water-rich inclusion populations.

(iii)   *Methane-water saturation method.* The approximate trapping pressure can be calculated for methane saturated water-rich inclusions by the empirically derived equation of methane solubility in aqueous NaCl solutions (Haas, 1978; Blount *et al.*, 1980) and the approximate trapping temperature.

A major problem of using all these approaches is to obtain reliable experimentally determined $P-\bar{V}-T-X$ properties of the systems concerned. The construction of isochores for *water-rich inclusions* is, therefore, approached by the addition of partial pressure of methane to that of the aqueous chloride solution (Zagoruchenko and Zhuravlev, 1970; Potter and Brown, 1977). Exactly the same approach can be applied to *methane-rich inclusions*, or

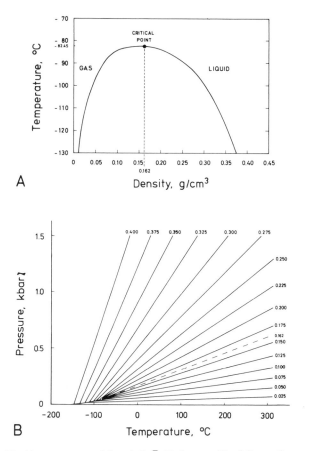

Figure 5.11   Liquid-vapour curve (*A*) and $P-\bar{V}-T$ diagram (*B*) of the methane system, after Zagoruchenko and Zhuravlev (1970).

alternatively, two other approaches can be used: (a) The isochore can be constructed using the $P-\bar{V}-T$ data of the CH_4 system from Zagoruchenko and Zhuravlev (1970), if the inclusions are considered to be filled entirely with methane, or (b) the isochore can be constructed using the equation of state from the $H_2O-CH_4-CO_2$ system (Jacobs and Kerrick, 1981) as the presence of dissolved salt in the gas-rich inclusions can be ignored.

Additionally, further problems arise in estimating the volumetric methane/water phase ratio in the inclusions at room temperature. Small inaccuracies cause considerable variations in composition and density. Therefore, the *Th*-isochore 'intersection' method of pressure evaluation (see (ii)) in comparison with the other approaches is most recommended.

As long as water-rich inclusions are saturated with methane during trapping, a pressure correction is not needed. Indeed, on the contrary, the application of a pressure correction would result in an overestimation of trapping temperatures if the presence of dissolved methane is neglected (Hanor, 1980). For example, the temperature value deduced along the dashed isochore of aqueous chloride solution at 1.5 kbar would erroneously be at 375 °C, 126 °C above the real trapping temperature of unmixed methane-saturated aqueous fluid (Figure 5.10B). Therefore, it is of vital importance to know the methane content dissolved in aqueous chloride solutions.

5.6 Case studies

This section presents some typical examples of very low-grade metamorphic fluids in metasedimentary rocks that have experienced different tectono-metamorphic histories. The intention is to describe fluid inclusions as closely representative as possible of the temperature maximum for a given metamorphic event as distinct from true retrograde fluid inclusions. These are discussed and compared with illite 'crystallinity' and coal-rank data from the same metasediments.

5.6.1 *External parts of the Central Alps*

Fluid inclusion studies in the earliest quartz generation from the external parts of the Central Alps indicate there is a close relationship between the regional variation of fluid composition and metamorphic grade. Earlier studies documented the presence of HHC-, CH_4- H_2O- and CO_2-bearing fluid inclusions in fissure quartz (Stalder and Touray, 1970; Touray *et al.*, 1970; Poty and Stalder, 1970; Poty *et al.*, 1974). However, on a regional scale the fluid inclusions display a zonation of compositions which parallels the metamorphic zones (Mullis, 1979, 1983a; Frey *et al.*, 1980a, b; Mullis *et al.*, 1983; see Figure 5.12). In the unmetamorphosed sediments, fluid inclusions contain 1 to > 80 mole% HHC. The adjacent low- and medium-grade anchizone fluid inclusions are characterized by 1 to > 90 mole % CH_4 (but with < 1 mole % HHC) whilst the high-grade anchi- to medium-grade epizone inclusions are dominated by water (70 to > 99 mole % H_2O, < 10 mole % CO_2 and < 1 mole % CH_4).

The fluid zonation directly correlates with the regional metamorphic mineral zonation (Frey, 1970, 1978) and also with both illite 'crystallinity' and coal rank (Frey *et al.*, 1980a, b; Breitschmid, 1982). If the fluid inclusion data are compared with the illite 'crystallinity' and coal rank in the external parts of the Central Alps a general trend is evident (Frey *et al.*, 1980a). The fluids evolve with increasing T and P from the HHC-zone through the methane-zone to the water-zone, whereas the illite 'crystallinity' index decreases from values of > 0.50 to < 0.25 °Δ2θ and reflectivity of coal rank (R_m) increases from < 1 to > 5% (Figure 5.13).

Occasionally, relatively high concentrations of N_2 (> 70 mole%) and

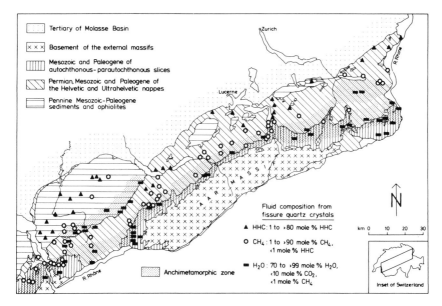

Figure 5.12   Fluid zones in the external parts of the Central Alps. The stippled area refers to the anchizone. Modified after Mullis (1979).

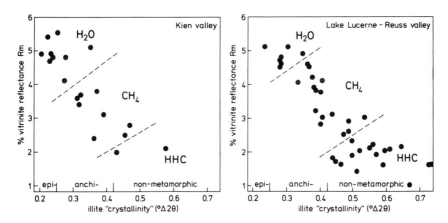

Figure 5.13   Fluid zones can be correlated with vitrinite reflectance and illite 'crystallinity' data along two cross-sections through the external parts of the Central Alps. (Modified after Frey *et al.*, 1980*a*.)

CH_4 (> 30 mole %) may occur in the volatile part of water-rich inclusions from the low- to medium-grade epizone. Sometimes CO_2 contents of up to 10 mole % are also observed. Such elevated concentrations can be directly related to variation in the host rock lithology, whereby N_2 and CH_4 are associated with fissure quartz in black shales and CO_2 in carbonate lithologies. Although the regional fluid zonation can be observed within a given nappe, unexpected relationships between different nappes can also be found (Mullis, 1979, 1983b). For instance, the southern part of the lower Helvetic nappes, the Glarus nappe and some pre-Alpine nappes contain higher grade metamorphic fluids than their underlying units. This implies fluid trapping before final tectonic emplacement of the nappes concerned. Since the peak of maximum pressure precedes the peak of maximum temperature during metamorphism of a tectonically thickened continental crust, a slow temperature decrease subsequent to the thermal maximum is normally postulated (England and

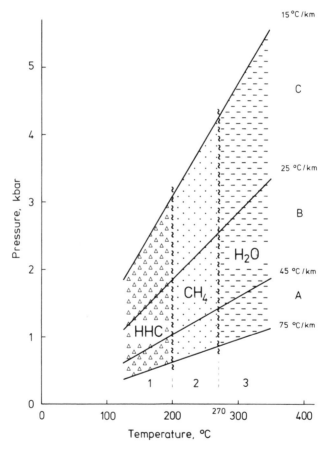

Figure 5.14 Fluid zonation shown in a P–T diagram. Based on the earliest fluid inclusion composition of fissure and vein quartz in sedimentary rocks from the non-metamorphic to the epimetamorphic zone. Metamorphic zonation: 1 = non-metamorphic zone; 2 = uppermost non-metamorphic zone, low and medium-grade anchizone; 3 = high-grade anchizone, low- and medium-grade epizone. Examples: A = Quebec Appalachians; B = External parts of the Central Alps; Terres Noires, French Alps; C = Porretta, Northern Apennines; Kodiak accretionary complex.

Richardson, 1977). Thus, fluids of the earliest syncinematically grown fibre quartz are thought to correspond to those developed during or very slightly after the temperature maximum. Temperatures and pressures derived from fluid inclusions of earliest fissure quartz may, therefore, be used for estimation of minimum to approximate $P-T$ conditions of the related metamorphic events. For the methane zone, these range from 200 to 270 °C and 0.9 to $\geqslant 2$ kbar, with geothermal gradients scattering between $\leqslant 25$ °C/km and 45 °C/km (Figure 5.14). In the eastern part of the studied area, pressure values increase up to > 3 kbar. Assuming that fluid pressures are equal to lithostatic pressures, the elevated values would indicate increasing overburden by parts of the Austro-Alpine nappes actually eroded. Pressure values may sometimes show differences of more than 1 kbar between two adjacent outcrops. Such differences may be caused by differences in time of trapping, by differences in $\bar{V}-X$ properties of the methane-rich fluids, by pressure drop during sudden enlargement of fissures or by the local development of overpressure.

5.6.2   Western Drauzug, Austrian Alps

Fluid inclusion studies on fissure quartz have been combined with illite 'crystallinity' measurements of some Permo-Triassic to Cretaceous sediments of the Western Drauzug in the Austrian Alps (Niedermayr et al., 1984). In the earliest fluid inclusions, only aqueous chloride solutions without CO_2, CH_4 or higher hydrocarbons were detected. The temperature of bulk homogenization to the liquid phase from different lithological settings ranges between 90 and 145 °C. As the metasediments sometimes contain dispersed organic matter, then HHC and CH_4 should have been generated during burial and initial metamorphism. The volatile-free aqueous fluids trapped within the fissure quartz together with their Th values are, therefore, considered to be indicative of the water-zone within the high-grade anchizone. These fluids are comparable with those from the Swiss Alps. Furthermore, illite 'crystallinity' measurements converge to similar low values as also recorded in the Swiss Alps. By analogy the minimum $P-T$ conditions of 270 °C and 1.5 to 2.0 kbar inferred for the Swiss Alps are interpreted to have been reached in this part of the Austrian Alps.

5.6.3   Mauléon region, Basses Pyrénées

In the Basses Pyrénées methane-bearing fluid inclusions in skeletal quartz have been described by Touray and Sagon (1967) and Touray (1970). Methane was detected as the main volatile species in gas-rich inclusions by mass spectrometry and by microthermometry (Th of CH_4 $L + V(L)$: -89 to -78 °C). Methane-bearing water-rich inclusions homogenize to the aqueous phase between 180 and 200 °C. The common occurrence of methane-rich and water-rich inclusions suggests methane immiscibility with minimum to approximate trapping temperatures of $190 \pm 10\,°C$. This corresponds to temperature conditions close to those of the diagenesis—anchizone boundary.

5.6.4 Terres Noires, French Alps

In the southern part of the French Alps Barlier (1974; see also Barlier et al., 1974) studied the thermal evolution of the Jurassic Terres Noires, using X-ray diffractometry of clay minerals, reflectometric measurements of carbonaceous matter and also microthermometric and gas-chromatographic analysis of fluid inclusions.

Across the area studied the content of chlorite and illite increases in the clay mineral fraction from the west to the east. At the same time there is a decrease in the expandable component of interstratified illite-smectite across the transect and kaolinite disappears towards the east. A parallel increase in metamorphic grade is documented from coal rank studies (reflectivity values varying from $\leqslant 2$ to $\geqslant 5\%\ R_m$) and from decreasing illite 'crystallinity' index measurements.

Barlier's fluid inclusion data reflect a fluid zonation similar to that found in the external parts of the Central Alps. HHC are recorded from fluid inclusion studies in the diagenetic zone (western part) whilst methane-bearing inclusions containing very small amounts of HHC occur in the high-grade diagenetic zone and in the low-grade anchizone. Water-rich inclusions are restricted to the high-grade anchizone and epizone in the eastern part.

Now, if the water- and methane-rich inclusions were genetically related, applying the Th-isochore 'intersection' method of pressure evaluation discussed in section 5.5 enables a minimum to approximate trapping pressure of 1.0–1.2 kbar at Th of $200\,°C$ to be calculated. As Barlier (1974) describes the inclusions as secondary this implies a retrograde trapping. Both the T and P values interpreted here indicate low-grade anchizonal metamorphism, although they must be considered as representative of minimum values for the peak of a metamorphic event.

5.6.5 Gaspé Peninsula, Quebec Appalachians

In the Taconic belt of the Gaspé Peninsula (Quebec Appalachians) Islam and Hesse (1983) and Hesse (pers. comm.) studied fluid inclusions in vein quartz from ac-joints and compared them with the thermal maturation zones established by pyrobitumen reflectance and illite 'crystallinity' studies of Islam et al. (1982). Methane-bearing water-rich and methane-rich fluid inclusions are characteristic for the late diagenetic and low-grade anchimetamorphic zone. The temperature of homogenization for water-rich inclusions (H_2O–$CH_4(H_2O)$) is $< 200\,°C$ in the late diagenetic zone and 200–$270\,°C$ in the low-grade anchizone. Methane-rich inclusions display Th values in the range -84 to $-81\,°C$ (L–V(L)) providing minimum trapping pressures of between 0.6 and 0.8 kbar at Th of 180–240 °C. In the high-grade anchizone, the fluid consists of an aqueous chloride solution containing < 1 mole % CH_4 or CO_2, with Th values of 150–200 °C. In the epizonal aureole surrounding the McGerrigle mountain pluton, fluid inclusions are enriched in CO_2, displaying Th values of up to 320 °C. A coherent correlation is apparent between the evolution of fluid and increasing metamorphism despite the very small number of localities studied in the anchi- and epi-metamorphic zones and the uncertainty of trapping time of fluids with respect to maximum metamorphic temperature. This low-temperature metamorphism is documented by increasing pyrobitumen reflectance from $\leqslant 2.7$ to $\geqslant 5.0\%$ Ro and decreasing illite 'crystallinity' from $\geqslant 0.43$ to $\leqslant 0.24\,°\Delta2\theta$.

5.6.6   Porretta, Northern Apennines, Italy

'Window-shaped' skeletal quartz crystals from Porretta in the northern Apennines were first described by Gambari (1868). Detailed studies by the author on quartz morphology, fluid inclusions and illite 'crystallinity' have been used to supplement coal rank studies of carbonaceous matter by Reutter et al. (1983).

Fibre quartz is overgrown by several generations of skeletal quartz, displaying more than 10 different fluid generations. Fibre quartz contains only methane-bearing water-rich inclusions, with homogenization temperatures into the liquid phase at $227 \pm 6\,°C$. The earliest generation of skeletal quartz, by contrast, displays methane-rich and water-rich inclusions with Th values into the liquid phase of $-114 \pm 4\,°C$ and $223 \pm 4\,°C$ respectively. Applying the different techniques of pressure estimation as described in section 5.5, the minimum fluid pressure is calculated as 2.1 to 2.3 kbar for a temperature of 225 °C. If the fluid pressure is assumed to be equal to the lithostatic pressure and the average rock density is 2.5 g/cm³, then the resulting rock pile thickness is 8.4–9.2 km. The average geothermal gradient for this thickness of sediments is 24 °C/km. According to the constraints discussed in section 5.5 these values must be considered as the minimum to approximate $P–T$ conditions of the peak of very low-grade metamorphism. However, these data are not in accordance with the coal rank reflectance values of 0.97 to 1.10% R_m and the illite 'crystallinity' values of 0.47–0.61 $°\Delta2\theta$ found in samples of the area studied which reflect diagenetic oil-window conditions. This discrepancy between the two data sets can be explained in terms of the kinetics of burial or very low-grade metamorphic reactions. During burial the evolution of organic matter is dependent not only upon the maximum temperature attained but also upon time–temperature relationships. The studied host rocks in the Porretta region are arkosic sandstones of the Modina-Cervarola tectonic unit. These were deposited during lower to middle Miocene times as loading from the northeastwards thrusting of Liguride nappes took place (see Reutter *et al.*, 1983).

Thus, fluid inclusion data, low vitrinite reflectance and tectonic history are consistent with a relatively short duration of heating, whilst the high fluid pressure indicates rock burial of at least 8.4 to 9.2 km. Subsequently, uplift must have been important and rapid during late Miocene time. If this interpretation is correct, then the 'crystallinity' of illite is not only dependent upon temperature but also upon duration of heating.

5.6.7 *Kodiak Islands, Alaska*

Recent fluid inclusion studies in the Kodiak accretionary complex which was underplated along the N. American Margin in late Cretaceous and early Palaeocene time, record fluid pressures of 2.5 to 3.5 kbar for temperatures of 215 to 290 °C (Vrolijk, 1985*a*, *b*; Vrolijk *et al.*, 1985; Vrolijk, 1987; Myers and Vrolijk, 1985). Calculated geothermal gradients lie in the range of 17 to 20 °C/km at vitrinite reflectance values of approximately 3% (Myers and Vrolijk, 1985).

Cogenetic methane-rich and water-rich fluid inclusions in fibrous quartz veins of the Kodiak accretionary complex record a gradual pressure drop during crystal growth by 20 to 45%. This decline in pressure is interpreted to indicate a shift from high, near lithostatic fluid pressure during fracture formation to a lower pressure as the fracture grew and filled. This indicates either a large local dilatancy, or fluid escape, transporting heat and solutes to shallow levels.

In the Maastrichtian Kodiak Formation, Myers and Vrolijk (1985) report prograde metamorphism from early methane saturated aqueous inclusions (methane zone) to methane undersaturated aqueous inclusions (probably water zone). The transition from the methane-saturated (pre-cleavage) to a methane undersaturated (syn-cleavage) system occurred at temperatures of between > 200 and 250–300 °C and at pressures > 3.2 kbar.

Both the Kodiak accretionary complex and the underplated Maastrichtian Kodiak Formation provide evidence of high fluid pressures and moderate trapping temperatures during active subduction and accretion. In addition, the Maastrichtian Kodiak Formation demonstrates fluid evolution concommitant with prograde metamorphism passing through the methane zone.

5.6.8 *Athabasca Basin, Alberta, Canada*

Fluid inclusions occur around detrital quartz within quartz overgrowths from Precambrian sandstones of the Athabasca Basin (Pagel, 1975; see also Pagel and Poty, 1984). The sandstones do not contain any organic matter except for graphitic material probably originating in the crystalline basement (Landais and Dereppe, 1985). The inclusions are composed of highly saline brines, commonly containing halite cubes and occasionally platelets of haematite. Neither hydrocarbons nor CO_2 were detected in any of the inclusions. Estimated maximum burial conditions are 5 km and 250 °C (Pagel, pers. comm.).

This example illustrates a potential shortcoming of the fluid inclusion technique. Even though these sediments are inferred to have experienced anchizonal metamorphism, the complete absence of hydrocarbons and CO_2 precludes the use of such fluid inclusions for refined geothermobarometry with C–O–H-bearing systems.

5.6.9   Trends in fluid inclusion salinities

Bulk salinities of very low-grade metamorphic fluids are generally low and rarely exceed 7 wt% NaCl equivalents. They usually increase in the non-metamorphic sediments which display salt concentrations of up to $\geqslant 25$ wt %. Fluids trapped in the proximity of evaporitic sediments are often enriched in dissolved salts (e.g. methane-bearing quartz from Col d'Allos with 24 wt% equivalent NaCl; Barlier et al., 1973).

Salt composition may vary widely as indicated by eutectic melting values. Nevertheless, CH_4-bearing aqueous inclusions of the CH_4-zone commonly contain NaCl and $CaCl_2$ dominant fluids. By contrast, brines of the unmetamorphosed zone tend to chloride mixtures enriched in $CaCl_2$ which probably also contain small amounts of sulphates, carbonates and other salts.

5.7   Discussion and implications

5.7.1   Methane immiscibility

A question of basic interest in geothermobarometry is how the composition of the adjacent inclusions in the host mineral evolves from a water-rich to a methane-rich fluid. This question can be answered by considering the process of methane–water unmixing. Methane saturation in aqueous solutions is < 3 mol% at 1.5 kbar and 250 °C (see Figures 5.8 and 5.9). Methane saturation decreases with decreasing P and T but with increasing salinity (Haas, 1978; Blount et al., 1980). Combining methane solubility data with geological evidence then raises four possible mechanisms that can be envisaged to explain unmixing phenomena.

(i) Unmixing during isobaric decrease in temperature
(ii) Unmixing during isothermal pressure decrease
(iii) Unmixing by input of salt-enriched fluids
(iv) Upward and downward movement of the gas–water contact mimicking unmixing phenomena.

All four mechanisms may operate individually or in combination with each other. In the example of fibre quartz growth from external parts of the Central

Alps (Mullis, 1975, 1976a, b, 1979) a slight decrease in temperature seems to control unmixing during early quartz growth stages. By contrast, repeated isothermal pressure decreases predominate during late quartz growth as a result of rapid enlargements of fissure systems. Additionally, vertical displacement of the gas–water contact may result in alternate trapping of methane saturated water-rich and water-saturated methane-rich inclusions. As a consequence, trapping of the two immiscible fluids may thus be possible without evidence of unmixing on the scale of the fluid inclusions within a given host.

5.7.2   Growth of fibre- and skeletal quartz and precipitation of quartz cements in sandstones

Fibre- and skeletal quartz are widespread in fissures from very low-grade metamorphic rocks (section 5.6). Both of these quartz types are useful for thermobarometry. Fibre quartz grows during syncinematic opening of tension gashes from a water- or gas–enriched fluid. By contrast, skeletal quartz seems to have been precipitated from a predominantly gas-enriched fluid.

Skeletal quartz morphology (section 5.2.1) indicates episodic rapid quartz precipitation from a gas-rich emulsion-like fluid, although the exact growth mechanisms are not yet known. Nevertheless, changes in pressure, temperature, or salinity during retrograde conditions may lead to methane oversaturation of aqueous chloride solutions (Figures 5.8 and 5.9) resulting in fluid unmixing and rapid quartz precipitation. In Alpine fissures, skeletal quartz thus formed (Mullis, 1975, 1976a, b); in gas reservoirs, intense quartz cementation along the water–gas contact is expected to be the consequence.

5.7.3   Thermobarometry

Simultaneously trapped or genetically related water-rich and methane-rich fluid inclusions in quartz can be used directly as a geothermometer and a geobarometer (section 5.5.2), such that minimum to approximate trapping temperatures and pressures can be deduced. This approach may be applied with particular effectiveness to the methane zone where the $P-T$ conditions may vary according to the time–temperature history of the rocks concerned.

In regional metamorphic terrains (such as the external parts of the Central Alps) the methane zone extends from 200 to 270 °C and 0.9 to $\geqslant 2$ kbar. Assuming that fluid pressure equals load pressure, calculated geothermal gradients then range between < 25 °C/km and 45 °C/km (Figure 5.14).

In rapidly buried or subducted sediments (such as the Modina-Cervarola unit in the Apennines or the Kodiak accretionary complex and the Maastrichtian Kodiak Formation), pressures may have reached values of 2 to $\geqslant 4$ kbar at temperatures of between 220 and 300 °C. Calculated geothermal gradients then become lower and range between 17 °C/km and 24 °C/km. By contrast, minimum fluid pressures in the Quebec Appalachians do not exceed 0.8 kbar at temperatures of between 180 and 240 °C. Retrograde fluid trapping or a small overburden at elevated geothermal gradients may explain this behaviour.

In summary, approximate temperature limits of the methane zone as deduced from earliest methane-bearing water-rich inclusions are ~ 200 to ~ 270 °C (exceptionally up to 300 °C in the Kodiak accretionary complex), whereas approximate trapping pressures may vary in a wide range from 0.8 to 3 kbar (Figure 5.14). Note that methane-rich inclusions trapped during retrograde conditions can be found at temperatures < 200 °C (Mullis, 1975, 1976a, b).

5.7.4 Origin of very low-grade metamorphic fluids

In the non-metamorphic zone higher hydrocarbons form as a thermal maturation product of organic matter during burial (HHC-zone). Further heating produces cracking of hydrocarbons and kerogen, forming wet and ultimately dry gas. The dominant fluid in the high-grade non-metamorphic and low- to medium-grade anchizone is methane, which is associated with essentially methane-saturated aqueous chloride solutions. Immediately prior to the transition from the methane to the water zone, CO_2 increases significantly in the CH_4-rich volatile part of all inclusions.

The transition from the methane zone to the water zone containing methane-undersaturated to methane-free aqueous chloride solutions seems to be controlled by several factors.

(i) The production of CH_4 becomes very small towards and after the end of the medium-grade anchizone (lower part of the metagenesis zone of the modified van Krevelen diagram; Figure 5.15). Aqueous solutions will therefore tend to be undersaturated with methane.

(ii) The amount of H_2O in the system increases as dehydration reactions and water-releasing recrystallization reactions (Voll, 1976) proceed. Simultaneous decrease in methane and increase in the amount of water thus both result in the dilution of methane.

(iii) The disappearance of methane may also be affected by redox-reactions, an inference supported by a significant increase of CO_2 in the volatile part. Fluid-graphite equilibria at these $P-T$ conditions cannot account for the chemical evolution of the fluid as organic matter has not yet reached the organizational state of graphite. Furthermore, the kinetics of the reaction between graphite and fluids are very slow as shown experimentally by Ziegenbein and Johannes (1980) and confirmed by calculations of fluid-graphite equilibria from fluid inclusion data (Dubessy, 1985; Ramboz et al., 1985). However, it is possible that reactions between C–O–H fluids and Fe-bearing silicates or oxides may cause oxidation of CH_4 to CO_2. Ultimately, CO_2 generated by such reactions may be consumed by the precipitation of

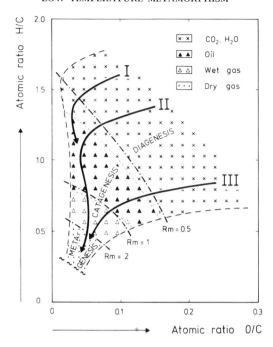

Figure 5.15 General scheme of kerogen evolution from diagenesis to metagenesis in the van Krevelen diagram. I to III are the principal types and evolution paths of kerogen. Vitrinite reflectance values for hydrocarbon-bearing reservoirs are shown. (Modified after Tissot and Welte, 1978, and Durand, 1980.)

carbonates, as observed in Alpine fissures typical of this zone which contain abundant calcite.

At higher metamorphic grades (low- to medium-grade epizone) CO_2, CH_4 and N_2 are occasionally detected in water-rich inclusions of the water zone. Such fluids may react with iron-bearing silicates or iron oxide minerals (Eugster and Wones, 1962). If C–O–H fluids at temperatures between 300 and 400 °C react with poorly-crystallized carbonaceous matter, additional equilibria with this material may contribute to the chemical evolution of the fluids (French, 1966; Eugster and Skippen, 1967; Ohmoto and Kerrick, 1977; Frost, 1979). As calculated by Holloway (1984), the fluid coexisting with graphite at temperatures below 400 °C and pressures above 0.3 kbar should consist either of CO_2–H_2O or CH_4–H_2O mixtures depending on oxygen fugacity. This could be the case in epizonal rocks such as black shales containing elevated concentrations of organic matter. By contrast, CO_2 incorporated in carbonates seems to predominantly originate from decarbonation reactions, whilst N_2 probably derives from ammonium-bearing minerals such as illite, biotite or K-feldspars.

In conclusion, C–O–H fluid evolution in very low-grade metamorphic rocks is primarily controlled by irreversible thermal cracking of kerogen in the HHC- and CH_4-zones. In the water zone, decarbonation reactions and equilibria between C–O–H fluids, iron-bearing silicates and iron oxide minerals determine the volatile composition. Equilibria between C–O–H fluids and dispersed carbon seems to play rather a subordinate role at this stage.

5.7.5 Correlation of fluid inclusion data with illite 'crystallinity' and coal rank

Fluid evolution from HHC-dominant to methane- and water-dominant compositions broadly correlates with increasing coal rank, illite 'crystallinity' and temperature (Figure 5.13). However, a very short duration of heating does not seem to significantly affect maturation of organic matter or illite 'crystallinity'. In marked contrast, fluid inclusions record very subtle changes in the ambient $P-T$ conditions. Therefore, fluid inclusion studies provide the best $P-T$ data currently available for rocks undergoing very low-grade metamorphism where only short or rapid crustal thickening is involved.

Acknowledgements

This chapter benefited through the criticism and advice of S. Burley, J. Dubessy and M. Frey. Stuart Burley patiently improved the English.

6 Radiogenic isotopes in very low-grade metamorphism

JOHANNES C. HUNZIKER
With a contribution on fission track dating by Anthony J. Hurford

6.1 Introduction

Work with non-stable isotopes in very low-grade metamorphic terrains not only renders information about the timing of metamorphic events, but also yields data which can quantify both the kinetics and the distances of reactions which took place during metamorphism. In addition, metamorphic temperatures can be approximated and palaeogeographic provenance problems can be solved. Only equilibrium parageneses yield chronometric information, while under disequilibrium conditions radiogenic isotopes are mainly used as tracers. The establishment of equilibrium conditions in a given system is achieved both through independent geochemical or crystallographic methods, and through coincidence of at least two independent radiometric methods.

In metamorphic terrains, the most widely used decay chains are K–Ar, Rb–Sr, and the spontaneous fission of ^{238}U exploited in the fission track method. The normal U–Pb, as well as Th–Pb, decay are used preferentially for higher-grade rocks and will only be mentioned briefly. Similarly, the U and Th disequilibrium methods, mostly in use for dating very young sediments, are not discussed.

Normally, work with non-stable isotopes is performed on discrete mineral phases. Potassic minerals usually also contain adequate quantities of Rb to allow determination by both K–Ar and Rb–Sr methods. The most commonly measured minerals are micas, feldspars, amphiboles, clay minerals and volcanic glass. The fission track method centres mainly on apatite, zircon and sphene. Whole rock samples, or enrichments of mineral phases, may also yield geologically meaningful information.

Under the conditions of very low-grade metamorphism, the slowness of reaction kinetics generally results in disequilibrium relations for mineralogical, geochemical and isotopic parameters (see also sections 2.2 and 3.2.4). Only under favourable conditions is equilibrium reached.

As reaction kinetics depend not only on the temperature attained, but also on the duration of a reaction, on deformation of the rock body, as well as on the grain size of the reacting particles, on fluid pervasivity and on fluid and mineral chemistry, to mention only the most important parameters, quantification becomes very complex, and nearly every example constitutes a case history of its own. In addition, experimental data are scarce, and, because of the large number of possible variables, are very difficult to compare with nature.

6.2   General problems of isotope studies in very low-grade terrains

The temperature range of very low-grade metamorphism, that is temperatures below around 400 °C, poses several specific problems.

(i) Grain sizes of authigenic minerals are generally very small. Here again the size reached depends not only on the reaction temperature, but also on the duration at elevated temperatures, the fluid chemistry and the porosity of the rock.

(ii) Distinction between authigenic or neoformed minerals and detrital or inherited minerals is very critical, as detrital minerals tend to react more slowly to the new environmental conditions than authigenic minerals; newly-formed minerals are always in equilibrium.

(iii) The slow evolution of both authigenic and inherited minerals to equilibrium with the metamorphic conditions, comprising not only recrystallization, but also restructuring and solid-state diffusion, results in the predominance of disequilibrium. Criteria must be found which facilitate the interpretation of isotopic data and the evaluation of these conditions.

(iv) The small size of the minerals may also pose severe problems for the retention of radiogenic daughter products, as the diameter of mineral grains plays an exponential role in the diffusion behaviour.

(v) Finally, the small grain size of minerals also poses special technical problems for mineral enrichment, separation and analysis. Sample crushing and grinding has to be performed very carefully to preserve original grain sizes, using either mechanical means with the teamer mill, grinding for 30 seconds, or ultrasonic treatment, or else repeated freezing and thawing.

Carbonatic samples, after eventual extraction of apatite, are decarbonatized in 5% acetic acid for calcite, or 10% hydrochloric acid for dolomite, and subsequently neutralized by rinsing with distilled water until they reach a pH of around 7, slightly on the basic side to prevent the suspension from flocculating.

For coarse-grained detrital minerals the usual heavy liquid, electromagnetic and shaking-table separation techniques are applied to coarse fractions preferentially enriched, often by a Wilfley table. For fine-grained clay minerals, illite to muscovite, settling techniques in distilled water are used for the fractions between 1 and 60 microns, and below 1 μm centrifuging is a common procedure.

The problem of adsorbed Sr may pose severe problems for Rb–Sr analysis of clay minerals. Leaching with acetic and dilute hydrochloric acids is widely applied (see Clauer, 1976; Clauer et al., 1986). Recently Morton and Long

(1980) have proposed leaching with ammonium acetate to purge loosely-bound Rb and Sr.

6.2.1   Inherited v. neoformed minerals

In isotope work on very-low grade metamorphic terrains, two different cases must be distinguished: (i) retromorphosed magmatic or metamorphic rocks; and (ii) prograde sedimentary rocks.

Retromorphosis of higher-grade metamorphic or even high-temperature magmatic rocks under conditions of very low-grade metamorphism generally results in unmixing into different mineral species. Sericitization of high temperature feldspars under very low-grade metamorphic conditions is one of the most common features. The neoformed micas are easily detected under the microscope, and, if separable, their isotope ratios and amounts can be measured. In the high-pressure regime, the original amphibole of mafic rocks may be transformed into glaucophanitic amphibole. However, these neo-formed minerals pose some dating problems, as it is a priori uncertain to what extent the primarily accumulated radiogenic daughter products are lost during such solid–solid reactions. In addition, these changes can also result in a loss of radioactive parent isotopes, thereby accentuating the effect. Bocquet et al. (1974) have reported glaucophane K–Ar ages from eo-Alpine glaucophane–schist facies rocks of the Western Alps, exhibiting pre-Alpine ages up to 1800 Ma. These anomalous ages are attributed to inherited argon problems.

Low-temperature alteration of volcanic rocks is very common, generally leading to anomalously low or high ages compared with the ages of extrusion. Only in rare cases can such altered ages be attributed to the time of metamorphism because exchanges tend to be incomplete.

If acidic volcaniclastic tuffs with high amounts of primary glass matrix are metamorphosed under very low-grade metamorphic conditions, glass can be completely transformed into illite during diagenesis and incipient metamorphism (Ahrendt et al., 1978, 1983a), thus representing ideal dating material, as no old lattice components are preserved.

The presence and admixture of detrital minerals to the neoformed paragenesis in prograde sediments create considerable problems, which have to be considered with care, but are not generally insoluble.

Authigenic illites are usually of the 1 Md disordered polytype during early diagenesis and slowly evolve to $2 M_1$ illites during prograde metamorphism. Unfortunately detrital illites and fine-grained muscovites can also be of these two polytypes. Depending on grain-size, lithology and composition of the pore fluids, the micas reach the $2 M_1$ stage between around 150 and 300 °C (Hunziker et al., 1986).

Detrital potassic white micas that have not reacted with their sedimentary environment are of the $2 M_1$ and 3T polytype, presumably depending on the presedimentary pressure conditions (Frey *et al.*, 1983).

During incipient metamorphism these two populations of potassic white micas progressively converge and afterwards cannot and need not any longer be treated separately. Before convergence, nevertheless, the two different generations need to be separated. A separation of the two polytypes, if present in the same rock, is occasionally possible by grain-size separation, smaller grain sizes tending to be more adjusted to their environment or *in situ* neoformations (Hunziker, 1986).

The presence of detrital feldspar in a sample, especially K-feldspar, can lead to complications, resulting in a general lowering of the radiometric ages due to the partially open system behaviour of K-feldspars during very low-grade metamorphism, sometimes outlasting a metamorphic phase down to very low temperatures and depending on pore-fluid chemistry (Hunziker, unpublished results).

Detrital amphiboles are usually easily detected and rarely occur together with neoformed amphiboles.

Detrital apatites and zircons are suitable for fission track dating. They provide cooling ages after the last metamorphic event. By means of track length meaurements insight into more complex thermal histories can be gained (see Gleadow *et al.*, 1983).

6.2.2   *Problems of disequilibrium in evolving mineral parageneses*

It is generally accepted that mineral growth during prograde and retrograde metamorphism occurs at the expense of pre-existing minerals. The open question for unstable isotopes remains: to what extent do such transformations fulfil the principal requirements for radiometric dating, that is absence of inherited radiogenic daughter products in the neoformations. Here again two cases must be distinguished:

(i)   A mineral neoformed through a process of crystallization or recrystallization, breaking the lattice bonds
(ii) A 'cleaning process' of an old lattice leading to a young mineral of the same species; the excess elements are expelled and/or new elements incorporated without breaking the old lattice bonds. Both cases may occur in the same rock.

Three examples may illustrate these two cases.

(i)   *Evolution of glauconite to biotite.* Frey *et al.* (1973) have shown, in a cross-section through the Helvetic Glarus Alps, how glauconite-bearing formations change during incipient metamorphism into stilpnomelane and biotite-bearing rocks (cf. section 2.7.7).

Stratigraphic control of the age of metamorphism in the Glarus Alps (Hunziker *et al.*, 1986) demonstrates that all these neoformations result in complete expulsion of pre-existing radiogenic daughter products from the glauconite.

In the Alps, generally, before the first occurrence of brown biotite, a green biotite is formed at the expense of the pre-existing brown pre-Alpine biotite. This recrystallization product normally also results in almost complete expulsion of radiogenic daughter products (Dempster, 1986). If this colour change from brown to green occurs under static conditions, generally the lattice bonds more or less survive (Frey et al., 1976) and often only part of the old radiogenic daughter products are expelled, leading to mixed pre-Alpine ages. Arnold and Jäger (1965) have described such an example, where the degree of rejuvenation goes hand in hand with the higher content of titanium unmixing from the old biotite lattice expressed in increasing sagenite content of the biotite, leading to mixed Hercynian/Alpine ages.

(ii) *Transformation of acidic volcanic glass into illite–muscovite.* In the very low- to low-grade metamorphic Rhenohercynian zone of the Variscides, Ahrendt et al. (1978, 1983a) demonstrated with syntectonically crystallized illites, that grew at the expense of Devonian acidic to intermediate glass tuffs, that this material is very suitable for age determination. These authors were able to show that the evolution of illite/muscovite from a glass matrix leads to reliable radiometric ages dating the time of phyllosilicate formation. In the specific case of the Rheinisches Schiefergebirge, as the whole rock sericite schist contains up to more than 7% potassium, an input of potassium from pore fluids into the glass tuffs during metamorphism has to be assumed.

(iii) *Evolution of illite to muscovite in mudstones-slates.* In an Alpine cross-section from the diagenetic Jura mountains to the epizonal southern Glarus Alps, Hunziker et al. (1986) showed that during the prograde changes in fluid composition and mineralogy the $< 2 \mu m$ illite population, with less than 10% mixed-layers illite/smectite component, slowly evolves from a sheet silicate with less than 7% K_2O to a content of 11% (Figure 6.1). Linked to this increase in potassium, the total layer charge increases from 1.2 to 2.0, parallel to the increase in 'crystallinity'. The authigenic 1 Md polytype during this evolution changes to a metamorphic $2 M_1$ polytype (Figure 6.2).

The $1 Md$–$2 M_1$ polytype change does not follow the theoretical mixing line, but two distinct trends. First, the radiogenic isotopes are preferentially lost with little change in bulk chemistry, e.g. argon is lost without noticeable change in the potassium content. This evolution line is referred to as the diffusion controlled branch. Above 150 °C the stacking order between tetrahedral and octahedral sheets is slowly changed along the restructuration branch—the low-temperature 1 Md polytype slowly evolves to a $2 M_1$ musco-vite. These changes are completed at higher anchizonal conditions. Depending on the chemistry of the interstitial pore waters and the duration of the thermal event, they may already be completed at considerably lower temperatures.

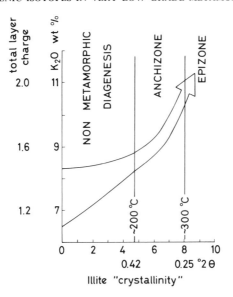

Figure 6.1 'Crystallinity' (Kübler index), total negative layer charge and potassium content of diagenetic to epizonal illites. Temperatures are derived from a regional metamorphic environment (modified after Figures 6 and 7 of Hunziker *et al.*, 1986).

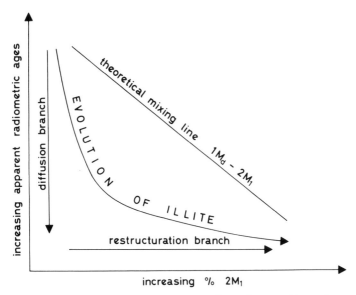

Figure 6.2 Correlation between decreasing apparent radiometric ages and increasing amount of $2M_1$ polytype during evolution of illite to muscovite. This evolution follows first a diffusion branch and then a restructuration branch (modified after Figures 20 and 22 of Hunziker *et al.*, 1986).

During this evolution, the radiometric ages for both Rb–Sr and K–Ar are progressively lowered from the age of detritus respectively of sedimentation, to the age of the new metamorphic overprint. All the changes in chemistry and mineralogy lead to the slow evolution of the mineral illite to the well-known muscovite or phengite. This continuous transformation is also seen in the shift of the dehydration peak (loss of hydroxyl water) in DTA curves (Hunziker et al., 1986, Figure 17), slowly shifting from 550 °C for diagenetic illites, to 750 °C for epizonal muscovites.

It is not at all clear a priori if all these continuous reactions which, according to stable isotope data, occur in the solid state without breaking of the tetrahedral and octahedral bonds, in other words without recrystallization, lead to complete expulsion of pre-existing radiogenic daughter products.

The extrapolation of measured K–Ar ages versus actual borehole temperatures of borehole samples from the Swiss Molasse Basin and metamorphic samples from the Glarus Alps (Hunziker et al., 1986) shows that for illite fractions $< 2 \mu m$ a temperature of 260 ± 20 °C during 10 ± 5 Ma led to complete resetting of K–Ar ages (Figure 6.3). Figures 6.1 and 6.2 show that the evolution of illite to muscovite for the fraction $< 2 \mu m$ under static conditions reaches the muscovite stage around 260 °C in the higher anchizone with over 85% $2 M_1$ polytype present.

Combining two different radiometric systems for age determination, e.g. Rb–Sr and K–Ar, on these samples results in concordant ages between the two methods from the higher anchizone upwards. This convergence is a strong argument in favour of meaningful geologic ages from this point onwards.

The resetting of a radiometric system is temperature- and grain-size-dependent, as shown in Figure 6.4. Coarser grains reach a zero age at higher

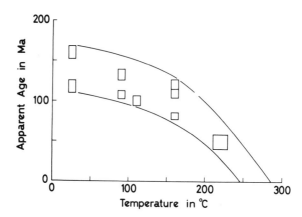

Figure 6.3 Borehole samples from the Swiss Molasse Basin and metamorphic samples from the Glarus Alps showing that for illite fractions smaller than 2 microns a temperature of 260 ± 20 °C during 10 ± 5 Ma leads to complete resetting of the K–Ar ages (modified after Figure 19 of Hunziker et al., 1986).

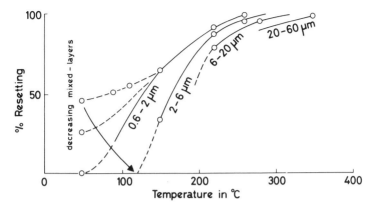

Figure 6.4 The resetting of radiometric ages is temperature- and grain size-dependent. Finer grains reach zero age at lower temperatures than coarser grains. Data from Hunziker *et al.* (1986).

temperatures than finer grains. In addition, the chemistry of the pore fluids plays an important role. For the 20–60 μm illites this temperature is already around 350 °C, the temperature proposed by Purdy and Jäger (1976) for the rejuvenation of Alpine K–Ar ages from potassic white micas. On the other hand, for grain diameters below 1 μm the temperature of resetting is considerably lower, thus allowing for interesting applications in hydrocarbon prospecting in sedimentary basins. Besides these static resetting mechanisms, dynamic recrystallization plays an important role in orogenic environments.

Black *et al.* (1979) have shown a close correlation between deformation and the response of the Rb–Sr system. Liewig *et al.* (1981), investigating polyphase sheet silicates from the *schistes lustrés* of the French Alps, were able to show that the K–Ar response of phengitic micas is in close correlation with the structural site, and suggest a two-step response of the micas to deformation.

Kligfield *et al.* (1986) were able to show a correlation between the degree of crenulation of potassic white micas and the amount of resetting of K–Ar and ^{40}Ar–^{39}Ar ages. A similar observation was already reported by Rickard (1965). By careful selection of end-members Kligfield *et al.* (1986) were able to tentatively date the beginning of the deformation in the Apennines to 27 Ma and the ending of subsequent pulses to 10 Ma.

6.2.3 *Retention of radiogenic daughter products in small mineral grains*

The admittedly small grain size of illite fractions potentially poses retention problems, and volume diffusion behaviour of argon at ambient or low temperatures of such small grains is theoretically to be expected, as the diameter of a mineral grain plays an exponential role in the diffusion behaviour. Grain diameter differences of 10^4 have to be taken into consideration between high-grade metamorphic and very low-grade metamorphic rocks.

Hammerschmidt and Wagner (1983) concluded from diffusion measurements that a decrease in diameter by two orders of magnitude did not affect retention properties of biotites down to $20 \mu m$ in diameter as long as temperatures stay around $100 \,^\circ C$ below the blocking temperature. The borehole data of Hunziker et al. (1986) allow the same conclusion, and the low temperature loss of the very fine fractions (see Figure 6.4) may be explained by the open system behaviour of mixed-layer illite/smectite admixtures in the diagenetic samples.

Illite/smectite mixed-layers easily exchange their hydroxyl ions at low temperatures of diagenesis. This is not the case for pure illites, where hydroxyl is only expelled at temperatures between 500 to $750 \,^\circ C$. $^{40}Ar-^{39}Ar$ incremental release data show clearly that Ar is expelled from the illite lattice at temperatures above $500 \,^\circ C$ together with hydroxyl water, in good agreement with the early work of Gerling et al. (1963), Brandt and Voronovsky (1964), Fechtig and Kalbitzer (1966) and Giletti (1974).

Various studies of mechanical grain size reduction of high-temperature micas and observed argon loss at low temperatures are reported in the literature. There are useful reviews in Fechtig and Kalbitzer (1966) and Lippolt (1971). All these experiments have one thing in common: a coarse-grained mica was ground down to micron and submicron size, and then the micas were submitted to both dry and a variety of hydrous ion exchange processes, resulting in argon and/or potassium losses at different temperatures above and below the assumed blocking temperature (Kulp and Engels, 1963).

There are various arguments against such an approach for prograde illitic muscovites.

(i) The separation process of illitic muscovite, if adequately performed (Clauer, 1976; Huon, 1985; Reuter, 1985; Hunziker et al., 1986), results in preservation of original grain-sizes as revealed by electron microscopy (Huon, 1985; Reuter, 1985; Knipe, pers. comm.), and, therefore, preserves the original grain boundaries.

(ii) Illitic muscovites were not ground down to their actual grain-size, but grew as idiomorphic crystals in these dimensions with completely sharp and discrete grain boundaries, and size depending on metamorphic conditions. Huon (1985) could demonstrate that the fraction $< 2 \mu m$ of the clay mineral fraction ($< 63 \mu m$ of mudstones) slowly decreases from around 30% in diagenetic environments, to below 5% in epizonal rocks. This decrease is counterbalanced by an increase of the $10-30 \mu m$ fraction trending from below 20% in the diagenetic zone to over 60% in the epizone. The mean grain diameter of the clay fraction in the diagenetic zone is c. $4 \mu m$ and in the epizone c. $15 \mu m$. This aspect is in sharp contrast to the completely ragged and torn, vertically embayed grain boundaries of mechanically reduced micas. Gerling et al. (1963) showed that mechanically ground micas changed their lattice distances and thus lost up to 30% of their argon.

(iii) In general, the tectonothermal history of the sample used in the grain-size reduction experiment is only incompletely understood. Fission track and fluid inclusion information about the lower-temperature cooling history is often not available, so that a post-metamorphic low-temperature exchange at the grain boundaries preferentially ground down into the fine fraction cannot be excluded.

(iv) Frank and Stettler (1979), Kligfield *et al.* (1986) and Hunziker *et al.* (1986) have shown with $^{40}Ar-^{39}Ar$ incremental release spectra of illites which yielded plateaux down to low-temperature release steps, that obviously very few diffusion problems exist for these well-crystallized illites down to $< 2\,\mu m$ fractions, and that therefore they are in equilibrium with their environment. A post-metamorphic leaking of argon at low temperatures would result in a staircase release pattern at low temperatures.

(v) Ahrendt *et al.* (1977, 1983a, b) have concluded in two cases, from the age coincidence between coarse-grained 500- and 300-Ma-old high-temperature muscovites and illitic micas in the same region, that expected argon losses of illitic muscovites are below detection limits even for 500-Ma-old minerals. The same conclusion was reached by Hunziker *et al.* (1986) from the age coincidence between fissure muscovite and illitic muscovite of the country rock, in the Gotthard massif, differing in diameter by five orders of magnitude.

6.2.4 *Summary*

Illitic micas carefully selected, separated into different grain size fractions and chosen according to strict mineralogic and granulometric criteria, can yield meaningful radiometric ages in low- and very low-grade metamorphic terrains, provided that at least two different radiometric methods are applied to facilitate the interpretation of the data.

Biotites in very low- to low-grade metamorphic terrains represent chronometers with few problems. They are generally suitable for age determinations with both K–Ar and Rb–Sr methods. As temperatures in these terrains rarely exceed by a large amount the assumed blocking temperature of $300 \pm 50\,°C$ (Dodson, 1979; Purdy and Jäger, 1976), biotites commonly date here an event closely related to the thermal peak of metamorphism.

Metamorphic amphiboles, due to their low potassium contents, pose problems of argon inheritance as already discussed, and therefore are not very reliable. In addition, Rb–Sr analysis rarely works on amphiboles because of the unfavourable Rb/Sr ratios so that no independent radiometric control is possible. Here, U–Pb offers a possibility, especially for older ages. Zircon, sphene and apatite fission track ages give useful information for both thermal maxima and cooling histories in very low-grade metamorphic terrains (see section 6.4.4).

6.3 Appropriate lithologies for non-stable isotope work in very low-grade terrains

According to the classical view (e.g. Coombs, 1954), metavolcanics, due to their generally high reaction rates, show the first signs of very low-grade metamorphic changes and are therefore widely used to determine the metamorphic grade that a region has attained in its geologic history. Today, work in very low-grade metamorphic terrains is performed on a vide variety of different rock types, some of which are more suitable for work with non-stable isotopes than others, due to their specific petrography and mineralogy.

6.3.1 *Limestone–marble*

Impure limestones/dolostones to marbles represent nearly ideal material for K–Ar and Rb–Sr work in the very low-grade metamorphic range. Unfortunately, due to the low K-content of these rocks, smectite and mixed-layer illite/smectite generally persist to higher temperatures than in metaclastites (see section 2.5.2). By means of sedimentological criteria the presence of detrital input of silicates can be either detected or ruled out. Authigenic illite–muscovites of high purity are easily extracted by decarbonatization of these rocks, either with 5% acetic or 2.5 normal hydrochloric acid and subsequent neutralization with distilled water. Hunziker *et al.* (1986) noted no effect of this treatment on the measured radiometric ages. Micas usually are aligned in the metamorphic schistosity, or form coatings on quartz grains recrystallized during incipient metamorphism. Unfortunately, calcite shows a rapid response to deformation, so that structural criteria for the timing of sheet silicate growth are often not applicable (Kligfield *et al.*, 1986). Apart from potassic white micas, phlogopite and amphiboles from impure limestones to marbles are suitable for non-stable isotope work in very low-grade terrains.

6.3.2 *Mudstones–slates–mica schists*

Low-grade metamorphism of metapelites leads to a great variety of sheet silicate parageneses including the illite–muscovite transition, and biotite- and chlorite-bearing rocks; often more than one type of sheet silicate is present in the metamorphic parageneses.

Harper (1967, 1970) has proposed the use of whole rock slates for K–Ar dating of tectonometamorphic events. Hunziker (1974), Bocquet *et al.* (1974) and especially Kligfield *et al.* (1986) have shown that whole rock slate dating poses problems, arising from fluid inclusions in quartz and other minerals, which may contain excess radiogenic argon, thus possibly yielding higher K–Ar ages as whole rocks than as discrete mineral phases. Excess radiogenic argon can easily be detected by measuring the argon contained in fluid inclusions of quartz or other K-poor to K-free minerals. To overcome these difficulties a separation of the sheet silicates by settling and centrifuging techniques has been applied in more recent work (Clauer, 1976; Thöni, 1981; Hunziker, 1979; Huon, 1985). Separation and enrichment in different grain sizes also leads to reduction or elimination of problems with feldspars and clay minerals (Hammerschmidt, 1982). The purity of the sample can be confirmed by X-ray diffractometry, but it is clear that the 99.9% pure mineral concentrates normally attained from high-grade rocks can rarely be prepared from very low-grade rocks. Normally illite/chlorite mixtures, or illite/quartz, are measured from the same locality, but with variable illite content to control the effects of the impurities on the radiometric age (Zingg et al., 1976).

Pelitic lithology is especially well suited for unstable isotope work in very low-grade metamorphism because of its predominantly sheet silicate mineralogy. Detrital zircons and apatites can be used for fission track work but the dimension of neoformations of these minerals is usually too small for track counting.

6.3.3   Metasandstones and metagreywackes

In metasandstones and metagreywackes during very low-grade metamorphism a generation of neoformed minerals evolves at the expense of the detrital paragenesis. Depending on the temperature, the duration of the thermal event, and the fluid chemistry of the circulating fluids, these neoformations remain small or become bigger; usually, nevertheless, in comparison to slates, exchanges in sandstones are facilitated by the higher fluid pervasivity. Only tedious and careful geochemical and X-ray work (Kübler, 1984) will allow a distinction between inherited and neoformed minerals, which develop normally in different $P-T$ conditions and/or in different rock chemistry.

In quartzitic metaclastites complications are sometimes encountered due to the disappearance of rock porosity prior to the climax of metamorphism. Detrital micas and K-feldspars can then behave as armoured relics. When, finally, these K-bearing phases recrystallize to neoformed mica, they may still contain radiogenic daughter products from a former cycle (Frank et al., 1977; Hammerschmidt, 1982).

Metasandstones and metagreywackes are generally good sources for apatites, zircons, and sphenes and thus represent very suitable material for fission track dating.

Finally, during diagenesis under favourable conditions, the percolating fluids lead to the generation of authigenic illites in the pore cavities of metaclastites, often with the concomitant disappearance of detrital K-feldspar (Rossel, 1982; Lee, 1984; Lee et al., 1985; Liewig et al., 1987). Sometimes several generations of illites can be distinguished by electron microscopy and multi-episodic histories can be reconstructed, thus opening a wide field of applications for the elucidation of tectonothermal histories of sedimentary basins.

6.3.4 Metavolcanics

Acidic volcanic rocks often show high Rb–Sr ratios and are therefore widely used for age determination, one special application being the bracketing of ages in absolute time scale work (see Harland et al., 1964, 1971). Within these rocks, and between volcanics and their country rocks, extensive exchanges can take place during incipient metamorphism, sometimes even through auto-metamorphic processes generated by the remaining thermal energy.

As these processes lead to exchange of both radiogenic mother and daughter products, the ages can become altered in both directions and thus are highly unreliable.

In basic volcanic rocks under hydrothermal conditions of very low-grade metamorphism, celadonite is sometimes formed, offering a unique possibility to date the formation of this mineral (Odin et al., 1986). The glass matrix of acidic to intermediate glass tuffs under very low-grade to low-grade metamorphic conditions, is transformed into illite and thus provides potential material for age determinations, as demonstrated by Ahrendt et al. (1978, 1983a) and as discussed above. Reuter (1985), in a more detailed study, emphasizes radiogenic ^{40}Ar retention problems in the smallest size fractions of these tuffs.

Finally, low-temperature but high-pressure conditions lead to the generation of alkali amphiboles and white micas in basic volcanic rocks. Both minerals can be used with caution to determine the age of the high-pressure phase. The amphiboles usually are very low in potassium content and the white micas very often show a high paragonite content, and thus are generally not suitable for Rb–Sr dating (Hunziker, 1974; Bocquet et al., 1974). Acidic to intermediate volcanic rocks show abundant apatite and zircon, and thus provide material for fission track dating, in basic rocks these accessory minerals are rather rare and large samples are required.

6.3.5 Mylonites

The pressure and temperature conditions for the generation of mylonites correspond to the temperature regime of very low- to low-grade metamorphism. Generally, mylonitization is a process of recrystallization of a rock under high shear stresses (Wise et al., 1984) that lead to highly ductile deformation during this recrystallization. Micas tend to recrystallize or simply rotate into the newly formed schistosity. The two cases can be distinguished by optical and geochemical means. K-feldspars tend to transform into micas (phyllonitiz-ation). Recrystallization of quartz and calcite offer unique opportunities for temperature estimations (Voll, 1976).

Every rock type can become a mylonite. Mylonitization is normally restricted to narrow zones that follow overthrusts, e.g. nappe overthrusts like the Glarus nappe, and also steep belts along fault zones, e.g. the Insubric line system. As the mylonitization process is a fast geological process, these rocks

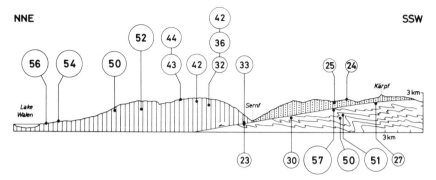

Figure 6.5 Schematic N–S profile across the Glarus Alps showing the anchimetamorphic and epimetamorphic (stippled area) overthrust block (hatched field), the anchimetamorphic parautochthonous Flysch below and the Glarus thrust plane in between. Potassic white micas extracted from the thrust plane mylonites yield K–Ar ages between 25 and 20 Ma. While the ages in the lower block are still controlled by the amount of detrital component of the sheet silicates, the K–Ar ages of the overthrust block still exhibit the imprint of the main metamorphic phase around 35 to 30 Ma before present.

are particularly suitable for dating tectonic movements. Hurford (1986a) combining Rb–Sr, K–Ar data on micas with fission track data on zircons and apatites, was able to determine a phase of movement on the Insubric line mylonites around 23 Ma before present, indicating an early Miocene uplift of over 9 km of the northern Pennine block, probably resulting from a major phase of backthrusting along the Insubric line. Hunziker *et al.* (1986) have used K–Ar and Rb–Sr ages in illitic–muscovitic K-white micas recrystallized in the fault gouge of the Glarus thrust to demonstrate that the Glarus thrust movements (Ruchi phase) occurred about 25 to 20 Ma ago (Figure 6.5), more or less contemporaneously with the backthrusting in the Insubric domain.

Thöni (1986) showed by the Rb–Sr thin slab isochron method that mylonite whole rocks, under certain circumstances, can be used to date the time of movement related to mylonitization. Here again, as with whole-rock slate dating, a higher temperature and a higher degree of reaction respectively are required for whole rocks than for mineral dating of mylonites, where careful textural and geochemical observations permit the selection of appropriate minerals in which exchange has gone to completion. Prograde rocks transformed into mylonites generally pose fewer inheritance problems than retrogressed rocks, in which high-temperature minerals may survive as clasts or as armoured relics and may therefore influence the radiometric age.

6.4 Possibilities and limitations of radiogenic isotopes

Non-stable isotopes applied in very low-grade terrains have their specific limitations and advantages. Generally a combination of different isotopic systems on differing but cogenetic materials leads to a higher degree of credibility of interpretation of the results.

6.4.1 K–Ar

In metamorphic terrains of all grades, the K–Ar method is ideal for dating the age of metamorphism, both in reconnaissance and in detailed work. As already pointed out in previous sections, a great variety of K-bearing phases (minerals, glass, and whole rock) are adequate for dating purposes, micas being the most widely-used minerals.

In very low-grade metamorphic terrains, neoformed or restructured micas generally only grow to very small grain sizes, depending on a variety of parameters, such as temperature, rock chemistry, degree of deformation, fluid pervasivity and fluid chemistry. One of the main problems (not only in K–Ar dating) is to determine the appropriate grain size in equilibrium with the metamorphic conditions and therefore adequate for dating the metamorphism, as mentioned previously. Decreasing grain size will usually yield decreasing ages, as the detrital micas are preferentially enriched in the coarser fractions. However, neoformations can also show this age trend, larger grains simply corresponding to longer periods of growth. Huon (1985), in a study on Palaeozoic schists of Morocco, could establish that generally finer and coarser fractions of the clay fraction tend to equilibrate to metamorphic conditions, while an intermediate fraction may still yield older ages. This author could show that while the smaller fractions have adjusted statically, the coarser fractions, being an obstacle in the new schistosity, tend to be reoriented, and thus recrystallized dynamically; the intermediate fractions not being rotated but being too large for a static equilibrium remained unaffected, at least in part, and yield older ages.

Careful analysis of the whole grain size spectrum by optical microscopy and X-ray techniques together with knowledge of the chemistry of detrital and neoformed micas in the rock type under consideration will enable selection of the appropriate grain size for age determinations. In the Alps, for example, the detrital micas are generally muscovitic, whereas the neoformations are more phengitic. As only a limited number of grain size splits is feasible, one is often confronted with mixtures or with partially adjusted samples.

To resolve problems of partial adjustment to metamorphic conditions, several grain sizes of the same sample are measured (Figure 6.6). In a $^{40}Ar/^{36}Ar$ v. $^{40}K/^{36}Ar$ diagram, all equilibrated minerals will plot on a straight line defining a single age—they are 'isochronic'—whereas the coarser and only partly equilibrated minerals will define a curve, with coarser grains yielding higher ages. This graphic treatment is only valid for monomineralic samples, as, obviously, different minerals of the same rock with different blocking temperatures will experience different degrees of equilibration to metamorphic conditions, and thus yield different age values. This method of

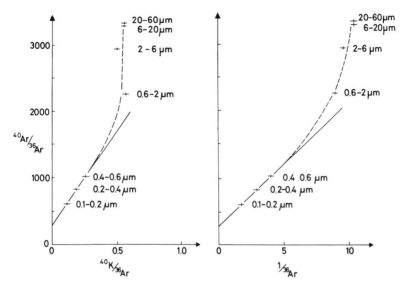

Figure 6.6 ^{40}Ar/^{36}Ar $v.$ ^{40}K/^{36}Ar and ^{40}Ar/^{36}Ar $v.$ $1/^{36}$Ar plots showing grain size populations with different degrees of equilibration. Data points of small grain sizes define straight lines while larger grain sizes still show a detrital influence leading to higher ages with increasing grain size.

measuring the K–Ar age of a variety of grain sizes of the same mineral leads to the establishment of the maximum grain size in equilibrium with the metamorphic conditions to which the rock has been submitted.

Another approach to reaching this conclusion is to apply two different radiometric systems to the same sample. Equilibration of the sample with metamorphism will result in concordant ages.

6.4.2 $^{40}Ar–^{39}Ar$

The ^{40}Ar–^{39}Ar method used in addition to the routine K–Ar method will lead to more detailed insight into the degassing properties of the minerals analysed. 39–40 total gas ages should generally correspond with conventional K–Ar ages; if this is the case, recoil effects do not have to be considered. In the case of 1 Md illites, however, this is not the case. Due to preferential ^{39}Ar loss of the samples during irradiation, the total gas ages tend to be up to 30% higher than conventional K–Ar ages and are therefore geologically meaningless (Hunziker *et al.*, 1986). The same phenomenon was described by Klay (1984) for glauconites.

A far better use of the ^{40}Ar–^{39}Ar method consists of incremental release age spectra. This unique tool for detecting only partial resetting or re-equilibration to metamorphic conditions is derived from the ^{40}Ar–^{39}Ar incremental release age spectra as shown by Hammerschmidt (1982) and by

Dallmeyer's work in Hunziker *et al.* (1986) and Kligfield *et al.* (1986). While 2M$_1$ illites in equilibrium with metamorphic conditions yield convincing plateau ages, partially reset illite fractions yield systematically increasing older apparent ages throughout incremental analysis (staircase patterns), typical of argon systems which have experienced partial post-crystallization loss of radiogenic ^{40}Ar by volume diffusion processes (see Turner, 1970, and Figure 6.7). With increasing resetting, either due to higher metamorphic temperatures, longer duration of the metamorphic event, higher degree of deformation and recrystallization, or smaller grain size, or a combination of these effects, the staircase pattern becomes increasingly flatter, until eventually a perfect plateau is reached. As this process is obviously transitional, quite convincing plateaux may turn out to still yield inadequately high ages with no geological meaning (Hammerschmidt, 1982; Hunziker *et al.*, 1986). Strict criteria for a plateau must therefore be applied (Dalrymple *et al.*, 1980).

As already pointed out, another pitfall may arise in the dating of 1Md illite. Here, conventional K–Ar ages are up to 30% lower than ^{40}Ar–^{39}Ar ages due to

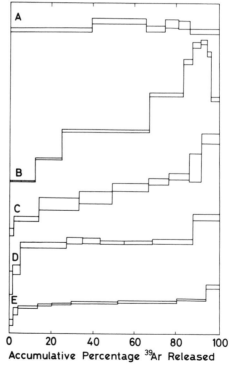

Accumulative Percentage ^{39}Ar Released

Figure 6.7 ^{40}Ar–^{39}Ar incremental-release age spectra of illites. (*A*) 1 Md illite with anomalously high plateau age, due to recoil problems in the reactor. (*B*)–(*D*) Staircase patterns reflecting partial post-crystallization loss of radiogenic argon (diminishing from *B* to *D*). (*E*) Nearly perfect plateau demonstrating that strict rules for plateau definition have to be applied.

significant recoil loss of ^{39}Ar during irradiation, yet with quite convincing plateaux (see Hunziker *et al.*, 1986, and Figure 6.7).

6.4.3 Rb–Sr

The Rb–Sr method is commonly used in metamorphic studies of all grades, for both dating purposes and measurement of exchange distances, as well as establishment of geologic environmental conditions. For diagenetic as well as for very low-grade metamorphic conditions, Clauer and co-workers have produced a large amount of Rb–Sr data, the bulk of which has been reviewed by Clauer (1979).

The main problem of Rb–Sr work in metamorphic terrains comes from the estimation of the initial ^{87}Sr/^{86}Sr ratio of a sample. Clauer's approach to this problem is based on the following principles. First, the neoformed minerals (clay minerals, micas, zeolites etc.) are separated according to grain sizes and enriched. Rb and Sr concentrations as well as Sr isotope ratios of an acid leachate of these minerals (preferentially obtained with dilute hydrochloric acid) as well as the mineral itself are measured. The age of the mineral is defined by a three-point Rb–Sr isochron, whereby the leachate containing the loosely-bound common Sr of the sample represents one point and the untreated mineral, which it is hoped contains the radiogenic component, represents the second point, the residue after leaching being the third point of the mineral–leachate isochron (Clauer, 1976). This technique presents problems related to the evaluation of the correct amount of leaching, so far estimated by trial and error and by measuring different leachates, solutions, HCl-insoluble residues and the mineral itself. Commonly the mechanism controlling the linear array in this sort of sample involves mixing between common and radiogenic Sr, so that it is not immediately clear whether the calculated age represents a geological event or not. Cross control with another decay chain, e.g. K–Ar, can help to answer this question.

Hydrothermal events very often lead to the generation of K- and Rb-rich minerals such as K-feldspars and micas. Fissure mineralizations can ideally be dated by the K–Ar as well as the Rb–Sr method, although here the problems of the initial ^{87}Sr/^{86}Sr ratios often cannot be solved adequately.

6.4.4 *Fission track dating*

The principles and techniques of fission track dating have been reviewed recently by Hurford (1986*b*) together with a description of the specific application to the dating of young samples. The method is particularly well suited to samples with moderate trace amounts of uranium, the limiting factor being the need for sufficient tracks—neither too many nor too few—from the spontaneous fission of ^{238}U. In practice, most use is made of apatite, zircon, sphene, and natural glass, spanning a time range of 1 to 400 Ma, although both

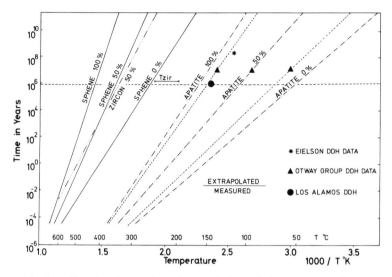

Figure 6.8 Annealing data for apatite, sphene and zircon. Laboratory annealing curves are extrapolated to geological time. The percentage track loss is shown on each curve. The annealing fan suggested for apatite is narrowed by inclusion of 100, 50 and 0% track loss points found in the Otway Group drill-holes (Gleadow and Duddy, 1981) and the 100% track loss points from the Eielson and Los Alamos deep drill-holes (Naeser, 1979). The probable closure temperature range for zircon, as given by geological evidence, is denoted T_{zir}. (Compiled from Fleischer et al., 1965; Wagner, 1968; Naeser and Faul, 1969; Krishnaswami et al., 1974; and Hurford, 1986a and unpublished data).

older and younger ages have been reported. Fission tracks are healed or annealed with heating, the temperature thresholds being substantially lower than the 'blocking temperatures' of minerals for the Rb–Sr and K–Ar isotopic systems. Figure 6.8 shows a compilation of annealing data from heating experiments and from geological evidence. Laboratory studies, which define the degree of track loss in various minerals at different times and temperatures, may be extrapolated by many orders of magnitude to geological time. In addition, the degree of track loss in apatite from deep drill-holes may be related to the measured borehole temperature and its probable duration in the past. If an apatite sample is held at 100 °C for 10^6 years, then all tracks will be lost. For zircon and sphene, geological evidence suggests effective closure temperatures of 200–250 °C, somewhat less than those predicted by annealing experiments (Gleadow and Brooks, 1979; Hurford, 1986a). Annealing of fission tracks in glass takes place at temperatures appreciably less than 100 °C (Fleischer et al., 1975) and care must therefore be taken in the interpretation of such data.

The combination of different isotopic systems and minerals from a single sample provides a useful first approximation of the time-temperature history of the sample. Figure 6.9 shows such a cooling curve for a single orthogneiss

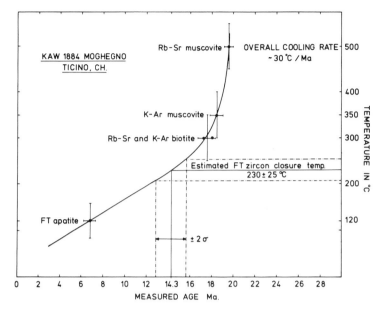

Figure 6.9 A cooling curve for a single 30-kg orthogneiss sample from the Central Swiss Alps showing measured mineral ages plotted against assumed system closure temperatures (after Hurford, 1986a and references therein).

sample from the Maggia Valley in the Central Penninic Alps of Switzerland (Hurford, 1986a) where the measured ages have been plotted against the assumed closure temperatures for each system (see Wagner *et al.*, 1977). This particular sample has cooled from amphibolite grade metamorphism some-time in the Mid-Tertiary, but the same procedure is applicable to any cooling sample.

Use of fission track dating to decipher the low-temperature thermal history of a sample has found particular application in sedimentary basin analysis. The concept of an annealing zone for apatite between 70–125 °C in which tracks are progressively lost over periods of 10^6–10^8 years, represents a more realistic model than the simple idea of a single closure temperature. In the Otway Basin, Victoria, Australia, Gleadow *et al.* (1983) have shown that the systematic way in which apatite ages decrease for drill-hole samples through a sedimentary sequence can be used to detect whether temperatures have decreased from an earlier maximum. Comparison of the position of the track annealing zone with the present-day drill-hole temperature profile, indicates whether temperatures are now at their maximum (Figure 6.10a) or if lowered ages are preserved at shallower levels and lower temperatures than those defining the present-day track annealing zone—in other words that the isotherms have sunk (Figure 6.10b). A second parameter for assessing thermal histories is the variation of confined fission track length distributions in apatite

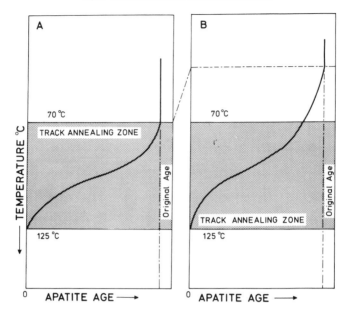

Figure 6.10 Hypothetical profiles of variation in apatite age with downhole temperature for 2 deep drill-holes. (*A*) shows a well where the temperatures are now at their maximum, and (*B*) a well where the temperatures have decreased from an earlier maximum. In (*B*) the reduced ages are preserved at shallower levels than the present day annealing zone. The broken line in (*B*) shows the earlier position of the top of the annealing zone. (Redrawn after Gleadow *et al.*, 1983).

with sample depth. Confined fission tracks do not themselves reach the polished and etched surface but intersect another track or fracture which allows the etchant to reach the confined track (Lal *et al.*, 1968; Green, 1981). The lengths of horizontal confined tracks may be measured under the microscope using a calibrated eyepiece graticule or an image analysis computer (see Hurford, 1986*b*). Fission tracks are produced throughout the history of the sample, each at a different time, according to the rate of spontaneous fission decay. Thus each track starts with approximately the same length but then experiences the subsequent temperature history of the sample, which may create varying degrees of shortening, dependent upon the intensity and duration of the heating. The present-day *overall* track length distribution therefore represents an integration of the temperature histories experienced by each track and thus records the overall temperature history of the sample. Figure 6.11 (after Gleadow *et al.*, 1983) shows three hypothetical apatite confined track length distributions for samples of differing thermal histories *A*, *B* and *C*, representing samples buried to different levels in the track annealing zone as, for example, with progressive burial in a sedimentary basin. In Figure 6.11 track length distributions are proposed for:

(*D*) a past thermal event, perhaps a mild metamorphism, with a bimodal

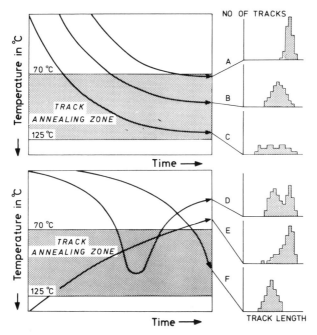

Figure 6.11 Temperature–time paths and resulting apatite track length distributions for rocks of varying thermal history. *A, B* and *C* show the evolution of temperatures for progressive burial to different levels in the track annealing zone. The lower 3 distributions show the patterns: *D*—a past thermal event; *E*—slow cooling; *F*—a recent heating event (Redrawn after Gleadow *et al.*, 1983).

    length distribution; the shortened tracks represent those formed before the thermal event, the longer tracks those formed since the event

(E) a slow-cooling event, such as slow uplift and/or the sinking of isotherms; here a greater proportion of longer, younger tracks are preserved, giving a right-skewed distribution

(F) a recent heating event, such as rapid burial or magmatism plunges the sample and its track length distribution similar to (*A*) into the annealing zone to give a similar distribution, but with each track shortened.

It is of particular significance that the apatite fission track annealing zone of about 70–125 °C spread over geological time is coincident with the temperatures for the maximum generation of liquid hydrocarbons (see Figure 6.12).

    Application of these analytical procedures to detrital apatites from sediments in the Canning Basin in Western Australia (Gleadow and Duddy, 1984) revealed two trends in apatite annealing: first the expected decrease in age and track length due to increase of temperature with burial; second, a further reduction in age and length associated with late Permian dolerite intrusions. The fission track data indicate that the thermal effect of the dolerite was much more widespread than would be expected from simple contact

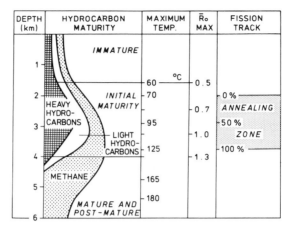

Figure 6.12 The comparison between hydrocarbon maturity, maximum temperature and fission track annealing properties shows the great potential of fission track studies for the evaluation of thermal histories of sedimentary basins. The fission track annealing zone of apatites coincides with the temperature for the maximum generation of liquid hydrocarbons (after Kanstler *et al.*, 1978, modified by Gleadow *et al.*, 1983).

metamorphic effects, suggesting the circulation of heated pore waters. Gleadow and Duddy conclude that such heating was sufficient to place many of the samples of the Canning Basin in the oil generation window.

6.4.5 *U–Th–Pb and Pb–Pb*

Uranium, thorium and lead isotope determinations are a powerful tool for solving age, provenance and mixing problems in low-grade terrains, as different decay chains can be measured in the same mineral or whole rock, and also in several minerals of the same paragenesis, or in several whole rocks. Minerals mainly used are zircon, monazite and sphene, but in principle all U-, Th-, and Pb-bearing phases, such as allanite, apatite, galena, pyrochlore, thorite, uraninite and xenotime can be measured, and even minerals with these elements in trace amounts, such as amphiboles, epidotes, feldspars and rutile. Mattinson (1981) applied the U–Pb method to such minerals containing uranium in trace amounts, to determine the age of blueschist metamorphism in the Franciscan. Gebauer and Grünenfelder (1976) proposed a low-temperature annealing model for zircons. This model is based on the observation that highly radiation-damaged (metamict) zircons will recrystallize, and consequently lose lead episodically at temperatures as low as 300 °C or even less. According to this model, an anchimetamorphic overprint of sediments will lead to concordant zircons, which may survive following amphibolite facies conditions.

Uranium mineralization and hydrothermal ore formation are often dated by the U–Th–Pb and Pb–Pb methods. Uranium ores, due to the high mobility of the uranium compounds down to ambient temperatures, pose their own problems, as measured ages are very often younger than expected and therefore result in discordancy between the uranium decay chains.

In this respect, Köppel (1968) found that whole rock analyses of rocks containing minute quantities of pitchblende will often give concordant ages, because of the low migration distances within the rock.

Besides the purely geochronologic aspects, the U–Pb decay chains measured in sulphides and whole rock samples may also serve as indicators outlining radiogenic halos around uranium deposits (Gulson and Mizon, 1980). Furthermore, lead isotopes play an important role in unravelling the genesis of sulphide ore bodies (Doe, 1970).

6.5 Economic importance

In view of the obvious potential of non-stable isotopes for solving problems in the field of economic geology, especially for the mining and oil industries, surprisingly few practical applications of isotope work have reached the academic literature. Most likely this is partly due to economic reasons, preventing publication of applied studies undertaken and financed by these industries. A few noteworthy exceptions concern the thermal evolution of sedimentary basins, problems of hydrocarbon maturation, transport and emplacement, and the complex problem of migration and formation of ore bodies.

6.5.1 *Thermal evolution of sedimentary basins*

A combination of fission track studies on apatites, zircons and sphenes (section 6.4.4, and Gleadow *et al.*, 1983) offers a new tool for the quantitative modelling of thermal histories of sedimentary basins, and opens up new aspects for the evaluation of hydrocarbon and geothermal energy resources.

Here especially, the fission track annealing zone of apatites between 70–125 °C spread over geological time coincides with the peak of liquid hydrocarbon generation and maturity (see Figure 6.12). As both processes, annealing of fission tracks in minerals, and generation and maturation of hydrocarbons, are a function of temperature and time, fission tracks in apatite contain a record of the heating to the oil window of potential reservoir rocks in sedimentary basins. According to Gleadow *et al.* (1983) the pattern of apatite fission track ages, together with detailed analyses of the distribution of track lengths, can give information not only on maximum palaeotemperatures but also on their variation through time. In this way, these authors managed to model time-dependent temperature paths for oil wells in the Otway Basin, Australia.

As discussed in section 6.3.3, neoformed illites in pore cavities of sandstones and greywackes also yield good information on the timing of tectonothermal events in sedimentary basins. Rossel (1982), Lee (1984), Lee *et al.* (1985) and Liewig *et al.* (1987) distinguish three different clay mineral parageneses during clay diagenesis in the Jurassic and the Rotliegend sandstones of the North Sea with at least three different stages of illite formation. The timing of such occurrences can ideally be established by radiometric illite analyses, thus providing valuable time marks for the tectonic pattern which governed the migration paths and the emplacement of oil and gas in the reservoir rock, as well as the displacement of the pore fluid in these rocks. It should be stressed, however, that with these brittle illites grown in pore cavities, especially great care must be taken to keep mechanical grain size reduction during separation as small as possible. These authigenic micas generally are in the micron size range, often even up to 50 microns. If such illites are mechanically ground down to the 0.1 micron size, the danger of argon losses is not negligible.

The quantitative modelling of thermal histories of sedimentary basins is also crucial for the evaluation of geothermal energy resources in such basins. One of the main answers that can be gained in this context is whether or not the actual measured temperatures at a certain depth represent maximum temperatures.

The growing demand for low-grade energy for heating purposes as a substitute for fossil fuels and electricity opens a wide field for the application of isotopes in low-grade surroundings, especially in sedimentary and volcano-sedimentary basins.

6.5.2 *Ore transport and formation of ore minerals*

Transport and formation of ore mineralization often occurs under hydrothermal, low- to very low-grade metamorphic conditions. Non-stable isotope work (besides stable isotopes) can help to characterize the mineralizing solutions, to determine the age of mineralization and the location and size of ore bodies.

Doe (1970) distinguished different genetic categories of sulphide ore bodies by means of lead isotopes. According to him, magmatogenic ores are usually uniform and their isotopic composition is close to that of the presumably related igneous source. Deposits of metamorphic origin generally show a greater isotopic variability if the host rocks are much older, while in sediments that are not much older than the ore body, isotopic compositions may approach the single-stage conditions, with model ages in close agreement with the time of metamorphism. Stratiform deposits in sediments of nearly equivalent age have uniform lead isotopic compositions, which show reasonably good agreement with the assumed age.

The processes of ore enrichment and deposition through geothermal brines can also be fingerprinted with U–Th–Pb isotopes as reported by Dickson *et al.* (1985). Gulson and Mizon (1980) report a dramatic increase in 206/207,

206/208 and 206/204 Pb ratios, within the radiogenic halos of uranium deposits, as mineralization is approached. Moreover, the quantification of this lead isotopic ratio increase leads these authors to estimations of the size of an uranium deposit.

Hydrothermal vein type mineralization often shows associated clay minerals, such as kaolinite, smectite, illite and chlorite, as well as mixed-layer clay minerals. According to Bonhomme *et al.* (1987) the clearest feature of these clays is their low titanium content. If the synchronous crystallization of clay minerals with the vein ores can be demonstrated, meaningful time marks for ore emplacement can be obtained by K–Ar and Rb–Sr analysis of the cogenetic illites. Bonhomme *et al.* (1986) have presented K–Ar results compatible with other age estimations for a series of vein mineralizations in the French Massif Central, comprising Au, U, Pb, Zn, Ba and F mineralizations. Diamond and Wiedenbeck (1986) have dated by the same method the gold–quartz veins of the Monte Rosa nappe. Clauer *et al.* (1985) were able to date the hydrothermal activity responsible for the uranium mineralization of the Athabasca Group by means of K–Ar analyses on the clay size illites in these rocks.

6.6 Conclusions

Isotope work in low- to very low-grade terrains has not yet reached the standards expected and routinely performed in work on high-grade or magmatic rocks. So far, every new data set potentially opens new and unexpected fields, thus rendering this sort of approach very interesting and exciting. Far more than for the more classical type of rocks, isotope work in very low-grade terrains is only reasonable after a combined multidisciplinary effort to solve the numerous problems resulting from the mostly very small grain dimensions, and the commonly predominating disequilibrium conditions. Microscopic work as a basis has to include not only optical microscopy, but all sorts of electron microscopical techniques, including microprobe analysis. Careful X-ray investigations, with both fluorescence and diffraction techniques, may help to clarify chemical and crystallographical problems and also with the selection of appropriate samples. Finally, isotopic work should preferentially make use of at least two independent non-stable methods and be combined with stable isotope determinations.

In low-grade terrains answers are sometimes still ambiguous and therefore often discredited by the more superficial potential user. However, the strong interest stemming from applied geology, where commercial criteria govern the application of a method as soon as a technique offers a positive answer, will it is hoped help to overcome these weaknesses in the near future.

This leads to some concluding remarks about present trends and future evolutions. Demands and techniques presently trend towards smaller and smaller samples. Many problems necessitate isotopic compositions of single grains, microfossils, overgrowths on grains and zones in zoned grains as well as series of cements. A laser system can vaporize a small volume of sample while leaving adjacent zones unaffected, thus avoiding time-consuming and often even impossible mineral separations. For small clay minerals, continuous centrifugation allows the use of great amounts of clay mineral suspension and a separation of grain sizes down to 0.1 microns in a rather short time, avoiding the long exposure of these small grains to fluids. Regarding isotope systems, the importance of the Sm/Nd system in low-grade terrains is still not clear. For very young sediments a great variety of disequilibrium systems are becoming increasingly popular. Answers provided from the K–Ar as well as from the Rb–Sr systems for the acid-soluble part of carbonate rocks are still not well understood and need further investigation.

Acknowledgements

I would like to thank T. Hurford for his fission track contribution. P. Bochsler, S. Burley, N. Clauer, M. Frey, S. Huon, B. Kübler and N. Liewig helped with their constructive criticism and discussions to considerably improve the quality of the manuscript. S. Ayrton kindly went through the pain of improving and correcting my continental English, and M. Frey is thanked for the tedious and meticulous editing of the manuscript, and for not giving up before the end.

7 Correlation between indicators of very low-grade metamorphism

HANAN J. KISCH

7.1 Introduction

The correlation between mineralogical indicators of burial metamorphism and incipient regional metamorphism ('very low-grade metamorphism') and with coal rank, was initiated some 25 years ago. Reviewing the correlation between the different mineralogical indicators at the current increased state of knowledge is more straightforward than it was some ten years ago: a reasonably consistent general correlation scheme has emerged from the large body of correlation data that has accumulated, as well as a more extensive documentation and better understanding of the chemical, mineralogical, lithological, and kinetic controls of the divergences.

The general pattern of correlation established earlier (Kisch, 1974; see Figure 7.1) has by and large been confirmed: a broad overlap between the ranges of occurrence—or approximate 'iso-metamorphism'—of (i) the 'diagenetic' or 'non-metamorphic' range of illite 'crystallinities' (with occurrence of illite/smectite mixed-layers), the zeolite facies, and sub-bituminous and bituminous coal ranks, and of (ii) the 'anchimetamorphic' zone as defined by Kubler (1967) on the basis of illite 'crystallinities' and other diagnostic phyllosilicate minerals (absence of kaolinite and illite/smectite mixed-layers), the prehnite-pumpellyite facies, and anthracitic* and low meta-anthracitic coal ranks, and of (iii) the 'epimetamorphic' zone of Kubler, the greenschist facies, and meta-anthracitic coal ranks, has been documented by published reports from many areas, and summarized in a number of reviews. One might attempt to further refine these relationships in the light of the increasing amount of data that recently have been, and still are forthcoming, particularly through the widespread application of the illite 'crystallinity' method. However, it is to be suspected that such attempts are subject to limits inherent in the relative inaccuracy and scatter of the parameters of incipient metamorphism, such as percentage expendable layers in mixed-layer clays, illite 'crystallinity', and metamorphic mineral assemblages as a result in local variation of chemical environment, and of vitrinite reflectance as a result of divergence of the maceral from pure vitrinite and of oxidation.

*Unless specified otherwise, 'anthracite' and 'meta-anthracite' are hereafter used in the sense of the ASTM coal classification, i.e. respectively 8 to 2% volatile matter (V.M.), corresponding to 2.5 to approximately 5% R_m or 2.7 to approximately 6% R_{max}, and < 2% V.M. corresponding to R_m > approximately 5% or R_{max} > approximately 6%.

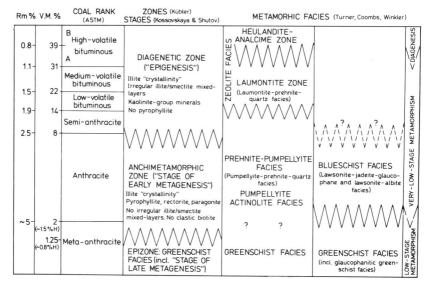

Figure 7.1 Schematic relationship of Kübler's zones, the 'stages of regional epigenesis and metagenesis', and facies of very low-grade metamorphism to coal rank. Slightly modified from Kisch (1974, Figure 1): coal ranks are given in the ASTM classification; reflectance values of Kisch (1980a, b) have been added.

On the other hand, several departures from the established correlation patterns of the parameters in some geological terrains that have come to light cannot merely be ascribed to such scatter of the data: they must reflect differences in the dependence of the evolution of different mineral parameters and coal rank on geological controls such as duration and re-occurrence of heating events, synmetamorphic v. postmetamorphic deformation, variation in P–T gradients or metamorphic facies series, presence and chemical composition of infiltrating or percolating solutions, etc. The understanding and evaluation of these differences in dependence on such controls may thus ultimately allow use of divergent correlations between parameters to reveal information relating to the differences in geological conditions and evolution of very low-grade metamorphism between different terranes. One of the purposes of this review is to contribute towards the understanding of the causes of these divergences.

The various diagnostic mineral parameters of incipient metamorphism and their controls have been described in several recent reviews (e.g. Kisch, 1983) and on the information furnished by divergences from the earlier established where possible referring to the above reviews. This review will mainly concentrate on more recent correlation data published in the last ten years, and on the information furnished by divergences from the earlier established general correlation pattern on the controls of the development of the parameters of burial diagenesis and incipient regional metamorphism.

7.2   Accuracy of determination and chemical controls of evolution

The causes of divergence from the general pattern of correlation are in part inherent in the different response of the parameters to the physical controls of metamorphism, and a closer understanding of these different responses is likely to become a major application of the study of divergences from the general correlation pattern.

However, much of the 'noise' in the correlations is due to lack of accuracy of determination of the various parameters, or scatter in their actual values. This scatter is largely due to chemical controls on the evolution of the various parameters. Such factors include the effects of composition of starting materials (smectite composition; glass v. plagioclase); oxidation conditions (for instance in stabilization of chlorite); supply of cations from destruction of clastic minerals or from interstitial solutions; SiO_2 supersaturation with respect to quartz; permeability and porosity of the rocks (affecting the supply of reactants and the removal of products); the a_{CO_2} of the solutions; change of composition of interstitial solutions with time; persistence of non-equilibrium mineral assemblages. These have been reviewed by Kisch (1983) and in some of the earlier chapters of this book; in this chapter they will be discussed only where directly relevant to the assessment of correlation—or lack of it— between the parameters.

Some of the inaccuracy is of a more pedestrian nature. For some of the parameters, the determination procedures are insufficiently standardized (particularly in the determination of illite 'crystallinity'), in others there is a problem in the selection of the material to be measured (particularly in vitrinite reflectance: recognition of oxidation, avoidance of 'pseudo-vitrinite' or 'semi-vitrinite' (Stach et al., 1982, pp. 233–235), etc.).

A more serious problem lies in the difference in the rate of development of the various parameters, particularly the difference between the rate of attainment of mineral equilibria on one hand, and the development of coalification on the other. The latter point has not been exhaustively dealt with in any of the preceding chapters, and will be briefly dealt with here.

The conventional 'classical' coalification model of Karwell (1956) is based on equations of first-order chemical reactions, rank increasing linearly with the duration of heating at constant temperature: heat at 120 °C, for instance, giving 0.6% R_0 after 10 Ma, 0.8 after 20 Ma, 0.95–1.0% after 30 Ma, 1.4% after 50 Ma, 2.5% after 100 Ma, and 3.5% after 200 Ma. In his model, Karweil did not only consider the duration of maximum heating, but added a logarithmic 'Z' scale for the addition of rank increments during successive periods of thermal history—including subsequent heating at temperatures lower than the maximum.

This approach has been elaborated by Lopatin (1971; see Lopatin and Bostick, 1974), who developed a temperature–time index based on the 'sum heat impulse' (SHI), the sum of the 'elementary heat impulses' (ΔT_n) (r^n), i.e. the time the coal resided in each 10 °C temperature interval (beginning with 50–60 °C), multiplied by the rank factor or temperature coefficient of the coalification rate for that temperature interval. This rank factor is based on the assumption that the coalification rate doubles for every 10 °C increase in temperature (beginning at 50 °C–heating below 50 °C does not result in formation of bituminous coal even during more than 100 Ma). The model was tested by comparison of the SHI values with coal ranks in the Münsterland 1 borehole; correlation turned out to be good. This approach has been popularized by Waples (1980), who proposed a numerical time–temperature index scale based on the Lopatin method.

Hood et al. (1975) took a somewhat simplified approach in relating coal rank (or LOM, level of organic metamorphism) only to maximum temperature and to the 'effective heating time'—the time the coal resided within 15 °C of the maximum temperature; their plot has been somewhat extended by Bostick et al. (1978, Figure 13—see Figure 4.23).

Gretener and Curtis (1982) submitted that not only are conversion rates below 50 °C so low that time has no effect, but at high temperatures (> 130 °C) the rate of the reactions is so high that time also plays no important role (at least, as far as passage through the oil window is concerned); time is taken to operate effectively between 70 and 100 °C. Termination of heating at a certain temperature level will result in a constant organic-matter conversion rate.

However, there has been disagreement with the prevailing view that geological time is a controlling parameter in organic metamorphism: Ammosov et al. (1975, 1977) and Neruchev et al. (1976; references from Sajgó, 1979) held 'that the duration of oil formation reactions is unimportant after a certain period of time'.

Lately additional evidence has been forthcoming that time may be of less importance in coalification than previously thought. Suggate (1982) tested Karweil's and Hood et al.'s models on New Zealand low-rank coals, and concluded that 'the generally-accepted influence of time is considered unproved and unlikely to be significant'.

In a study of sedimentary organic samples from six liquid-dominated geothermal systems in western North America, Barker (1983) has shown that the reflectance data are strongly temperature-dependent. Only for heating times of less than 10^4 years the curves diverge towards lower R_m values (also Barker and Elders, 1981). Barker concludes that in these systems, 'after about 10^4 yr, reaction duration has little or no influence on metamorphism of organic matter...(and) that vitrinite reflectance can be used to determine the maximum temperature reached in hot sedimentary basins of moderate longevity' (but see comment by Bostick and reply by Barker, 1984).

Price (1983) holds that, contrary to the accepted models, geological time is not a controlling parameter of organic metamorphism, times of 1 Ma or more having no influence on maturation reactions. For different sedimentary basins currently at or near maximum geothermal gradients, with sediment burial

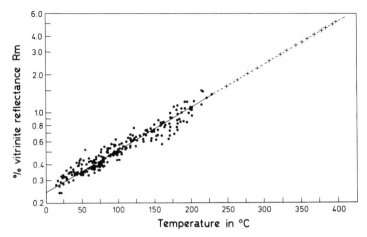

Figure 7.2   Plot of vitrinite reflectance v. burial temperature for burial times (times since attainment of 80% of present-day, maximal temperatures) range from 200 000 years to 240 million years. Solid line is from regression analysis of data from nine sedimentary basins. Burial times (times since attainment of 80% of present-day, maximal, temperatures) in these basins range from 200 000 years to 240 million years. Dashed line is an extrapolation for five different basins. (Modified from Price, 1983, Figure 19).

times ranging from 2 to 240 Ma, he showed that there is a strong R_0-T correlation (see Figure 7.2) but no correlation of R_0–burial (heating) times for any temperature interval. Price claims that the only solid documentation of the role of time as a controlling parameter of organic metamorphism—in the works of Karweil (1956), Lopatin (1971) and Connan (1974)—is entirely based on areas with moderate to extreme palaeogeothermal gradients, and that the levels of organic metamorphism attributed to geological time by these authors are better explained by high palaeogeothermal gradients.

Price considers that 'there is no evidence from natural systems that... maturation reactions have first-order reaction kinetics... and therefore no basis for the time dependence which has been assumed for these reactions' (p. 27): laboratory studies show that these reactions are characterized by multiple-order reaction kinetics (but see discussion by Kohsmann and reply by Price, 1985).

Barker and Pawlewicz (1986) have plotted a scatter diagram of maximum burial temperature (T_{max}) v. R_m for samples with a wide range of heating duration, temperatures for given values of R_m being some 30–35 °C lower on their regression line than on that of Price and Barker (in Price, 1983, Figure 19); they argue that most of the data scatter ($r^2 = 0.7$) can be explained by problems in determining R_m, 'and can be minimized... by sampling and analysis in a single laboratory, illustrated by the excellent correlation ($r^2 = 0.9$) between T_{max} and R_m found by Barker and Price (in Price, 1983)'. They conclude that functional heating duration has a limited influence on thermal maturation, and that vitrinite reflectance can be directly calibrated as a maximum geothermometer, implying that the chemical reactions must stabilize.

These views imply that for any burial temperature there is a 'quasi-equilibrium' rank, which, once stabilized, will increase only slightly or not at all with continued heating at the same temperature, and that the rank will appreciably increase only when this temperature is exceeded.

This writer feels that the duration of the heating necessary for reaching such a state of 'quasi-equilibrium' rank is itself likely to be a function of temperature, i.e. that it should be longer for attainment of equilibrium at low than at high temperatures. This would mean that at low temperatures the coalification reactions are 'slower' than at high temperatures.

Little is known about the effect of duration of heating on mineral reactions. However, reactions such as mixed-layering of smectite and the development of illite 'crystallinity' are to a large extent controlled by chemical factors such as have been listed above. Such factors are not likely to be very strongly dependent on temperature. It seems therefore to this writer that the *rate* of attainment of mineral equilibrium is not very strongly dependent on temperature, at any rate less so than the attainment of 'quasi-equilibrium' coal rank. This would be of obvious relevance to the relation between mineral equilibrium and coal rank as a result of brief periods of heating: for short durations of burial coalification would 'lag' behind temperatures and behind mineral reactions at low temperatures, whereas at high temperatures the opposite would be the case.

In the following discussion of the relation between the low coal rank associated with the disappearance of discrete smectite or the appearance of laumontite in rapidly accumulated sequences on one hand, and the associ-ation of illite 'crystallinity' with high rank in terrains subjected to strong heating by post-kinematic intrusive bodies on the other, we will see that the divergences from 'regional correlations' found offer some support for this view.

7.3   Illitization and chloritization of smectite

The most prominent changes in the clay mineralogy of sedimentary rocks upon burial diagenesis are the progressive disappearance of smectite and kaolinite, and the emergence of illite and chlorite as the major phyllosilicates at the onset of anchimetamorphism.

The discontinuous nature of the illitization of smectite through illite/smectite mixed-layers upon burial has been documented in a large number of papers. In particular, the decrease in percentage smectite layers from about 70 to about 20 percent takes place within a comparatively short depth interval. Two major 'clay dehydration stages' can be distinguished (Perry and Hower, 1972), the first corresponding to the disappearance of discrete smectite (with $< 25\%$ illite layers) and the second to the reduction of the percentage expandable layers from about 35 to 20%. At the latter stage,

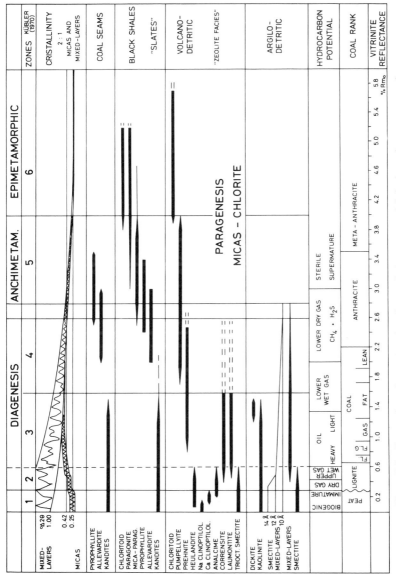

Figure 7.3 Comparison between mineral associations, potential hydrocarbon zones, and vitrinite reflectance. After Kübler *et al.* (1979, Table 3). Note: the German coal-rank nomenclature is given.

regular stacking order tends to appear, first of the IS (allevardite or K-rectorite) type, and when the percentage of smectite layers has decreased to about 10%, ISII stacking order (Kalkberg type) appears.

In his earlier review, this author (Kisch, 1983, pp. 312–314 and Table 5-III) has attempted to correlate the depth and temperature intervals of this reduction in expandable layers with the geothermal gradients, both based on measured borehole temperatures. However, since then the palaeothermal gradients in several Mesozoic and Palaeogene basins have been shown to have been much higher at the time of diagenetic alteration than at present (e.g. Price 1983, pp. 16–24). The correlation with borehole temperatures thus must be abandoned except for geothermal areas and for young sedimentary basins in which the present temperatures can reasonably be taken to be maximal.

The author will not consider the temperatures for the smectite–illite reaction as obtained from hydrothermal experiments: alteration and fluid data in experimental samples show major divergences from those in the Gulf Coast and Salton Sea—for instance formation of smectite at 200 °C and 300 °C, 500 bars. These differences can be argued to be due to kinetic effects associated with the short duration of the experiments, leading for instance to higher silica activity, lower pH, and more oxidizing nature of the experimental fluids.

7.3.1   Coal rank associated with illite/smectite (I/S) mixed-layering

In the following discussion of the relation of illitization of smectite to coal rank we will treat these two stages separately.

The conversion of discrete smectite (17 Å peak after ethylene-glycol (EG) solvation) into I/S mixed-layers with less than 70% smectite layers in argillaceous shales, mudstones and siltstones takes place usually in association with coal ranks of 0.45 to 0.65% R_0—data from Dunoyer de Segonzac, 1969, and Durand and Espitalié, 1976; Heling and Teichmüller, 1974; Foscolos et al., 1976; Monnier, 1982, and Rybach and Bodmer, 1980; Nadeau and Reynolds, 1981; Kisch, 1981a; Héroux et al., 1981; Dypvik, 1983; and Robert, 1985, pp. 287–291; see also compilations by Kübler et al. (1979; see Figure 7.3), Kübler (1980, Figure 7), and Héroux et al. (1979, Figure 1). Local association with slightly lower ranks, for instance 0.4% R_0 in the Miocene of the Gulf Coast has been ascribed to very rapid subsidence and failure of the coalification to keep pace with the increase in temperature (Heling and Teichmüller, 1974). Higher coal ranks of 0.7% R_0 have been found for the transformation in bentonites and carbonate rocks (section 7.3.2), and in areas with high geothermal gradients (section 7.3.3).

The coal ranks associated with the formation of illite/smectite mixed-layers with low percentages of expandable layers and the disappearance of kaolinite are shown in Table 7.1. Contents of about 25% expandable layers are associated in several areas with a range of 1.1 to 1.5% R_0 (Foscolos et al., 1976; Héroux et al., 1979, Ivanova et al., 1980; Spears and Duff, 1984), but association with much lower ranks of 0.8% R_0 and even less has been documented (Środoń, 1979; Dypvik, 1983; Pollastro and Barker 1986).

7.3.2 Lithological and mineralogical controls on illite-smectite mixed layering

(i) *Illite-smectite mixed-layering in tuffaceous and calcareous rocks.* The above data come from argillaceous shales or 'argilo-detrital' mudstones and siltstones. In other rock types, such as tuffaceous rocks and calcareous rocks, the relationships may be different due to a delay or 'lag' in the illitization of smectite. In tuffaceous and other pyroclastic rocks such a 'lag' in the stage of illitization of smectite compared to shales has been noted by Środoń (1979), Hoffman and Hower (1979), Ivanova *et al.* (1980), and others. The survival of discrete smectite and I/S mixed-layers is particularly evident in zeolite-facies sequences developed in pyroclastic and volcanoclastic rocks (Boles and Coombs, 1975, p. 164; Kisch, 1981*b*, p. 353; Utada and Vine, 1984); in the altered vitric tuffs of the Hokonui Hills, discrete smectite-rich I/S mixed-layers persist together with heulandite into the lowest units in association with coal ranks of about 1.2% R_0 (Kisch, 1981*b*, Table 1; see Table 7.5).

The ordering of mixed layers in bentonites has been reported to take place both prior to and at a lower illite content (50%) than in adjacent shales (60%) (Schultz, 1978; Nadeau and Reynolds, 1981), as is also apparent from the low rank of associated coals (0.7% R_0 according to Pevear *et al.*, 1980). Elsewhere ordering appears in association with the same content of 25–30% expandable layers in shales and in pyroclastic rocks, but in the latter both at higher ranks ('coking to lean coals' or $R_0 > 1.15\%$—Ivanova *et al.*, 1980).

The preservation of the montmorillonite content of the mixed-layer minerals upon burial by fine grain size and high carbonate content (e.g, Árkai and Vizclán, 1975) is apparent in the persistence of discrete smectite in Mesozoic calcareous rocks (Persoz, 1982, pp. 31–32), up to 3000 m below its disappearance in Tertiary Molasse in association with *c.* 0.6% R_0 (Monnier, 1982), and in the high contents of expandable layers in I/S mixed-layers in carbonate rocks with stated coal ranks (e.g. Gill *et al.*, 1977; Bertrand *et al.*, 1983).

(ii) *Availability of K for illitization of smectite.* Illitization at low temperatures may be inhibited by lack of K, as has been demonstrated experimentally by Blank and Seifert (1976). The importance of the content of clastic mica and K-feldspar in the depth of persistence of smectite and the content of illite layers in I/S mixed layers has been demonstrated by Teodorovich and Konyukhov (1970), Hower *et al.* (1976), Eslinger and Sellars (1981), and Bruce (1984). Much of the K required for the illitization of smectite seems to be derived from the decomposition of detrital mica. This is evidenced by the deterioration of illite 'crystallinity' with depth concurrent with the disappearance of discrete

Table 7.1 Relation between coal rank and attainment of illite/smectite mixed-layers (I/S) with low percentages of expandable layers and disappearance of kaolinite

Age of formation; area	Percentage smectite layers (sm) in I/S; disappearance of kaolinite; borehole temperatures	Coal rank R_0 (%) unless otherwise indicated; calculated temperatures. Reference if different from first column
Shales, mudstones, and siltstones		
Mesozoic and Tertiary		
Logbaba wells, Palaeocene–Cretaceous of the Douala Basin, Cameroun (Dunoyer de Segonzac, 1969)	Disappearance of kaolinite (c. 1900 m depth, 85 °C) I/S with c. 25% sm (c. 3500 m depth, 135 °C)	0.7–0.8 (Durand and Espitalié, 1976) c. 2.7
Lower Cretaceous shales, NE British Columbia (Foscolos et al., 1976)	Disappearance of kaol. ('usually absent') (early mesodiagenesis) I/S with 25% sm (boundary meso/telodiagenesis)	1.0–1.5 1.5
Compilation by Héroux et al. (1979; see also Kubler, 1980, Fig. 7)	Disappearance of kaol. (lower part of diagenesis zone 3) I/S with 25–10% sm Complete disappearance of sm layers (base of diagenesis 4)	0.9–1.4 1.1–1.4 c. 2.3
Jurassic–Cretaceous of the Sverdrup Basin, N.W.T. (Foscolos and Powell, 1979a,b; Powell et al., 1978)	'First dehydration': I/S with c. 55 to c. 35% sm + verm (<0.2 μm fraction) (d_{001} from >14 Å to <12 Å) 'Second clay dehydration' (d_{001} from c. 11.5 to 10 Å)	0.5–0.55 > 2 (below total depth reached)
Sandstones of the Cretaceous Kootenay Fm at Mt Allan, SW Alberta (Hutcheon et al., 1980)	Kaol + dolom→chlor + calc (265 m below top of Kootenay Fm)	1.55 (Hughes and Cameron, 1985) 180°–230 °C from mineralogic equilibria
10 wells in Jurassic and Cretaceous of the North Sea (Dypvik, 1983)	I/S with <30% expandable layers— 80 to 100 °C	Above 0.65 to 0.8

Lower Cretaceous bentonites, E British Columbia (Spears and Duff, 1984)	I/S with 13–31% sm and rectorite-type ordered interstratification; kaol.	Medium-volatile bituminous (equiv. to 1.1 to 1.5%)
Well in Cretaceous-Tertiary of Green River Basin, Wyoming (Pollastro and Barker, 1986)	I/S with 30% sm layers; transition from random interstr. to allevardite-type order I/S with 10–15% sm; transition from allevardite-type to Kalkberg-type order	0.7 R_0; 120 °C from R_0 1.3 R_0; 175 °C from R_0
Palaeogene sandstones, E part of Santa Ynez Mountains, Calif. (Helmold and Snider, 1981)	I/S → ill-chlor assemblage	2.4 R_0; 235 °C from R_0
Upper Palaeozoic		
Upper Permian coal measures, Bowen Basin, E Queensland (Kisch, 1968)	Incipient chloritization of kaolinite (Bluff) Disappearance of kaolinite; common illite + chlorite + ankerite and/or siderite (Baralaba)	V.M. equiv. to 1.75–1.8 V.M. equiv. to 1.8–1.85
2 wells in Carboniferous of the Upper-Silesian Basin (Środoń, 1979—see Fig. 6.5)	I/S with <25% sm I/S with 15% sm (change from IS to ISII type ordering) I/S with 10% sm (ISII type) Illitization of kaol. becomes evident Only minor kaol.	0.80 0.85 2.2 >1.2 1.7 to 2.2
Carboniferous of Donetz Basin (Ivanova et al., 1980)	IS with 25–30% sm; tendency towards regular interstratification	Coking and lean coal (equiv. to >1.2)
Upper Carboniferous of the well Saar 1 (Teichmüller et al., 1983)	Base of kaolinite zone at 1450 m depth	1.1
Carboniferous of Askrigg block and Stainmore trough, N Pennines, England (Smart and Clayton, 1985)	I/S with 30–35% sm; less than half partial IS ordering I/S with c. 25% sm I/S with c. 15% sm; beginning of ISII ordering	0.7 R_{max} 1.1 R_{max} 2.5 R_{max}

Table 7.1(*Contd.*)

Age of formation; area	Percentage smectite layers (sm) in I/S; disappearance of kaolinite; borehole temperatures	Coal rank R_0 (%) unless otherwise indicated; calculated temperatures. Reference if different from first column
Lower Palaeozoic		
Ordovician sediments overlying the Ballantrae Ophiolite Complex, Scotland (Bevins *et al.*, 1985, p. 73)	Kaolinite	CAI 1–2.5 (Bergström, 1980) (equiv. to < 0.8 to 1.2–1.5)
Silurian N of B.O.C. and of Pentland Hills (Bevins *et al.*, 1985, p. 73)	Kaolinite	CAI 2 and 2.5 (equiv. to 0.85–1.3 and *c.* 1.2–1.5)
Carbonate rocks		
Ordovician platform carbonates and shale, St Lawrence Lowlands, Quebec (Bertrand *et al.*, 1983)	Neuville section: 40–60% I/S; discrete smectite in marly ls, mean ill. peaks 1.6 to 2.1° 2θ	mean 0.84 (11)
	Joliette section: 23–32% I/S; kaol. in ls; mean ill. peaks 1.0 to 1.3° 2θ	mean 1.76 (5)
	Montréal section: 7–10% I/S; appreciable kaol.; mean ill. peaks 0.6 to 0.8° 2θ	mean 1.64 (4)

smectite in diagenesis zone 2 and the upper part of the mixed-layer diagenesis zone 3 of Kubler (1980; Kübler et al., 1979) in association with coal ranks in the range 0.3 to 0.8% R_m (Kübler et al., 1979—see Figure 7.3; Kubler, 1980, Figure 7; Héroux et al., 1979, Figure 1; Héroux et al., 1981, Figure 15; Monnier, 1982; also Hutcheon et al., 1980), in which diagenesis erases the 'crystallinity' of the inherited micas due to alteration or neoformation of disordered 'illitic' layers (Kubler, 1980, pp. 11–12; Héroux et al. 1981; Monnier, 1982).

On the other hand, improvement of illite—or, more correctly, illite + I/S mixed-layer 'crystallinity'—with depth through all or most of the mixed-layer zone has been reported from sections in the Upper Cretaceous of the Douala Basin, Cameroun (Dunoyer, 1969, Figure 15), and from profiles of the Zechstein, Rotliegendes, and Carboniferous of East Germany (Blank and Seifert, 1976); see also Kübler et al. (1979), see Figure 7.3; and Kubler (1980, Figure 7).

The above-mentioned deterioration of illite 'crystallinity' in the upper part of the mixed-layer zone does not hold for calcareous rocks: in the upper zones of diagenesis the 'crystallinity' is better than in marl and shales (Kubler, 1967; Kübler et al., 1979, p. 356; Dunoyer, 1970; Bertrand et al., 1983, Figure 5; Persoz, 1982, p. 35). In eight wells in the Swiss Plateau and the SE border of the Jura, Persoz (1982, Figure 15) has shown that the 'crystallinity' of illite (EG-solvated) as measured on the same marly lithologies in the Jurassic improves with increase in maximum burial depth from 1 to 7 km throughout diagenetic zone 3 of Kubler. A similar improvement of 'crystallinity' with depth in the mixed-layer zone is documented in profiles in Ordovician carbonate rocks from Quebec by Bertrand et al. (1983).

Some of the K required for the illitization of smectite may be derived from the destruction of the smectite itself (Eslinger and Sellars, 1981) or by selective 'cannibalization' of smectite layers of I/S (Boles and Franks, 1979; Pollastro, 1985).

Dissolution of smectite has also been proposed as a mechanism for the conversion of smectite to illite by Nadeau et al. (1984, 1985), who showed that random illite/smectite mixed-layers can be regarded as physical mixtures of elementary smectite and 'illite' particles (respectively 10 Å and 20 Å thick), and that regularly interstratified I/S and illite consist mainly of 'illite' particles 20–50 Å thick and > 50 Å thick, respectively. Dissolution of smectite and coarsening of the illite particles would therefore give first 'random mixed layer', subsequently 'regular mixed layer', and finally 'illite' without detectable expandable components, consisting of particles 20 to 160 Å thick.

In K-poor, but Na-rich environments illite/smectite mixed-layers may be replaced by a paragonite/muscovite mixed-layer (Frey, 1969b, 1970) rather than by illite. In such environments illite/smectite mixed-layers may persist into the lowest grade part of the anchizone, locally together with paragonite/muscovite mixed-layer, for instance in Jurassic marls of the Helvetic nappes of central Switzerland (Breitschmid, 1982), in association with ranks of 1–4% $R_{m(0)}$.

Availability of K from solutions is a function of permeability (Heling, 1974; Howard, 1981). However, Velde and Nicot (1985) have argued that the imposition of elements such as H^+ and K^+ through the solution is more likely to occur in rocks with few clay phases, whereas in an invariant assemblage a_{H^+} might be buffered by the clay-mineral assemblage.

Morton (1985) even argues that after sudden 'punctuated' diagenesis in the Oligocene Frio Formation, initiated by a change in pore-water chemistry resulting in rapid formation of diagenetic illite at shallow burial, subsequent pore-water changes inhibited further clay reaction, even upon 7000–8000 ft of additional post-Oligocene burial.

(iii) *Inhibition of illitization by release of cations other than K.* Strong inhibiting effects of the release of Ca^{2+} and Mg^{2+} on the illitization of smectite have been demonstrated experimentally by Blank and Seifert (1976), Roberson and Lahann (1981), and Howard and Roy (1985). This effect is demonstrated by the delay of the 'second dehydration step' in the Sverdrup Basin, N.W.T., to the zone where cracking of liquid hydrocarbons to gas occurs, i.e. between 1.0–1.2 and 1.4% R_0 (Powell *et al.*, 1978). Ca release during albitization of plagioclase could help to explain the delay of attainment of 65–90% illite layers until coal ranks as high as 3% R_0 in bentonite from Chuckanut, Washington (Pevear, 1983).

Similar Ca-release as a result of albitization could explain the association of smectite and illite-poor illite/smectite mixed-layers in the 'classical' zeolite facies sequence in the Triassic of Southland, New Zealand, with coal ranks ranging from 0.60 to as high as 1.33% R_0 (Kisch, 1981*b*), and the persistence of illite/smectite mixed-layer to a coal rank of 2.4% R_0 in the laumontite-bearing eastern part of the Santa Ynez Basin (Helmold and Snider, 1981).

(iv) *Effect of the composition of the smectite and of reducing environments upon illitization.* There is a growing body of evidence indicating that compositional varieties of smectite react differently to increase of temperature upon burial. Boles and Franks (1979) ascribed the apparent increase in Fe and Mg in unreacted smectites with depth to the (preferential) illitization of the more aluminous smectites the smectites with high octahedral $(Fe + Mg)/Al$ resisting conversion to illite until temperatures high enough to produce ordering are attained. Whereas illite/smectite forms an ordered mixed layer phase at 50–30% smectite layers, which remains stable to relative high temperatures, Velde and Odin (1975) showed that glauconite solid solution diminishes rapidly as temperature increases, without ordering at 30% smectite layers. Fe reduction as a factor enhancing smectite to illite transformation has been stressed by Eslinger *et al.* (1979) and Velde (1985*b*).

7.3.3 Chloritization of smectite: occurrence of corrensite

Smectite may be converted to chlorite. This transformation usually goes through an intermediate stage of chlorite/smectite mixed layers, which often show ordered interlayering. Corrensite is a 1:1 regular chlorite/smectite or chlorite/vermiculite mixed layer with a low-angle 28 Å diffraction peak; the two varieties have been referred to as respectively low-charge and high-charge corrensite (the latter does not expand with glycerol or ethylene glycol).

Corrensite is common in basic and intermediate pyroclastic and volcanoclastic rocks, particularly in the laumontite zone of the zeolite facies (e.g. Kübler, 1973; Lippmann and Rothfuss, 1980). It is an almost ubiquitous constituent of the laumontite-bearing Grès de Taveyanne (Kubler et al., 1974, pp. 465–466; Stadler, 1979; Kisch, 1980b, p. 768), associated with coal ranks from 0.9 to 2.5% R_m (see Table 7.5): on the basis of a study of these and similar rocks Lippmann and Rothfuss (1980) went so far as to conclude that laumontite does not occur in corrensite-free rocks!

Regular and irregular chlorite/smectite and chlorite/vermiculite mixed-layers occur throughout the zeolite-facies sequence of the North Range Group of Southland, southern New Zealand (Boles and Coombs, 1975, p. 164; 1977, p. 993; Kisch, 1981b, p. 353), particularly in laumontite-bearing and other illite-poor samples, in association with coal ranks ranging from c. 0.6 to at least 1.2% R_0 at the base of the sequence (Kisch, 1981b, Table 1). The laumontite-bearing units of the disturbed belt of Montana '... contain either corrensite... or another chlorite/expandable phase' (Hoffman and Hower, 1979).

Temperature data from deep wells in the Neogene of northern Honshu (Iijima and Utada, 1971), and geothermal areas in northern Honshu (Seki et al., 1983; Liou et al., 1985a) and Iceland (Kristmansdóttir, 1979, p. 364)—in all of which corrensite is associated with laumontite and/or wairakite—corroborate Kübler's (1973) conclusion that corrensite appears in pyroclastic rocks in 'normal conditions of diagenesis' at 90° to 100 °C, and at 200 °C in low-pressure hydrothermal conditions.

According to the general schemes of Kübler et al. (1979; see Figure 7.3) and Héroux et al. (1979), its distribution in diagenesis is almost identical to that of laumontite: they appear in diagenesis zone 2 at an equivalent coal rank of 0.5–0.6% R_0, and persist throughout diagenesis zone 3 (up to a coal rank of 1.6% R_0). Similar relations are found for the Grès de Taveyanne by Kisch (1980b).

Their persistence through diagenesis zone 4 (up to a coal rank of 2.6% R_0) is indicated as uncertain or minor by Kübler et al. (1979–see Figure 7.3) but as certain by Héroux et al. (up to a rank of about 2.9 to 3.1% R_0); this seems to be corroborated by reflectances of 2.1–2.8% and 2.6% measured close to the occurrences of laumontite + prehnite + pumpellyite + corrensite in the Grès de Taveyanne of the Diablerets nappe (see Table 7.5). No corrensite was found in any of the Grès de Taveyanne with prehnite and pumpellyite without laumontite—which are associated with anchimetamorphic illite 'crystal-linities' and anthracitic coal ranks (Stalder, 1979; Kisch, 1980b).

In addition to the references given by Kisch (1983, pp. 324–325) a number of references document the occurrence of low-charge corrensite in the higher-grade part of the 'diagenetic' zone: in the Tournaisian-Viséan carbonate rocks of northern France (Dunoyer et al., 1968, Figure 3; Kübler, 1973, p. 550), and in the northern part of the Dinant syncline (Dandois, 1981, p. 312), in the Triassic–Jurassic of the Maliaque zone of central Othrys (Holtzapffel and Ferrière, 1982); low-charge corrensite has subsequently also been found in the overlying Upper-Cretaceous and Tertiary flysch of central and eastern Othrys by Kisch (unpublished data).

Robinson and Bevins (1986, p. 109) mention regular mixed-layered illite/smectite and chlorite/smectite as dominating the clay-mineral as-semblages in the diagenetic zone of the Welsh Borderland, characterized by an illite-'crystallinity' range of 0.83–0.32 $°\Delta 2\theta$.

More specific indications of associated rank are given by Chudaev (1979), who describes montmorillonite and random chlorite/montmorillonite mixed-layer in association with long-flame coal ($R_0 < 0.7\%$) from Kamchatka, and ordered corrensite-like mineral in association with gas and fat coal (corre-sponding to 0.7 to 1.2% R_0), which has disappeared from the deeper series in association with coking coal ($R_0 > 1.2\%$). Monnier (1982) similarly describes appearance of corrensite near the transition between smectite and illite/smectite mixed-layers—which is associated with coal ranks of 0.6 to 0.65% R_0—and Pollastro and Barker (1986) describe the occurrence of corrensite in association with the strong increase in percentage illite layers in illite/smectite mixed-layers from 20 to 75%, and in association with vitrinite reflectances of 0.7–0.8% (ibid., Figure 5).

In the Muschelkalk of northwestern Germany corrensite appears to be restricted to the Göttingen area (see also Lippmann, 1956), associated with 'diagenetic' illite 'crystallinities' of $Hb_{rel} = 273$–498 (Brauckmann, 1984, pp. 37–38, Figures 5 and 6), which correspond to reflectances of less than 2% (ibid., Figure 4).

A rare example of gradual decrease in percentage expandable layers in chlorite/smectite mixed-layers with depth of burial is given from the Paleocene sandstones of the Gibraltar road section in the central part of the Santa Ynez basin, California, by Helmold and van de Kamp (1984). The percentage expandable layers decrease from about 80% at $2\frac{1}{2}$–3 km to 0–20% at about 6 km estimated maximum burial depth, in association with a rank increase from 0.3 to ca. 2% R_0, laumontite appearing at about 1.3% R_0 (5670 m maximum burial depth) in association with about 0–20% expandable layers.

There are some reports of 'anomalous' occurrences of corrensite in the anchizone, particularly in calcareous rocks: e.g. in Tertiary limestones at Dérochoir, Massif de Platé (Aprahamian et al., 1975, p. 103, 108), in Triassic dolomitic sandstones of the cover of the Aiguilles Rouges Massif (ibid., p. 106,

108), and in association with anchi-epimetamorphic illite 'crystallinities' in the limestones and dolomites of Arudy, in the 'Châinons calcaires' of the Pyrenees (Kubler, 1967a, Figure 9); for additional occurrences see Kübler (1973, p. 550).

However, many of the reports of occurrence of corrensite in the anchimetamorphic and epimetamorphic zones appear to deal with regular chlorite/vermiculite mixed layers (high-charge corrensites) rather than with regular chlorite/smectite mixed layers (low-charge corrensites). Examples include their persistence from high-grade 'diagenetic' into anchimetamorphic and epimetamorphic terranes of the Upper Palaeozoic of the NW Moroccan Meseta (Piqué, 1975; 1982, p. 494, Figures 3 and 8), and into the anchimetamorphic zone in Middle-Cambrian greywackes (Wybrecht et al., 1985), their occurrence in the anchimetamorphic Pelagonian zone of eastern Othrys (Holtzapffel and Ferrière, 1982), and (high-charge corrensite according to Hauff, pers. comm.) in association with high-grade anchimetamorphic illite 'crystallinities' in the Ossa Unit (Pelagonian allochthon) of the W. margin of the Olympos window (Kisch, 1981a).

7.3.4   Illitization and chloritization of smectite in geothermal areas and under the influence of high thermal gradients

The most consistent divergence from the predominant relationship between illitization and chloritization of smectite and coal rank is found in areas that underwent strong, particularly short-lived heating during or after burial.

This is particularly evident along contacts of hypabyssal bodies. For instance, in the Carboniferous of the Alston Block of the northern Pennines, England, Smart and Clayton (1985) found high vitrinite reflectances of mainly $2\frac{1}{2}$–4% R_{max} for a distance of about 1.5 times the thickness of the Whill Sill, whereas—with the exception of a very strong drop where sample sites are very close (i.e., c. 50 m) to the contact—the sill had little if any effect on the percentage smectite layers in I/S.

Such a 'lag' of mineral alteration behind coalification is also evident in areas that have been affected by comparatively short-lived thermal events with high thermal gradients, and to a lesser extent should also exist in other areas affected by high thermal gradients.

However, in the case of the disappearance of discrete smectite this 'lag' is comparatively minor: even in the axis of the Anambra Basin in the Benue Trough (Nigeria), with a high geothermal gradient—as apparent from an unusually high reflectance gradient of 1.5% R_0/km—smectite persists only to a rank of 0.7% R_0 (Robert, 1985, p. 297); in the southern part of the Tulameen coal field, B.C. (Pevear et al., 1980), the transformation of smectite to I/S mixed-layers with 55% illite layers and an IS superlattice (K-rectorite) under the effect of a thermal event with steep thermal gradients is associated with a similar rank (0.7% R_0), somewhat lower than the 0.8% found associated with I/S mixed-layers with 60% illite layers by Środoń (1979).

Table 7.2 Relation between attainment of illite/smectite mixed layers (I/S) with low percentages of expandable layers and disappearance of kaolinite in black shales and underclays in areas with possibly high palaeogeothermal gradients

Age of formation; area	Percentage smectite layers in illite/smectite mixed-layer (I/S); disappearance of kaolinite; presence of pyrophillite	Coal rank R_0 (%) unless otherwise indicated; (number of samples); calculated temperatures. Reference if different from heading
Wells in Jurassic Terres Noires, southern French Alps (Dunoyer 1969)		
Well Grand Lubéron (Dept. Vaucluse) 4050–4350 m	I/S: deepest kaol.; ill. peaks 17–13 mm (6)	2.4–3.0 R_m (2) (Kisch, 1974, Tab. II)
4700–5100 m	deepest I/S; ill. peaks 11–6 mm (5)	3.4–4.0 R_m (2) (Kisch, 1974, Tab. II)
Well Montagne de Lure (Dept. Basses Alpes) 2400–3100 m	I/S: deepest kaol.; ill. peaks 16–11 mm (7)	2.3–2.7 R_m (2) (Kisch, 1974, Tab. II)
3100–3600 m	deepest I/S; ill. peaks 16–13 mm (5)	2.5 R_m (Robert, 1985, p.296)
Well Aurel (Dept. Drôme) 1500–1800 m	I/S: deepest kaol.; ill. peaks 14–12 mm (4)	3.0 R_m (Kisch, 1974, Table II)
	(NB: limit diagenesis/anchizone: 5.5 mm)	3.3 R_m (Robert, 1985, p.296)
Jurassic Terres Noires in southern French Alps (Barlier et al., 1974)		
	(Mineralogy of clay fraction)	
Zone A	largely 25 to 45% irregular I/S up to 25% kaol. illite crystallinity predominantly > 8.5	1.7 to 2.5 R_0 (18); 155–185 °C (curve of Vassoyevich et al., 1970)
Zone B	predominantly < 10% irregular I/S no kaol. detected illite crystallinity predominantly 8.5–3 (limit diagenesis/anchizone: 4.5)	2.9 to 4.2 R_0 (8–includes samples from Carboniferous and Lower Cretaceous); 200–255 °C (curve of Vassoyevich et al., 1970)
Upper Carboniferous of Westphalia (Stadler, 1963; Teichmüller et al., 1979, see Fig. 6.11) Well Münsterland 1, 3000–3300 m depth to base of kaolinite zone in shales		2.5 to 2.9 R_m (Lensch, 1963)
Underclays in eastern Pennsylvania anthracite fields (Hosterman et al., 1960)		
Southern field, Pottsville quadrangle		
N of Pottsville	little kaol.; no pyro	5.1 to 5.4 R_{max} (2) (Kisch, unpubl.)
Sharp Mountain (SE of Pottsville)	kaol; pyro common	5.3–3.6% V.M. (eq. to 3.5–4.2 R_{max}) (1*, pp. 159–160)

Western Middle field

Shamokin quadrangle, W pt. of field Bear Valley	kaol; no pyro	2.6 to 3.0 R_{max} (7) (Hower, 1978) 8.5–6.2% V.M. (eq. to 2.7–3.2 R_{max}) (1*, pp. 127–128)
Cameron Mine		7.4–5/9% V.M. (eq. to 2.9–3.2 R_{max}) (1*, pp. 129–131)
Ashland quadrangle, E part of field Centralia Mine	minor kaol; pyro	5.2 R_{max} (1) (Kisch, unpubl.); 4.1–3.0% V.M. (4–5 R_{max}) (1*, pp. 49–50)

Eastern Middle field

Hazleton quadrangle, Lattimer Mine	little kaol; pyro common	3.5–2.4% V.M. (eq. to $4\frac{1}{2}$–$5\frac{1}{2}$ R_{max}) (1*, pp. 97–98)

Northern field

Nanticoke quadrangle, Wanamie Mine	kaol; no pyro	4.0 R_{max} (1) (Hower, 1978; 6.2–4.8% V.M. (eq. to 3.1–3.7 R_{max}) (2*)
Wilkes–Barre West quadrangle, Sugar Notch Mine	kaol; local pyro	4.5, 5.0 R_{max} (Kisch, unpubl.); 5.1–3.6 V.M. (eq. to 3.6–$4\frac{1}{2}$ R_{max}) (1*, pp. 111–112)
Wilkes–Barre East and Pittston quadrangles, interstate 81	kaol; local pyro	5.1 to 5.2 R_{max} (3) (Kisch, unpubl.)

Notes

1* Approximate R_{max} equivalents of whole-coal analyses in Cooper *et al.* (1944), US Dept. of Mines, Tech. Paper **659**.
2* R_{max} equivalents of analyses of float-and-sink fractions in Sanner, W.S. (1967), US Bur. Mines, Rept. Invest. **7086**; see also Kisch (1974, Table II, # 4).
(1968), US Bur. Mines, Rept. Invest. **7086**; US Bur. Mines, Rept. Invest. **6989**, and Brady, G.A. and Griffiths, H.H.

In the case of attainment of advanced stages of illitization the lag is much more evident: in the above-mentioned Anambra Basin, I/S with 10–13% expandable layers is reached only at a rank of 2.5–3.5% R_0, and in the well Nagele 1 (the Netherlands)—with an even higher reflectance gradient—I/S persists to a coal rank of 3.5% R_0 (Robert, 1985, pp. 299–300).

The association of I/S with 80–90% illite and common Kalkberg-type ordering in Mancos shales (similar to that found at or within the calculated 200 °C isotherm 3 km from the Cerillos, New Mexico, pluton) up to 5 km from the Tertiary intrusives at Crested Butte, Colorado, with anthracite (Nadeau and Reynolds, 1981, pp. 255–256) constitutes according to this writer another example of the 'lag' in illitization in thermal aureoles.

The persistence of I/S to anthracitic coal ranks (3 to 4% R_m) in the Jurassic Terres Noires of the French Alps (see Table 7.2) has also been ascribed to a regional thermal anomaly by Robert (1985, p. 302).

In a study of the relationship between the regional zoning of mineral catagenesis and of coal rank, in the Mesozoic coal-bearing formation of South Yakutia, Zhelinskii (1980) found divergence over large areas, the isovols intersecting the zones of mineral alteration; these 'disagreements' could be correlated with increases in the palaeogeothermal gradient, which underwent major increases under the effect of Cretaceous magmatism and related hydrothermal activity. In the most affected areas, coking-fat coal rank (c. 1.2% R_0) is already reached at the limit between early and late catagenesis, and gas coal rank (c. 0.8–0.9% R_0) in the zone of early catagenesis, these ranks normally occurring in the lower parts of the lower subzone and the upper subzone of late catagenesis, respectively.

The above studies clearly indicate that under the effect of steep palaeogeothermal gradient the illitization of smectite, and the degree of clay-mineral alteration in general, 'lags' behind the progress of coalification, resulting in association of a low degree of clay mineral metamorphism with higher than usual coal ranks. A similar effect of vicinity of late intrusions will be noted on the relationship between illite crystallinity and coal rank (cf. section 7.7).

7.4 Disappearance of kaolinite

Similarly to smectite, the stage of disappearance of kaolinite upon burial is strongly dependent upon compositional factors. However, since, contrary to smectite, kaolinite shows only very minor compositional variation, and is replaced by 'discontinuous' reactions, this compositional dependence is easier to establish: whereas various compositional variations find their expression in a wide range of smectite compositions, which affect the stage of their replacement, but are difficult to determine; such factors result only in variations in the stage of replacement of kaolinite, and in extreme cases in its replacement by minerals and mineral assemblages other than or in addition to illite and/or chlorite, such as pyrophyllite, pyrophyllite +

muscovite/paragonite mixed-layers, paragonite, and dioctahedral chlorite.

Data on the coal rank at which kaolinite disappears are summarized in Table 7.1.

Kisch (1969, Figure 3) has shown that, in terms of coal rank, the grade of burial diagenesis of disappearance of kaolinite differs greatly in different lithologies, particularly striking being its persistence to anthracite rank in coals, kaolinite-coal tonsteins, and underclays.

Kaolinite disappears from normal shaly mudstone ('argilo-detrital') rocks between approx. 0.9% and 1.6% R_0 in the lower part of diagenesis zone 3 (cf. Kübler et al., 1979—see Figure 7.3, Héroux et al., 1979, Figure 1; Kubler, 1980, Figure 7)—approximately in the same range as I/S mixed layers with 25–10% expandable layers and ordered interstratification. This is in agreement with the 'usual absence' of kaolinite in the rank range 1 to 1.5% R_0—'late mesodiagenesis'—in the Lower Cretaceous shales of NE British Columbia (Foscolos et al., 1976; see Table 7.1), and the evidence of illitization of kaolinite at $R_0 > 1.2\%$ in the Carboniferous of Upper Silesia (Środoń, 1979). Zen and Thompson (1974; also in Ghent, 1979) place the limit between these assemblages at 19% V.M.—approximately equivalent to 1.6% R_0 (but mistakenly place this stage in the anchimetamorphic zone, well beyond the zone of mixed-layer clays).

It may therefore be summarized that kaolinite in argilo-detrital shales tends to be replaced in association with the range of 1.2 to 1.9% R_0, but that in minor quantities it may persist to at least 2.2% R_0.

As in the case of illitization of smectite (Velde and Nicot, 1985), the stage of replacement of kaolinite by illite and/or chlorite is strongly dependent on the concentration of cations such as K, and Mg and Fe, respectively. In assemblages with many phases in equilibrium, the assemblage may be expected to have a low variance at constant P–T conditions, the activity of the cations consequently being buffered by the assemblage. Such a low-variance assemblage is kaolinite–chlorite–dolomite (or ankerite)–calcite + illite.

The similarity of the ranks (1.8 and 1.55% R_0) at which kaolinite is replaced by chlorite in the presence of ankerite/dolomite or siderite in the Permian of the Bowen Basin (Kisch, 1968) and the Cretaceous sandstones of the Kootenay Formation of SE British Columbia and SW Alberta (Hutcheon et al., 1980; see Table 7.1), and to the approx. 1.6% R_m corresponding to the disappearance of dolomite and ankerite to form chlorite and calcite ('chlorite isograd') in the Salton Sea geothermal area at 180–190 °C (Muffler and White, 1969; McDowell and Elders, 1980)—according to the regression curve for geothermal systems by Barker (1983, Figure 2)—suggests that we may indeed be dealing here with a buffered reaction. The grade of disappearance of kaolinite in defined assemblages thus appears to be a close approach to representing a true indicator of P–T conditions of metamorphism rather than of cation activities imposed through the solution.

As in the case of illitization of smectite, the process appears to be inhibited by high carbonate content. For instance, in the marls of the Helvetic nappes of central Switzerland kaolinite persists up-grade in association with illite 'crystallinities' characteristic of the highest grade of 'diagenesis' and coal ranks of 2.0 to 2.3% R_0 (Breitschmid, 1982, pp. 362–3; Frey *et al.*, 1980, Figure 5).

In Na-rich shales or 'slaty' black shales kaolinite disappears between 1.2 and 2.1% R_0 (Kübler *et al.*, 1979; see Figure 7.3), generally being replaced by allevardite (from about 2% R_0*), pyrophyllite (from about 2.4% R_0), or muscovite–paragonite mixed-layer (from about $2\frac{1}{2}$% R_0) and paragonite (in the anchizone**) rather than by illite and/or chlorite. In the Jurassic black shales ('Terres Noires') of the subalpine belt of the W. French Alps, kaolinite persists both in wells (Dunoyer, 1969, Figures 70, 71; Kisch, 1974, Table II) and eastwards (Barlier *et al.*, 1974) until ranks of about 2.5 to 3.3% R_0 (see Table 7.2).

From coal seams kaolinite is shown in Kubler's scheme to disappear at 1.3 to 1.5% R_0 (Kübler *et al.*, 1979—see Figure 7.3; Kubler, 1980, Figure 7). These ranks are almost certainly much higher. Countless occurrences (cf. Kisch, 1969) show that kaolinite in coals, tonsteins, and underclays may persist well into anthracite ranks (ASTM classification, with < 8% V.M. and > 2.5% R_0). Examples include the underclays of the Pennsylvania anthracite region (see Table 7.2) in which kaolinite persists without pyrophyllite in association with anthracite with up to at least 4% R_{max} (particularly in the W. parts of the Western Middle and Northern fields), and is locally partly replaced by pyrophyllite (Hosterman *et al.*, 1970) in association with anthracites with $3\frac{1}{2}$ to $5\frac{1}{2}$% R_{max} (Kisch, unpublished data).

Alternatively, the 'lag' in mineral diagenesis in the Jurassic Terres Noires and the Pennsylvania anthracite region that is apparent not only from the persistence of kaolinite, but also in the high coal rank associated with the onset of the anchizone (see section 7.5.5), could be ascribed to a regional thermal anomaly ('hyperthermie') that has been claimed for both terranes (Damberger, 1974; Robert, 1985, pp. 295–301).

7.5 Illite 'crystallinity' and coal rank

Since the publication of this author's earlier review of the relationship between the anchizone and anthracitic coal ranks a large amount of data has become available on coal rank and illite 'crystallinity' from a wide variety of geological terranes.

For the sake of briefness, these correlation data have been compiled in tabular form (Table 7.3).

---

* The appearance of 'allevardite' in Na-rich shale is correlated by Héroux *et al.* (1979, Figure 1) with a rank of 0.7–0.9% R_0: this rank is much too low, and corresponds to the appearance of K-rectorite rather than of allevardite.
** Not between 1.7 and 2.3% R_0 as indicated by Héroux *et al.* (1979, Figure 1).

7.5.1 *Limits of the anchimetamorphic zone*

The anchimetamorphic zone has been defined by Kubler (1967a) on the basis of limiting values of the half-height width of the 10 Å illite/muscovite diffraction peak, corresponding to Weaver's (1961) 'sharpness-ratio' (SR) values of 12.1 and 2.3 delimiting the zones of very-weak to beginning metamorphism (see Kubler, 1967a, Figure 3; 1968, Figure 3) these limiting values (in mm) were valid only for certain instrumental conditions. Subsequent authors have adopted different instrumental conditions, and therefore obtained divergent limiting peak widths (in mm); many of these authors do not appear to have calibrated their peak widths and limiting values against those of Kubler.

Kisch (1980a, b) expresses the peak widths in $^\circ\Delta2\theta$ rather than in millimetres, since this eliminates at least the differences due to varying chart speeds. However, even if converted to $^\circ\Delta2\theta$, the limits of the anchizone in the $-2\,\mu m$ fraction as given by various authors show appreciable differences: the low-grade and high-grade limits range respectively from 0.64° to $0.35^\circ\Delta2\theta$ and from 0.41° to $0.21^\circ\Delta2\theta$ (Kisch, 1983, Table 5–IV). One of the causes appears to be variation in goniometer scanning rates; Kisch (1980a, p. 275) has shown that the same peaks are by some 0.04 to $0.05^\circ\Delta2\theta$ narrower upon slow (e.g. $1/4^\circ\,2\theta/\text{min}$) than upon fast (e.g. $1^\circ\,2\theta/\text{min}$) scanning rates. However, many of the differences cannot be explained by this cause only (Kisch, in preparation).

In the tables, the peak widths are given as in the source of the data, but for evaluation this writer has added either equivalent $^\circ\Delta2\theta$ values (in brackets when computed by this writer), or the limits of the anchizone from the original paper.

7.5.2 *Use of Hb_{rel} illite 'crystallinity' values after Weber*

A number of authors, particularly in Germany, are using the Hb_{rel} illite crystallinity values after Weber (1972a, b). In the original paper by Weber (1972b, Figure 14), the limits of the anchizone as measured on polished slates, equivalent to Kubler's limits of 4 and 2.5 mm, were Hb_{rel} 150–155 ($Hb = 5\,\text{mm}$, equivalent to $0.25^\circ\Delta2\theta$) and Hb_{rel} 105 ($Hb = 3.5\,\text{mm}$, equivalent to $0.17^\circ\Delta2\theta$): the corresponding 2–6 μm fractions showed slightly smaller, and the $-2\,\mu m$ fractions in thin and thick preparates respectively somewhat and much larger Hb_{rel} values for the diagenetic/anchimetamorphism limit (Weber, 1972a, Figures 3–5).

In contrast, Ludwig (1973, p. 92) found that the 2–6 μm fractions showed *appreciably broader* 10 Å peaks than the corresponding slate plates. Using standards of Kubler, Ludwig (1972, 1973) found Hb_{rel} values of 181 ($Hb = 5.8\,\text{mm}$, equivalent to $0.29^\circ\Delta2\theta$) for 2–6 μm fractions and 222 ($Hb = 7.1\,\text{mm}$, equivalent to $0.35^\circ\Delta2\theta$) for $-2\mu m$ fractions for the low-grade limit of the

Table 7.3 Coal ranks associated with illite 'crystallinity'—regional correlations

Terrain; age or formation; area; tectonic unit	Illite crystallinity— $2 \mu m$ fractions unless otherwise indicated. Values as given; (number of samples); illite crystallinity zone	Coal rank $R_m(\%)$ unless otherwise indicated; (number of samples). Reference if different from heading
Helvetic zone, central Swiss Alps Jurassic to Eocene (Frey et al., 1980—limits of anchizone 7.5 and 4 mm)		
(a) Kien Valley section		
Flysch of 'Zone des Cols'; Wildhorn nappe, frontal part	8.2 to 19 mm (3)	2.1 to 2.8 (2.5 to 2.9 R_{max}) (3)
Intermediate flysch	6.7 to 9.3 mm (3)	2.0 to 3.1 (2.1 to 3.2 R_{max}) (3) mean 2.5 (2.7 R_{max}) — 2.65 (3.0 R_{max})
	limit diag.-anchizone	
Niesen nappe; Wildhorn nappe, southern part; Gellihorn nappe	4.1 to 7.1 mm (7) limit anchi-epizone	3.6 to 5.55 (3.9 to 6.15 R_{max}) (9) — 5.0 (5.7 R_{max})
Doldenhorn nappe; autochthonous sedimentary cover of the Aar massif	3.1 to 4.9 mm (6)	4.7 to 5.4 (5.2 to 6.2 R_{max}) (6)
(b) Lake Lucerne— Reuss Valley section (also Breitschmid, 1982)		
Klippen nappe; Drusberg nappe; Axen nappe, northern part	7.1 to 27 mm (21)	0.9 to 3.0 (0.9 to 3.3 R_{max}) (21)
North-Helvetic flysch	6.9 to 11.9 mm (8)	2.5 to 3.9 (2.8 to 4.7 R_{max}) (8)
	limit diag.-anchizone	mean 3.2 R_m (3.6 R_{max}) — 3.4 (4.0 R_{max})[1]
Axen nappe, southern part	4.7 to 7.4 mm (7) limit anchi-epizone	4.0 to 4.9 (4.3 to 5.8 R_{max}) (7) — 5.5 (6.5 R_{max})
Autochthonous sedimentary cover	3.7 to 5.3 mm (5)	4.5 to 5.1 (5.3 to 6.7 R_{max}) (5)
(c) Glarus Alps		
Säntis-Drusberg nappe	7.7 (1)	3.1 to 3.2 (3.5 R_{max}) (2)
Axen and Mürtschen nappes	4.1 to 6.9 mm (7)	4.5 to 5.1 (4.9 to 6.7 R_{max}) (8)

Lower-Tertiary flysches associated with the Grès de Taveyanne (Kisch, 1980b—limits of anchizone 0.37 and 0.21°Δ2θ)

Diablerets, Taveyanne, Kien Valley, middle Kander Valley (Diablerets-Gellihorn and Wildhorn nappes)	0.47°Δ2θ (one exception): diag.	0.85 to 2.3 R_{max} (14)
Seedorf (parautochthonous)	0.25 to 0.60°Δ2θ (5); [2]diag.-anchi.	2.1 to 2.9 R_{max} (3)
Reichenbachfall (Aalenien of Wildhorn nappe)	0.38 to 0.40°Δ2θ (2): high-diag.	2.7 R_{max} (1)
La Tièche section, N part	0.41 to 0.45°Δ2θ (4): high-diag.	3.3 R_{max} (1)
La Tièche section, S part	0.23 to 0.33°Δ2θ (4): anchi.	3.8 to 4.0 R_{max} (2)
Rosenlaui, Reichenbachfall, Oberalp-Asch (all parautochthonous), Reichenbachfall (Wildhorn nappe)	0.21 to 0.35°Δ2θ (11): anchi.	3.3 to 3.9 R_{max} (6)

Carboniferous zone, Rhone valley (Kisch, 1980b)

Chalais, Réchy, Grône-Réchy, Aproz, Haute Nendaz	0.15 to 0.20°Δ2θ (11):2epi.—all except 2 contain parag.	2.0 to 4.4 R_{max} (4) (Réchy, Grône-Réchy, Nendaz) 6.20 to 7.80 (8) (Kübler et al., (1979 Table 4)

French Alps–Mesozoic of external zones (Dunoyer, 1969; Artru et al., 1969; Abbas, 1974; Dunoyer and Abbas, 1976—limits of anchizone 5.5 and 3.5 mm)

Norian-Rhaetian of Briançonnais zone, Lac de l'Ascension, S of Briançon	$7\frac{1}{2}$ to $3\frac{1}{2}$ mm (33), mean 5.1 mm: low-anchi	Dogger of Champcella, 10 km to the SW 3.55–3.85 (4) (Chateauneuf et al., (1973)
Terres Noires of Dauphiné zone in Barcelonnette window	5.0–5.6 mm (2): low-anchi	Lias-Dogger—4.5 (Robert, 1971)
Trias-Rhaetian of Dauphiné zone at Barles (in front of Digne thrust)	$7\frac{1}{2}$ to $4\frac{1}{2}$ mm (42), mean 6.1 mm: high-diag.	Underlying Carboniferous-3.5 (Robert, 1971)

Hercynian of the Rheinische Schiefergebirge, West Germany (Devonian and Carboniferous)

NE Rheinische Schiefergebirge (Wolf, 1975, Table 2, slightly modified—limits of anchizone 105 and 155 Hb_{rel})

N of Latrop anticline (Devonian)	111–130 Hb_{rel}:'high-anchi.	5–6 R_{max}
East-Sauerland main anticline (Eifelian)		
–W of Altenbüren fault	100–120 Hb_{rel}: high-anchi; anchi. -epi.	5–7 R_{max}
–E of Altenbüren fault	100–130 Hb_{rel}: high-anchi; anchi. -epi	5–6 R_{max}
SW of Meggen, S of Elspe sync. (Eifelian)	121–130 Hb_{rel}: mid-anchi.	4–5 R_{max}
N Wenne Valley (Givetian)	131–145 Hb_{rel}: low-anchi.	5–6 R_{max}
Middle Wenne Valley, E of Eslohe (Givetian)	111–145 Hb_{rel}: anchi.	5–7 R_{max}
NE end of Elspe syncline, Wenne Valley (Upper Devonian)	131–145 Hb_{rel}: low-anchi.	4–5 R_{max}

Table 7.3(*Contd.*)

Terrain; or formation; area; tectonic unit	Illite crystallinity— $-2\,\mu m$ fractions unless otherwise indicated. Values as given; (number of samples); illite crystallinity zone	Coal rank R_m (%) unless otherwise indicated; (number of samples). Reference if different from heading
NE part of Lüdenscheid syncline, Ruhr Valley (Namurian A)	$> 160\,Hb_{rel}$; diag.	3–$4\,R_{max}$
Möhne Valley (Namurian B)	131–$145\,Hb_{rel}$; low-anchi.	3–$4\,R_{max}$
Stavelot-Venn massif (NE part) (Kasig and Spaeth, 1975, Fig. 2; Kramm *et al.*, 1985, Fig. 2 and 4—see Fig. 6.9) Devonian and Carboniferous 'mantel-schichten'		
Inde syncline and Aachen anticline (6–11 km NW of Venn overthrust)	260 to 450 Hb_{rel} (8), mean curve 330 to 400 Hb_{rel}; diag.	2.4 to 3.1 R_{max} (10) (Teichmüller and Teichmüller, 1979b) 0.34 to 1.2 (Hollerbach *et al.*, 1982, Table I)
Within 3 km NW of Venn overthrust	240 to 500 Hb_{rel} (7), mean curve 240 to 400 Hb_{rel}; diag.[(3)] Zone A–8.5 to 13mm (diag.–anchi boundary at 8.5 mm) (Fieremans and Bosmans, 1982, Table 1 and Fig. 1)	2.5 to 3.6 R_{max} (Teichmüller and Teichmüller, 1979b) Fig. 1; Kramm *et al.*, 1985, Fig. 1
NE part of the Cambro-Ordovician 'Kernschichten'	120–250 Hb_{rel} (17), mean curve do:anchi	5.7 to 6.1 R_{max} (3) (Teichmüller and Teichmüller, 1979b) Fig. 1; Kramm *et al.*, 1985, Fig. 3)
SE part of the Cambro-Ordovician 'Kernschichten' and the 'Mantel-schichten' to the SE	90 to 160 Hb_{rel} (35), mean curve 100 to 140 Hb_{rel}; anchi to anchi-epi Zone C–3 to 4 mm (anchi-epi boundary at 3.5 mm) (Fieremans and Bosmans, 1982, Table 1 and Fig. 1)	5.7 to 8.1 R_{max} (8), mainly 5.7 to 6.5 R_{max} (Teichmüller and Teichmüller 1979b, Fig. 1; Kramm *et al.*, 1985, Fig. 3) 4.7 to 4.8 (Hollerbach *et al.*, 1982, Table I)

Hercynides of South Wales
Carboniferous of the South Wales Coalfield (Gill et al., 1977)

Eastern part	SR 3.0–4.0 in terrigenic rocks	Low-volatile steam coal
'Metadiagenetic' zone II[4]	2.5–3.2 in carbonate rocks	(10–20% V.M.)
	<35% expandables in I/M m-l	
Western part	SR 4.0–6.0 in terrigenic rocks	Anthracite
'Anchimetamorphic' zone III[4]	3.2–4.0 in carbonate rocks	(<10% V.M.)
	<20% expandables in I/M m-l	

Carboniferous of the Pembroke Coalfield (Robinson et al., 1980—limits of anchizone: SR 2.3 and 12)

Eastern part	SR 1.6 to 4.5; mainly 0.33 to 0.49° $\Delta 2\theta$: high-diag. and low anchi.	Anthracite: 10 to 5% V.M. mean of 6.5% V.M.

Hercynian of Southern Spain
Palaeozoic of Zone of Almadén (Saupé et al., 1977—limits of anchizone 5.5 and 3.5 mm)

M Devonian NW of Almadén	6.1 mm (1): high-diag.	3.7 R_{max} (fossils), 6.8 R_{max} (metabituminite)
Silurian of Almadén mine	4.0–6.1 mm (6): anchi	6.1, 9.9 R_{max} (roof of deposit) 8.2, 12.4 R_{max} (dolomite nodule)
U Ordovician of Almadenejos	4.4–5.1 mm (3): anchi	3.4, 4.8 R_{max} (graptolite) 6.1 R_{max} (metabituminite)

Variscan of N Czechoslovakia
Carboniferous of the Ostrava-Karviná coal basin and the Nízký Jeseník Mountains (Králík, 1984)

Petřkovice member (lowest member in coal basin)	Hb_{rel} 400–160, S.I. 1.78–2.20: diag.	1.32–2.08% R_o
3 boreholes in Ostrava-Karviná coal basin	mean microcrystal size 29.65 to 46.41 nm: high-diag	35.4 to 17.6% V.M.
Borehole in Nízký Jeseník Mts	mean microcrystal size 154.19 nm: low-epi	$d_{002} = 34.0$ Å
2 boreholes in Nízký Jeseník Mts	mean microcrystal size 174.8 and 177.9 nm: epi	$d_{002} = 3.37$ Å

Alleghenian metamorphism in the Appalachians
Pennsylvanian of the Narragansett Basin, Massachusetts and Rhode Island (Murray et al., 1979; Rehmer et al., (1979)

Portsmouth, Cranston, Bristol, and Somerset, R.I., areas	anchi (7.25 to 4.6 mm or 0.28 to 0.18° $\Delta 2\theta$)	Anthracite to meta-anthracite 4.6 to 5.1 R_{max} (Lyons and Chase, 1981, Table II)

Table 7.3(*Contd.*)

Terrain; age or formation; area; tectonic unit	Illite crystallinity — $-2\,\mu m$ fractions unless otherwise indicated. Values as given; (number of samples); illite crystallinity zone	Coal rank R_m (%) unless otherwise indicated; (number of samples) Reference if different from heading
Mansfield, Mass, area (N part of Narragansett Basin)	diag. (> 7.25 mm or > 0.28° $\Delta 2\theta$)	Anthracite to meta-anthracite 5.6 to 7.0 R_{max} (Lyons and Chase, 1984, Table II)
E Pennsylvanian Appalachians (Kisch, unpublished data — limits of anchizone 0.37 and 0.21° $\Delta 2\theta$)		
Delaware Water Gap-Columbia area, N.J. (Ordovician)	0.16 to 0.20° $\Delta 2\theta$ (5): high-anchi. to epi.	CAI > 5 (Harris *et al.*, 1978)[6]
Delaware Water Gap-Stroudsburg area (Silurian and Devonian)	0.22 to 0.36° $\Delta 2\theta$ (10): anchi	CAI 4.5 to 5 (Harris *et al.*, 1978)
Jacksonville-Pleasant Corners area (Ordovician)	0.31 to 0.37° $\Delta 2\theta$ (4): low-anchi.	CAI around 5 (Harris *et al.*, 1978)
Potters Mills area and Juniata Valley (Ordovician and Silurian)	0.32 to 0.46° $\Delta 2\theta$ (5): high-diag. and low-anchi.	Ordov: CAI 4 to 4.5 (Harris *et al.*, 1978) Sil.: CAI 3 to 3.5 (Harris *et al.*, 1978)
Susquehanna Valley (Silurian to Middle Devonian)	0.31 to 0.47° $\Delta 2\theta$ (13): high-diag. and low-anchi.	CAI 4 to 4.5 (Harris *et al.*, 1978)
Anthracite fields (Pennsylvanian)		
Trevorton area, Western Middle field	0.83° $\Delta 2\theta$ (1): diag.	Semi-anthracite (V.M. > 8%)
Shamokin area, Western Middle field	0.36 to 0.54° $\Delta 2\theta$ (4): diag.	2.3 to 2.8; 2.6 to 3.0 R_{max} (7) (Hower, 1978)
Anthracite fields except the above	mainly 0.41 to 0.82° $\Delta 2\theta$ (17): diag. several with kaol. + pyro. and/or par.	3.5 to 4.9: 4.0 to 5.8 R_{max} (15) (Hower, 1978; mainly S and E Middle fields) 4.5 to 5.4 R_{max} (Kisch, unpubl. data)
Appalachian Basin (Hosterman and Whitlow, 1983) Devonian black shales	'Crystallinity factor' 0.11[5] 'Crystallinity factor' 0.10	CAI 2.5 to 3 (Epstein *et al.*, 1977) CAI 1.5 to 2 (Epstein *et al.*, 1977)
Ouachita orogenic belt of Oklahoma-Arkansas (Guthrie, Houseknecht and Johns, 1986)		
Jackfork Fm (Pennsylvanian)	7.5 mm (0.60° $\Delta 2\theta$) or SR $2\frac{1}{2}$-3 = $2\frac{1}{2}$-3 R_m 4 mm (0.31° $\Delta 2\theta$) or SR $5\frac{1}{2}$-6 => 5 R_m	$\log CI = 1.02 - 0.05\ R_m$ $(n = 18)$ $\log SR = 0.29 + 0.05\ R_m$ $(n = 23)$

Stanley Fm (Missisippian)	7.5 mm (0.60°Δ2θ) or SR $2\frac{1}{2}$–3 = $1\frac{1}{2}$–2 R_m 4 mm (0.31°Δ2θ) or SR $5\frac{1}{2}$–6 = 4 R_m Local pyrophyllite Local mixed layers	log CI = 1.02 – 0.10 R_m ($n = 18$) log SR = 0.23 + 0.13 R_m ($n = 23$) > 2.7 < 1.5

Post-Taconic (? Acadian) orogeny in the Appalachian belt of SW Gaspé, Quebec (Duba and Williams-Jones, 1983a)

3 Traverses across L Devonian Fortin Group	-zone with 0.43–0.62° Δ2θ -zone with 0.33–0.42° Δ2θ -zone with <0.32°Δ2θ	1.54 to 2.77 (3) 3.55 (1) 3.58 to 5.19 (10)

Taconic zone of the Quebec Appalachians
Cambro-Ordovician continental margin sequences of external domain around Quebec City (Ogunyomi et al., 1980)

Middle Ordovician Quebec Promontory nappe (lowermost tectonic unit); Lower Ordovician Pointe-de-Lévy nappe	Zone I 'middle diagenesis': 5.5–8.5 mm or 0.72–1.10° Δ2θ	1.0–1.5 (asphaltic pyrobitumen)
Cambro-Ordovician Bacchus nappe; Cambrian Chaudière nappe (highest tectonic unit)	Zone II 'late diagenesis': 3.2–5.5 mm, 0.42–0.72°Δ2θ (black shales up to 8.0 mm, 1.04°Δ2θ)	1.5–2.8 (asphaltic pyrobitumen)
St. Hénédine nappe, north	Zone III 'anchimetamorphism': 1.8–3.2 mm, 0.23–0.42°Δ2θ	2.8–3.8 (asphaltic pyrobitumen)
St. Hénédine nappe, south	Zone IV 'epimetamorphism'; <1.8 mm, <0.23°Δ2θ	≥ 3.8 (asphaltic pyrobitumen)

Cambro-Ordovician deep-water shales, Gaspé Peninsula (Islam et al., 1982)

External domain of Taconic belt	Late diag.: IC > 0.43°Δ2θ Anchizone: IC 0.24–0.43°Δ2θ (epizone is an aureole around a pluton)	< 2.7 (asphaltic pyrobitumen) 2.7–5 (asphaltic pyrobitumen) ≥ 5 (asphaltic pyrobitumen)

Ordovician platform carbonate of the St Lawrence Lowlands (Bertrand et al., 1983)

Neufville area	12–15 mm, mean 13.48 mm (34), 1.6–2.1°Δ2θ, mean 1.75°Δ2θ (34): diag. c. 40–60% m-l	0.7–1.0, mean 0.84 (11)
Joliette area	7–10 mm, mean 8.24 mm (25), 1.0–1.3°Δ2θ, mean 1.07°Δ2θ (25): diag. c. 20–35% m-l	1.6–1.9, mean 1.765 (5)
Montréal area	5–7 mm, mean 5.66 mm (60), 0.6–0.8°Δ2θ, mean 0.736°Δ2θ (60): diag. c. 5–10% m-l	1.3–1.8, mean 1.64 (4)

Table 7.3(*Contd.*)

Terrain age or formation; area; tectonic unit	Illite crystallinity— $2\,\mu m$ fractions unless otherwise indicated. Values as given; (number of samples); illite crystallinity zone	Coal rank R_m (%) unless otherwise indicated; (number of samples). Reference if different from heading
Paratectonic Caledonides of the British Isles (L. Palaeozoic)		
Upper Ordovician and Silurian of the Lake District	mid-anchi: 99 to 150 Hb_{rel} (40), mean Hb_{rel} 130 (Bevins *et al.*, 1985, p. 65)	CAI = 5 (Bergström, 1980)[6] CAI 4–5 (Bevins *et al.*, 1985)
Southern Uplands accretionary complex	anchi and epi: 90 to 160 Hb_{rel} (317) mean Hb_{rel} 110 (Bevins *et al.*, 1985, p. 69)	CAI 5 or more (Bergström, 1980)
Ordovician of Ball Hill-Abington area, northern belt of Southern Uplands	anchi (Hepworth *et al.*, 1982, p. 532)	3.91 to 8.22 R_{max}
Southeastern Ireland	anchi: 107 to 162 Hb_{rel} (17) (Bevins *et al.*, 1985, Fig. 7)	CAI = 5 (Bergström, 1980)
Paratectonic Caledonides of western Sweden (L. Paleozoic)		
Cambro-Silurian of the Jämtland Supergroup of Jämtland (Kisch, 1980*a*—limits of anchizone 0.37 and 0.21° $\Delta 2\theta$)		
Zone A		
Brunflo area (autochthonous)	> 0.5° $\Delta 2\theta$ (3): diag.	CAI 3.5–4 (Bergström, 1980)
Rest of zone A—Östersund-Krokom (autochthonous and parautochthonous)	⩾ 0.32° $\Delta 2\theta$ (17): diag. and low-anchi	1.4 to 2.9 R_{max} (7)
Onset of anchizone	appr. 0.37° $\Delta 2\theta$	CAI 4.5 to 5 (Bergström, 1980)
Zone B—Norderö-Näldsjön (parautochthonous)	0.27 to 0.35° $\Delta 2\theta$ (12): low- and mid-anchi	2.5–2.8 R_{max} 3.8 to 4.3 R_{max} (4) CAI = 5 (Bergström, 1980)
Zone C—Mattmar-Alsensjön	predom. 0.19 to 0.15° $\Delta 2\theta$: high-anchi	
Zone D—Mörsil-Järpen	predom. ⩽ 0.21° $\Delta 2\theta$: epi	predom. 3.7 to 4.3 R_{max} (3) (one exception)

Cretaceous of western Canada

Lower Cretaceous Buckinghorse Formation shales of NE British Columbia (Foscolos et al., 1976, Tables 2 and 5)

2440 m depth	19 mm	SR 1.4	2.22
2645 m depth	20 mm	SR 1.5	2.20
2850 m depth	15 mm	SR 1.8	
3078 m depth	13 mm	SR 2.2	2.53
Boundary diagenesis-anchizone	12.8 mm	SR 2.3	

Kootenay Formation (Hutcheon et al., 1980)

Mt Allan section, SW Alberta	470–670 Hb_{rel} (9): diag.	1.3 to 1.7 R_{max} (Hughes and Cameron, 1985)
(0 to 400 m below top of Kootenay Fm)		
Elk Valley section, SE B.C.	370–680 Hb_{rel} (10): diag.	From 0.82 at top to 1.6 R_{max} at 370 below top of Kootenay Fm
(0 to 600 m below top of Kootenay Fm)		

Notes

(1) According to Kisch's interpretation of the data of Frey et al. (1980) and Breitschmid (1982, sections Fig. 13 and 14; glycolated 10 Å widths), the diagenetic-anchizone transition in the North-Helvetic flysch of the Seedorf area is associated with a somewhat lower reflectance of 3.1% R_m or 3.5 R_{max}.

(2) According to Frey et al. (1980, p. 188)—basing themselves on the (narrower) glycolated 10 Å peak widths—this N Helvetic flysch zone is in the transition range from diagenesis to the anchizone; according to Breitschmid (1982, Fig. 18) in the low-grade anchizone. Reflectances measured in the North-Helvetic flysch zone of the area range from 2.5 to 3.9% R_m (6), mean 3.1 (Frey et al. 1980, Table 1b; Breitschmid 1982, Fig. 14).

(3) Diag.-anchi according to the Teichmüller et al. (1979) classification (Kramm et al., 1985, p. 318)

(4) Gill et al. (1977, p. 687) base their delimitation of the anchimetamorphic zone on its association with anthracites after Kisch (1974, 1975)

(5) The 'crystallinity factor' = height of the 001 illite peak/area of the 001 illite peak

(6) Correlation between conodont alteration index (CAI) and vitrinite reflectance and fixed-carbon after Epstein et al. (1977, Fig. 11)

Conodont	Vitrinite	
CAI	reflectance	fixed carbon (%)
1	<0.8	<60
1½	0.7–0.85	60–65
2	0.85–1.3	65–73
3	1.4–1.95	74–84
4	1.95–3.6	84–95
5	+3.6	+95

anchizone, and $125(Hb = 4.0$ mm, equivalent to $0.20\,°\Delta 2\theta$) for the high-grade limit in both size fractions.

In view of their calibration with Kubler's limits, these Hb_{rel} values will be used here. Kemp et al. (1985) have somewhat wider limits for the anchizone: their limiting Hb_{rel} values (2–6 μm fractions) are 170 and 80–100 ('transition zone').

Divergent definitions of anchimetamorphism exist. In 1979 Teichmüller, Teichmüller and Weber, on the base of coal rank and illite-'crystallinity' studies in the Mesozoic and Palaeozoic of Westphalia, proposed to re-define the onset of the anchizone primarily by a rank of 4% R_{max} (or 3.5% R_m), associated with Hb_{rel} of 500–350 ($-2\,\mu$m fractions, thick slides). This proposal is unfortunate in more than one way: not only has the anchizone been, and should remain, defined in terms of illite 'crystallinity' (see Chapter 1) but the coal rank-illite crystallinity relationships in Westphalia are atypical, as we will demonstrate below. In the following we will therefore use the Hb_{rel} boundary values for the anchizone of Weber (1972b) and Ludwig (1972, 1973) rather than those of Teichmüller et al. (1979).

Some authors have defined the range of illite-'crystallinity' values of 'anchimetamorphism' on the basis of its postulated association with the prehnite-pumpellyite facies and/or vitrinite reflectance (e.g. Árkai, 1973, p. 81; 1977, Figure 6a–d, 7a–b; Árkai et al., 1981, p. 272 and Figure 5a–d), rather than by calibration with Kubler's zones. The limits of resulting local 'anchimetamorphic zones' may show appreciable shifts from those of Kubler; in fact, Árkai's (1983) Figure 4 shows the NE Hungarian 'anchizone' to straddle the boundary between the anchizone and the epizone in the sense of Kubler. Correlation between parameters of incipient metamorphism is not sufficiently constant to justify such postulates: the terms anchimetamorphism and anchizone should be used only with the criteria established by Kubler, first and foremost illite 'crystallinity'.

7.5.3   General relationship between illite 'crystallinity' and coal rank

In view of the agreement that is emerging on the relation between illite 'crystallinity' and coal rank in a range of terrains, the following discussion will be brief on the main relationship, and long on the exceptions.

In a large number of terrains, including the Helvetic zone of the Swiss Alps (see Figure 5.13), the Taconic zone of the Quebec Appalachians, and the para-tectonic Caledonides of Jämtland, western Sweden, the limit between the diagenetic and the anchimetamorphic zone is associated with coal ranks ranging between 2.3 and 3.1% R_m (or 2.5 and 3.4% R_{max}) (see Table 7.3). In a recent compilation of partly unpublished data on illite 'crystallinity' and coal rank, Robert (1985) found that in the Lacq and Pau regions, north of the Pyrenees, the onset of the anchizone corresponds to about 2.4% R_0, in the Ordovician to Silurian of the Illizi Basin (Algerian Sahara) to even smaller values of 1.5 to 1.9% R_0. In their compilations of relations between indicators of very low-grade metamorphism, Kübler et al. (1979—see Figure 7.3) and Héroux et al. (1979, Figure 1) correlate the limit between the zones of diagenesis (or catagenesis) and anchimetamorphism respectively with 2.6–2.8 and 2.5–3.0% R_m. Data supporting this limit are obtained from diagenetic terrains without anchimetamorphism (including the Ordovician platform carbonates of the St Lawrence Lowlands, Quebec, and the Cretaceous of western Canada), and from anchimetamorphic terrains without diagenetic zones (including the Pennsylvanian of the Narragansett Basin, Massachusetts and Rhode Island, and the paratectonic Caledonides of the Lake District, N. England, the Southern Uplands of Scotland, and south-eastern Ireland). It is of interest that the anchimetamorphic zones of all these terrains grade into higher-grade metamorphic zones of medium-pressure ('Barrovian') type.

With regard to the limit between the anchi- and epizone there are far fewer relevant coal rank data: sections across this limit are available from the Helvetic zone of the Swiss Alps (Frey et al., 1980), the Jämtland Caledonides (Kisch, 1980a), and the Taconic zone of Quebec Appalachians (Ogunyomi et al., 1980) (see Table 7.3). In the former area (see Figure 5.13) the limit is associated with 5 to 5.5% R_m (or 5.7 to 6.5% R_{max}), in the latter two with somewhat lower values of respectively approx. 3.7–4.3% R_{max} and 3.8% R_0. In the compilation by Kübler et al. (1979—see Figure 7.3) the limit correlated with a comparable value of 4% R_m.

The value of 5% R_0 associated with the anchi-epizone transition in the Cambro-Ordovician of the Gaspé Peninsula (Islam et al., 1982) is not considered here, since the epimetamorphic zone is considered a thermal aureole around a Devonian pluton.

7.5.4   Illite 'crystallinity' and coal rank in areas of post-kinematic igneous activity

The most conspicuous departure from this relationship is in exposures that underwent contact metamorphism, e.g. Ahrendt et al. (1977) along dolerite dyke in Southwest Africa; Zingg et al. (1976) along andesite effusives in the Sesia zone; and Teichmüller et al. (1979) along an olivine–basalt dyke in the Carboniferous of the Ruhr district. In these cases anthracite or natural coke and meta-anthracite are associated with respectively diagenetic and anchimetamorphic illite 'crystallinities'. It is obvious in these cases that the short-lived heating affected the illite 'crystallinity' much less than it did the coalification.

Furthermore, there is an impressive body of evidence from areas with proven or suspected post-orogenic deep intrusive bodies that here the same divergency holds, although to a somewhat lesser extent than in contact metamorphism. Most of this evidence comes from areas with coal-rank maxima (e.g. Deutloff et al., 1980) in the Palaeozoic and Mesozoic of Westphalia, West Germany: the Bramsche and Vlotho massifs, both characterized by large intrusive bodies of probably late-Cretaceous age at depth, and the area south of the Lippstadt dome around the magnetic anomaly of Soest-Erwitte where the Devonian shows evidence of strong post-orogenic heating (Hoyer *et al.*, 1974). Coalification maps showing these anomalies have been published by Bartenstein *et al.* (1971), Wolf (1972, 1975), Deutloff *et al.* (1980), and Brauckmann (1984).

The data on the Bramsche massif show that in the roof of the massif meta-anthracite ranks are reached in association with poor illite 'crystallinities' that are still diagenetic or near the boundary of diagenesis and anchimetamorphism; on the W and SW margins of the Bramsche massif anthracitic coal ranks are associated with diagenetic, and meta-anthracitic ranks with diagenetic to diagenetic/anchimetamorphic illite 'crystallinities' (as shown in well Münsterland 1 below *c.* 4500 m depth—see Figure 7.4) (Teichmüller *et al.*, 1979). For the depth range 1843–4700 m in the well Münsterland 1 high palaeogeothermal gradients of 80–87 °C/km have been calculated by Buntebarth (in Teichmüller and Teichmüller, 1986, Table 1), while the well bottom at 5700 m

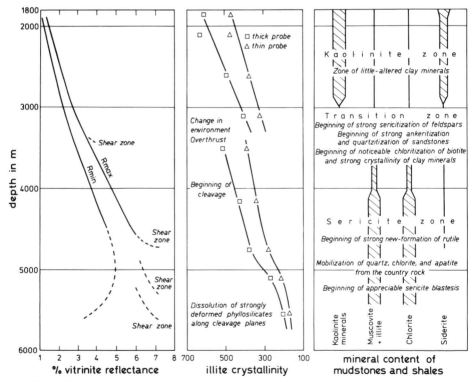

Figure 7.4 Increase in coal rank, illite 'crystallinity' (Weber index, fraction < 2 μm), and change in mineral content in Carboniferous and Upper-Devonian mudstones and shales in the well Münsterland 1. (Modified from Teichmüller *et al.*, 1979, Figure 18).

depth is estimated to have been subjected to a palaeotemperature of 300 °C or more (Teichmüller et al., 1979, pp. 227–228).

Teichmüller et al. (1979, pp. 261–262) show that samples from the well Alfhausen Z1 (roof of the Bramsche massif) have much higher, meta-anthracitic, coal ranks than those from the more distant well Ibbenbüren UB 150, for roughly similar illite 'crystallinities'. Brauckmann (1984, Table 1) has shown that in the Muschelkalk from the Vlotho area, and to a lesser extent in the Bramsche area, anchimetamorphic illite 'crystallinities' are attained in association with appreciably higher coal ranks than in the North German plain. These anomalous relationships are ascribed to the short-lived heating effect by the Bramsche and Vlotho plutons (Stadler and Teichmüller, R., 1971; Wolf, 1975; Teichmüller et al., 1979; Deutloff et al., 1980).

A very strong increase in reflectance with depth, and a 'lag' in mineral diagenesis similar to that in Münsterland 1 is also shown by the well Nagele 1 (Netherlands): 'diagenetic' illite 'crystallinities' persist in the Namurian in association with coal ranks of up to 6% R_0, taken by Robert (1985, p. 301 and Figure 197) to indicate a strong and very brief 'hyperthermie'.

However, some of the 'lag' of illite 'crystallinity' in such areas has been ascribed to the high content of organic matter. For instance, Rehmer et al. (1979; also Murray et al., 1979) ascribe the anomalously high coal rank associated with diagenetic illite 'crystallinities' (cf. Table 7.3) to poorer crystallization of illites in association with coals than those not so associated. Similarly, Deutloff et al. (1980, pp. 334–335) ascribe the 'lag' in improvement of illite 'crystallinity' (but not that of chlorite 'crystallinity'!) behind coal rank in the Rhaetian and Liassic of the roof of the Vlotho Massif to the high content of bituminous and coaly constituents.

Similarly, the relationships of the Lippstadt dome have to be compared to the context of the regional relationships in the NE Rheinische Schiefergebirge, where the onset of the anchizone ($Hb_{rel} = 150$–155 in 2–6 μm fractions) is associated with coal ranks of 4% R_{max}, which is higher than in most areas, though adopted as the diagenesis/anchimetamorphism boundary by Teich-müller et al. (1979). However, in the Belecke and Warstein anticlines, on the SW margin of the Soest-Erwitte magnetic anomaly, diagenetic and diagenetic/anchimetamorphic illite 'crystallinities' are associated with mar-kedly higher reflectances of 5–7% R_{max} (Wolf 1975—see Figure 7.5, Teichmüller et al., 1979). This effect is ascribed to heating by a post-kinematic pluton (Hoyer et al., 1974; Wolf, 1975; Teichmüller et al., 1979), which could also explain the recovery and incipient recrystallization of quartz and ore minerals in the Ordovician sediments of Soest-Erwitte and in the deformed quartz veins, which do not show such recrystallization phenomena in the East-Sauerland anticline (Teichmüller et al., 1979, pp. 257–259).

A similar discrepancy found in the bore Konzen 1, on the SE margin of the Cambro-Ordovician of the Stavelot-Venn Massif, where anchimetamorphic illite 'crystallinities' are associated with R_{max} values of 10%, is also interpreted

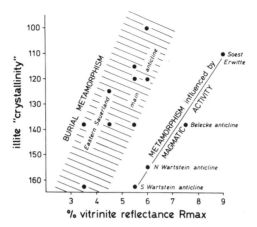

Figure 7.5 Relations between coal rank (R_{max}) and illite 'crystallinity' (Hb_{rel}) in the Palaeozoic of the eastern Rheinische Schiefergebirge and in adjoining areas affected by intrusive activity. (After Wolf, 1975, Figure 2).

as a result of relatively short-lived heating (Kramm *et al.*, 1985, p. 324).

A similar although weaker discrepancy may be apparent in the 16–20 km wide epimetamorphic aureole seen around the Devonian McGerrigle Mountains pluton in the Cambro-Ordovician flysch of the Gaspé Peninsula: here the anchi-epizone limit (IC = 0.24 °Δ2θ) is associated with a coal rank of 5% R_0 (Islam *et al.*, 1982).

In a recent compilation of illite-'crystallinity' and coal-rank data, Robert (1985) noted somewhat smaller, but still appreciable 'lags' of mineral diagenesis in the Ahnet Basin, Algerian Sahara (anchizone reached at 4–5% R_0 or more); the Douala Basin, Cameroon (anchizone not reached at 3.5% R_0); and the Jurassic Terres Noires (onset of the anchizone at approximately 4% R_0). Robert ascribes such intermediate lags of mineral diagenesis to 'hyperthermic anomalies' of regional extent, related to rifting or arc formation.

These data constitute a convincing body of evidence that upon post-kinematic intrusive activity the development of illite 'crystallinity' lags behind that of coal rank, as compared with the regional relationships.

The effect described above should be distinguished from the retrograde deterioration of illite 'crystallinity' around hydrothermal convection systems. Islam *et al.* (1982) have so interpreted anomalously low illite 'crystallinity' in the Cambro-Ordovician flysch of the Taconic belt of the Gaspé Peninsula, Quebec, within 6 km of the Devonian McGerrigle Mountains pluton, within the epimetamorphic and higher-grade aureole (IC ≤ 0.24 °Δ2θ, R_0 ≥ 5%) as caused by late-stage hydrothermal fluids. Similar anomalies are found in halos around cupriferous alteration centres—interpreted as hydrothermal convection systems—in the Upper Ordovician and Lower Silurian Matapedia Group of SW Gaspé (Duba and Williams–Jones, 1983b): the illite peaks tend to narrow somewhat towards the centres, but sediments immediately adjacent (less than some 100 m) to some of the centres contain mont-morillonite, a mixed-layer I/S, or have an anomalously poor illite 'crystallinity'.

Roberts and Merriman (1985, p. 621; also Merriman and Roberts, 1985) have noted the poor illite 'crystallinity', and the frequent occurrence of one or more of rectorite, corrensite, and pyrophyllite in pelites in adjacent to igneous intrusions in North Wales. However, they regard these minerals as older than the regional metamorphism, which they survived due to the induration of the rocks by the contact metamorphism, and *not* to subsequent hydrothermal activity.

Characteristically, in these areas of intrusive activity, pyrophyllite—which appears in regional terrains only in the anchizone—appears in association with anthracite, but with anomalously poor, diagenetic or diagenetic to anchimetamorphic, illite 'crystallinities'. One might surmise that the associ-ation of anthracite ranks with diagenetic illite crystallinities and pyrophyllite might be regarded as prima facie evidence for high geothermal gradients or magmatic heating. Such an association occurs widely in the anthracite area of eastern Pennsylvania, where anthracite and meta-anthracite ranks are associated with illite 'crystallinities' that appear to be uniformly diagenetic, with a common appearance of pyrophyllite with kaolinite in underclays (Hosterman *et al.*, 1970; see also Kisch, 1974, Table II). The high coal rank in this area has never been satisfactorily explained, and it is conceivable that heating by a deep intrusive body (the 'Scranton anomaly') has contributed to the anomalously high coal rank. The relationship between tectonic structure and the regional coalification pattern in the west-central part of the Southern Anthracite field—e.g., ranks being highest towards the centres of the two troughs of the Minersville synclinorium, and rank increasing in progressively younger coals along the plunge of the fields—are typical for large-scale post-deformational heating rather than for pre- or syn-deformational burial-related coalification pattern (Damberger, 1974).

Some areas appear to show a coal rank of about 4% R_{max} in association with the transition diagenesis-anchimetamorphism, somewhat higher than in most of the marginal orogenic areas studied. This is the case in the Hercynian of Westphalia, both in the northeastern Rheinische Schiefergebirge (Wolf, 1975; Teichmüller *et al.*, 1979), and in the north-eastern part of the Stavelot-Venn Massif (Kramm *et al.*, 1985). It is believed that these areas were metamor-phosed under the effect of high temperature gradients (low-pressure type of metamorphism).

7.5.5   *Kinetic effects in the development of illite 'crystallinity': synkinematic versus postkinematic heating*

Earlier in this chapter, we have given data indicating that coalification proceeds relatively rapidly with time. However, a major additional cause for the 'lagging' of illite 'crystallinity' behind coal rank in areas with short-lived, static heating compared with areas of synkinematic 'regional' very-low-grade metamorphism appears to lie in kinematic effects on the development of illite 'crystallinity'. Various studies (particularly Weber, 1976) show that the marked increase of illite 'crystallinity' through the anchizone corresponds to recrystallization.

Teichmüller *et al.* (1979, pp. 259–262) assume a threshold or activation temperature for the recrystallization of illite, below which illite 'crystallinity' increases slower than coal rank, unless assisted by deformation: for this reason the time relation between deformation and heating is an important consideration in the correlation of illite 'crystallinity' with coal rank. In areas with syncrystalline deformation this activation temperature is lowered, the lowering being more marked as the phyllosilicates are finer-grained (i.e., did not undergo previous recrystallization) and the deformation is stronger. In the Rheinische Schiefergebirge the synkinematic recrystallization is estimated to have begun at temperatures somewhat below 300 °C, at Hb_{rel} values around 220 ($-2\,\mu$m fraction, thick slides)—the onset of the anchizone using the boundary values of Ludwig (1972, 1973) rather than those proposed by Teichmüller *et al.* (see earlier discussion on p. 258).

Such effects are also apparent in the variation of illite 'crystallinity' with the intensity of deformation, for instance across folds. Using IR methods to detect differences in illite 'crystallinity', Flehmig and Langheinrich (1974) and Nyk (1985) found higher 'crystallinities' in the hinges than in the limbs of folds in the Harz Mountains and the Meggen area, Rheinische Schiefergebirge (see Figure 2.4); X-ray methods do not seem to detect this difference. Kalkreuth (1976) similarly found higher illite 'crystallinities' in the axial zones of tightly sheared anticlines in Carboniferous slates, without a concomitant variation in coal rank (Bostick, 1979, p. 27).

Vicinity to syn-metamorphic thrust zones also appears to enhance 'crystallinity'. Thus, Clauer and Lucas (1970) noted an improvement in illite 'crystallinity' from anchimetamorphic to anchi-epimetamorphic (and a concomitant decrease in I_{002}/I_{001}) in the 'zones de chevauchement' in the Schistes de Steige of the Alsace (abnormal contacts with the Schistes de Villé), and Aldahan and Morad (1986) report high-diagenetic and anchimetamorphic illite 'crystallinities' in the Dala Sandstones of central Sweden, but epimetamorphic values in the vicinity of a late Proterozoic thrust zone (where the sutured quartz contacts and development of almost linear triple junctions also reflect crystallization).

However, when the temperatures during deformation are so low that the threshold temperature is not reached—i.e., in the case of cataclastic deformation after uplift and cooling—mechanical thrusting ('cold working') can *reduce* the 'crystallinity' of illite by inducing lattice defects. Such a local decrease of illite 'crystallinity' by local post-crystalline ('mylonitic') thrusting has been described by Teichmüller *et al.* (1979, p. 213; see also Ahrendt *et al.*, 1978) from the Rheinische Schiefergebirge, for instance in the highly 'cry-

Table 7.4 Decrease of the illite crystallinity in the Solmsthaler Schichten with increasing mylonitization (from Teichmüller *et al.*, 1979, p. 213)

Rock type	Hb_{rel}
Phyllitic slate	165
Phyllitic slate, some layers mylonitized	184
Phyllitic slate, extensively mylonitized	234

stalline' (due to earlier syncrystalline deformation) Solmsthaler Schichten which, in their phyllonitic form—due to overprinting at lower temperatures— have a poorer illite 'crystallinity' (see Table 7.4).

Thus, syncrystalline—but not postcrystalline—shearing appears to lower the temperature of the recrystallization of illite, without enhancing coalification: when the heating is synkinematic—as is common in regional-metamorphic terrains—high illite 'crystallinities' should be expected in association with comparatively low coal ranks. However, when considering this effect, we should keep in mind that, contrary to coalification in the bituminous coal range, which is unaffected or even retarded by pressure, shearing may promote the increase of maximum reflectance and of reflectance anisotropy in anthracite and meta-anthracite. Such effects are in evidence in the local shear zones in the Münsterland 1 borehole (Teichmüller *et al.*, 1979), and in the Narragansett Basin (Raben and Gray, 1979c). Shearing effect may thus enhance anthracitization and graphitization (see Teichmüller, this volume).

In areas of heating without deformation—particularly areas of magmatic heating—where the temperatures did not reach the threshold temperature for recrystallization of illite, illite-'crystallinity' increase requires more time than coalification. Examples are the Bramsche and Vlotho Massifs and probably the Belecke anticline (see section 7.5.4): illite 'crystallinity' lags behind coal rank. Only at relatively high temperatures (e.g. the well Soest-Erwitte 1/1a, where temperatures are believed to have exceeded 350 °C) illite 'crystallinity' also increased rapidly, accompanied by recovery and incipient recrystallization of quartz in the sediments and of previously deformed quartz veins.

In areas around magmatic intrusives without synkinematic recrystallization illite 'crystallinity' thus lags behind coalification on both accounts: (i) the greater sensitivity of coalification to short-lived heating, and (ii) the absence of enhancement of illite 'crystallinity' by syn-crystalline deformation.

7.6 Zeolite and prehnite–pumpellyite facies and coal rank

In an earlier paper (Kisch, 1969) this author has attempted to establish the coal rank at which the diagnostic assemblage laumontite + quartz appears, and the

Table 7.5. Coal rank associated with laumontite occurrences

Terrain; age or formation*; area; tectonic unit	Diagnostic minerals; I/S mixed-layers**; depth and borehole temperatures	Coal rank $R_m(\%)$ unless otherwise indicated (number of samples)***; calculated temperatures
Tertiary and Mesozoic of California		
Upper Miocene of Los Angeles Basin (5 wells) (McCulloh et al., 1978)	top of lau at 2.7 to 4 km depth and 125° to 175 °C	0.36 to 0.47 (Bostick et al., 1978, Tables 4–6)
Miocene of Tejon area, San Joaquin Valley (Castaño and Sparks, 1974)	top of lau at 2.8 km depth and 96 °C	0.44
Cretaceous of Great Valley sequence, Cache Creek area (Castaño and Sparks, 1974)	top of lau at 10 km depth	0.63; 130 °C by analogy to Tejon area
Palaeocene sandstones of Santa Ynez Mountains (Helmold and van de Kamp, 1984)		
E part of basin (Wheeler Gorge)	incipient albitization (4570 m burial) complete albitization (5425 m burial) appearance of lau (5670 m burial)	0.5; 110 °C from R_0 0.9; 165 °C from R_0 1.0; 173 °C from R_0
Central part of basin (Gibraltar Road)	complete albitization (4310 m burial) appearance of lau (5670 m burial)	0.75; 150 °C from R_0 1.3; 190 °C from R_0
W part of basin (Point Conception)	incipient albitization (2150 m burial) complete albitization and first appearance of lau (2515 m burial)	0.35; 77 °C from R_0 0.5; 110 °C from R_0
Jurassic Coloradite Formation, Cedros Island, Baja California	heu and lau in underlying Gran Canon Formation	1.71 or 1.54 (TAI) (Boles and Landis, 1984, p. 519)
Helvetic zone of the French and Swiss Alps, Grès de Taveyanne (U Eocene–L Oligocene)		
SW part of Thônes syncline, Haute-Savoie (parautochthonous Clefs and Manigod units)	heu, lau (Sawatzki, 1975, Plate 5) I/S m–l + sm (ibid., p. 338)	0.53 to 0.84 (14) (Kübler et al., 1979, Table 4 and Fig. 2)

Locality	Mineral assemblage / zone	Data (references)
SW part of Thônes syncline, Haute-Savoie (parautochthonous Clefs and Manigod units)	heu, lau or lau, pum; corr I/S m–l, sm	0.9–1.0 (3)[†] (Stalder, 1979, Table 1)
Pernant (Arve Valley), Platé Massif, Haute-Savoie (Morcles nappe)	lau (Martini, 1968, Plate I); traces of expandable minerals (Martini, 1972)	1.5 (1) (Kisch, 1974, Table III and p. 100) 1.7* (Martini, 1972, p. 261)
Sougey (Giffre valley), Platé Massif, Haute-Savoie (Morcles nappe)	lau (Martini, 1968, Plate I); traces of expandable minerals (Martini, 1972)	2.1 to 2.3 (4) (Kübler et al., 1979, Table 4)
Champsaur	lau, pr, pum, corr	2.4, 2.5 (2)[†] (Stalder, 1979, Table 1)
Platé Massif (Morcles nappe)	lau, pum, ± pr, ± corr	1.3 to 3 (6)[†] (Stalder, 1979, Table 1)
Diablerets (Diablerets nappe)	lau, pum, ± pr, corr lau, pr, pum lau, pum, corr	2.1, 2.8 (2)[†] (Stalder, 1979, Table 1) 2.6 (2) (Kübler et al., 1979, Table 4) 0.9 to 1.3 (4) (Kisch, 1980b, Table 2)
Taveyanne (Diablerets nappe)	lau, pum, corr	1.25 (1) (Kisch, 1980b, Table 2)
Lower Kien Valley	lau, pum, corr	0.85, 1.2 (2) (Kisch, 1980b, Table 2)
Balmen, middle Kander Valley	lau, pum, corr	1.65, 1.7 (2) (Kisch, 1980b, Table 2)
Tertiary of Japan		
Miocene Green Tuff and Palaeogene coal fields (Hayakawa et al., 1979)	fresh glass zone clinoptilolite-mordenite zone analcime zone albite zone	<0.3 (2 coal mines) 0.3 to 0.6 (4 localities) 0.6 to 0.9 (3 localities) 1.9 to 4.4 (3 localities)
Paleogene coal measures (Iijima, 1978)	clinoptilolite-mordenite zone (zone II) analcime-heulandite zone (zone IIIa) analcime-laumontite zone (zone IIIb)	up to 0.6 (3 sections) up to 0.78 to 1.03 (4 sections) >0.78 to 1.03 (4 sections)
Hamayuchi drill hole in folded Tertiary of Hokkaido (Iijima, 1978) and three wells in Neogene of Niigata Plain (Shimoyama and Iijima, 1976)	clinoptilolite-mordenite zone (zone II) base at 85–90 °C analcime zone (zone III) base at 120–125 °C	lignite and sub-bituminous coal (up to 0.6) coking bituminous coal

Table 7.5 (*Contd*)

Terrain: age or formation*; area; tectonic unit	Diagnostic minerals; I/S mixed-layers**; depth and borehole temperatures	Coal rank R_m(%) unless otherwise indicated (number of samples)***; calculated temperatures
Palaeogene Munakata coal field, Kyushu (Miki and Nakamuta, 1985)	clinoptilolite-heulandite type 2 heulandite type 1 laumontite	up to 0.8 (3 sections) 0.75 to 1.2 (3 sections) > 1.1 (2 sections)
Mesozoic of Siberia		
Jurassic and Cretaceous Tarbagatai coal measures, W Chita Oblast	heu and lau	BD grade, equiv. to 0.5–0.6 (Buryanova and Bogdanov, 1967)
Cretaceous of Lena Coal Basin, northern Yakutia (Zaporozhtseva et al., 1963)	heulandite stilbite, heulandite, very little laumontite abundant laumontite (only zeolite)	B–D grade, equiv. to c. 0.5–0.6 D grade, equiv. to c. 0.55–0.72 G–Zh grade, equiv. to 0.72–1.2
North Range Group (Triassic), Southland Syncline, southern New Zealand (Coombs, 1954; Boles and Coombs, 1955, 1977)		
NW part of North Range (Taringatura Hills) North Peak Formation (NR5)	heu, lau, pr, pum (Coombs, 1954)	0.55 to 0.95 (7) mean 0.79 (Kisch, 1981b, Table 1)
SE part of North Range (Taringatura Hills) Stag Siltstone (NR6–8)	heu, lau, pum; highest unit containing lau	0.63 (1)[+] (Kisch, 1981b, Table 1)
Gavenwood Tuffs (NR2) Fairplace Formation (NR1)	heu, lau, pr, pum lau, pum; deepest occurrences of heu, pr	0.81 to 0.88 (3)[++] (Kisch, 1981b, Table 1) 0.82 to 1.25 (3) mean 1.00 (Kisch, 1981b, Table 1)
Beers Track, Hokonui Hills Stag Siltstone (NR6–8) North Peak Formation (NR5)	lau, heu, pr, pum; no analc lau, heu, pr, pum; no analc	0.75 (1) (Kisch, 1981b, Table 1) 0.52 to 0.92 (6) mean 0.7 (Kisch, 1981b, Table 1)
Fairplace Formation (NR1)	lau, heu, pr, pum; no analc	1.06 to 1.33 (3) mean 1.23 (Kisch, 1981b, Table 1)

Upper Permian of the Sydney Basin, New South Wales

Illawarra Coal Measures, Illawarra area, lower part	lau, abundant in underlying Gerringong Volcanics	1.0 to 1.2 (6)[+] (Kisch, 1966, pp. 415–416; Kisch, 1969, Table I)
Newcastle Coal Measures, Lake Macquarie area, middle part	ana	0.72 to 0.85 (9) (Kisch, 1966, p. 417; Kisch, 1969, Table I)

Upper Carboniferous and Lower Permian of the Tamworth Trough, New South Wales

'Glacial Stage' Beds, Seaham area, Lower Hunter Valley	lau, ? clinopt, heul	about 0.9 (Kisch, 1966, pp. 417–418; Kisch, 1969, Table 1)
400 ft above base of Permian, Werrie Basin	heu, minor ana; lau abundant in underlying Currabubula Fm (Carbonif.)	0.80 (Kisch, 1966, pp. 418–419; Kisch, 1969, Table I)

Paratectonic Caledonides of the British Isles

Top of Lough Nafooey Group (Tremadocian), Southern Mayo Trough, Ireland	lau; lower in section pr-pum without zeol	CAI = 3.5 in overlying Arenig ls (Bevins et al., 1985, p. 74)
Bail Hill Volcanic Group, Northern Belt of Southern Uplands, Scotland	thoms, pr, pum, analc (Hepworth et al., 1982, pp. 531–532)	2.2 R_{max} (graptolite) (Kemp et al., 1985, p. 337)
Silurian in margin of Welsh Basin	zeolite facies	CAI = 2.5 and up (Bevins et al., 1985, p. 62)
Silurian of shelf area of Wales and English Midlands (E of Church Stretton Fault)	probable zeolite facies	CAI 1.5 to 2.5 (Bevins et al., 1985, p. 62)

[+] one oxidized sample omitted
[++] two oxidized samples omitted
*) The age or formation given is that of the rocks containing the laumontite.
**) Reference to the source of mineral data is given only if different from or additional to the source of the coal-rank data.
***) Reflectance values calculated from V.M. yields in original source are marked[+].

assemblage analcime + quartz is being replaced by albite. He found that this change was associated with a wide range of coal ranks in various areas, varying from 'long-flame to gas coal' (i.e. high-volatile B bituminous coal with approximately 0.72% R_0) in the Cretaceous of the Lena Basin, through gas-flame coal (high-volatile A bituminous coal with approximately 0.85–0.90% R_0) in the Permo-Carboniferous of the Sydney Basin, N.S.W., to gas coal (high to medium-volatile bituminous coal with approximately 1.1% R_0) in the Werrie Basin of New South Wales.

   The lower limit of the laumontite + quartz assemblage was extrapolated to be associated with coals of at least ess coal (low-volatile bituminous) rank. The onset of the prehnite–pumpellyite facies has been shown in subsequent papers to be associated with somewhat higher, anthracitic rank (Kisch, 1980b, p. 773).

   In view of the recent reports of the wide overlap between heulandite- and laumontite-bearing assemblages—for instance in the type area in the Triassic of Southland, New Zealand (Boles and Coombs, 1975, 1977—see also Figure 3.21), it is not surprising that a wide area of overlap is expressed also in a wide range of coal ranks at which both minerals appear. However, in view of the many chemical controls on the appearance of diagnostic burial-metamorphic mineral assemblages, it is surprising that the available data point to a rather well-defined boundary between the zeolite facies and the prehnite–pumpellyite facies in terms of coal rank.

   In this section, some new information on the correlation between mineral facies and coal rank will be presented, and the 'state of the correlation' discussed.

7.6.1   Zeolite zoning and coal rank

In the following, we will consider as zeolite facies all assemblages containing analcime, clinoptilite, heulandite or laumontite, with or without prehnite and pumpellyite.

   Table 7.5 summarizes the available information on coal ranks associated with laumontite-bearing sequences. In this table, the data on coal ranks associated with analcime occurrences presented by Kisch (1969, Table I and Figure 7) are not repeated; it should be noted that all of these were associated with coals having more than 37% V.M. (or less than about 0.84% R_0). In the 1969 table, some coal ranks were obtained by downward extrapolation from coal-rank data higher in the sequence, assuming an average coal rank/depth gradient (for instance for the base of the analcime zone and the top of the laumontite zone in the Currabubula Formation, Werrie Basin); these extrapolated data are subject to major error and have now been omitted.

   The data presented in the table show that the appearance of laumontite, where documented, is associated with ranks ranging from as low as 0.36–0.47% R_0 (notably in the Miocene of California) through 0.5% (W part of Santa Ynez Basin), about 0.55% (Lena Coal Basin), 0.63% (Cache Creek), less than about 0.55–0.8% (North Range, Thônes Syncline), less than about 0.9%

(Lower Hunter Valley), c. 0.9–1.0% R_m (Sydney Basin; E part of Santa Ynez basin; Palaeogene of Japan) to 1.3% R_m (central part of Santa Ynez Basin).

The low rank of appearance of laumontite (0.36 to 0.47% R_0) in the Miocene of California is likely to be related to the brief duration of burial: the 'effective time' of burial (within 15 °C of its present maximum temperature) in the Los Angeles Basin is at most 2 million years, and it may thus be assumed that upon continued burial and coalification the onset of laumontite would become associated with higher coal ranks.

Assuming that the 'effective time' of burial in the Great Valley sequence at Cache Creek and in the Tejon area were similar, probably not more than 5 Ma, the higher vitrinite reflectance (approx. 0.63% R_0) and thus the higher calculated subsurface temperature associated with laumontite formation in the former sequence (130° v. 96 °C) could be related to load pressure in that section as against fluid pressure in the Tejon sequence. Similarly, the higher burial depths and palaeotemperatures of laumontite formation in the eastern and central portions of the Santa Ynez Basin are ascribed to overpressuring due to low permeability of turbidite sandstone encased in a shale lithosome, elevating P_f close to P_t, with a resulting 'dramatic' shift to higher temperatures in the equilibrium curve for the model reaction laumontite = anorthite + 2 quartz + 4 H_2O as pressure is raised from hydrostatic to lithostatic conditions (Helmold and van de Kamp, 1984, p. 267).

It may also be significant that the appearances of laumontite in the Tertiary of California in association with coal ranks of 0.4 to 0.5% R_0 occur without prior development of heulandite, by concurrent albitization of plagioclase (Castaño and Sparks, 1974; Helmold and van de Kamp, 1984). However, there is less reason to ascribe the appearance of laumontite in association with ranks in the range 0.4 to 0.5% R_0 to special causes such as brief duration of burial in view of the sequences containing both heulandite and laumontite, e.g. the Grès de Taveyanne of the Thônes Syncline, the North Range Group of Southland, and including the Hunter Valley and Kyushu occurrences. These span a wide range of ranks: from less than 0.55–0.8% R_m to as high as 1.25% R_m, covering most of the 'oil window', a range similar to that found for the heulandite zone in Tertiary of Japan (from about 0.6 to 1.2% R_m). In some isolated instances, for instance in the Coloradito Formation of Baja California (Boles and Landis, 1984, p. 519) the ranks associated with such 'transitional' heulandite- and laumontite-bearing sequences is reported to be even higher.

These coal rank correlations indicate that laumontite may appear with or without heulandite at grades of metamorphism much lower than usually thought, and may occur in the same sequences as heulandite over a wide range of coal ranks. This corroborates and broadens Boles and Coombs' (1975, 1977) findings of a breakdown of heulandite to laumontite or prehnite—the plagioclase associated with the laumontitized tuffs invariably being albitized (1975, p. 168)—over a stratigraphic interval of at least 3 km in the Murihiku Supergroup of the Hokonui Hills, southern New Zealand, controlled by such factors as P_{H_2O} and activity of ions in stratal waters, rather than by depth of burial. These authors have noted that heulandite in the Hokonui and Taringatura sections mainly forms from glass, and its silica content is a function of the silica content of the parent glass. They suggest that laumontite cements high in the Hokonui section could form from quartz-equilibrated waters at temperatures in the range 50–100 °C, similar to those at which clinoptilolitic heulandite formed in a high a_{SiO_2} environment controlled by glass; they express doubt as to 'the existence of a true stability field of heulandite relative to laumontite plus quartz' (1977, p. 998).

Excluding the low ranks associated with the Californian occurrences, which might be ascribed either to brief 'effective times of heating' or to formation of the laumontite as a result of albitization of plagioclase rather than by replacement of laumontite, the sequences containing laumontite without heulandite and/or analcime (in the Grès de Taveyanne, the base of the North Range Group, the Illawarra area of N.S.W., the Lena Coal Basin, and the paratectonic British Caledonides) appear to be associated with ranks ranging up from about 0.7–1.0% R_m.

In his earlier review Kisch (1974, p. 100 and Figure 1) tentatively placed the boundary between the laumontite zone of the zeolite facies and the prehnite-pumpellyite facies in the high-rank part of the low-volatile bituminous range at a V.M. yield of about 15%, equivalent to approx. 1.9% R_m, noting the scarcity of correlation data in the boundary region. Similarly, when writing in 1980b, he noted the absence of determinations of semi-anthracitic (and of low-volatile bituminous) ranks in association with with the Taveyanne greywackes '... leaving a rather unsatisfactorily wide range of uncertainty regarding the coal rank associated with the prograde disappearance of laumontite' (Kisch, 1980b, p. 773).

This gap is gradually being filled by low-volatile bituminous and semian-thracitic reflectances up to 2.5% R_m^* measured in association with 'Grès mouchetés à laumontite', for instance in the Platé Massif, Champsaur, Diablerets, and the Kander Valley; these are invariably accompanied by pumpellyite ± prehnite and corrensite, and with diagenetic illite 'crystallin-ities' and absence of illite/smectite mixed-layers with high contents of expandable layers.

Further evidence that laumontite persists in association with low-volatile bituminous and anthracitic ranks has been forthcoming from the paratectonic Caledonides of the British Isles. Kemp et al. (1985, p. 337) have concluded that the boundary between the zeolite facies and the prehnite–pumpellyite facies is at about 2.5% R_{max}, and state that '... coincidentally, Kisch (1974) defines the boundary between the zeolite and the prehnite–pumpellyite facies as vitrinite reflectance 3.0% $R_{0\ max}$'.

---

* Kisch (1980b) has measured lower reflectances of 0.9 to 1.3% in association with the laumontite-bearing Grès de Taveyanne at Diablerets, Taveyanne, and the lower Kien Valley, in association with similar diagenetic illite 'crystallinities' (see Table 7.5 and Kisch, 1980b, Table 1).

On the basis of the available data, the laumontite zone of the zeolite facies—including the laumontite–prehnite–pumpellyite-bearing assemblages,—seems to be associated with coal ranks ranging up to about 2.6% R_m, i.e. with the high-volatile bituminous A to semi-anthracitic ranks, and diagenetic illite 'crystallinities'.

7.6.2 Coal rank and prehnite–pumpellyite facies

Coal ranks associated with the prehnite–pumpellyite facies in the Grès de Taveyanne and the paratectonic Caledonides of the British Isles are shown in Table 7.6. Árkai's (1983) correlation of coal rank with the 'anchizone' of NE Hungary has been included in the compilation since this 'anchizone' has been equated 'a priori' with the prehnite–pumpellyite facies. Conversely, it should be realized that many papers that purportedly give coal rank data for the prehnite–pumpellyite facies in fact give data for anchimetamorphic terrains without prehnite–pumpellyite assemblages on the basis of similarity of the illite 'crystallinities' with other areas which do have prehnite–pumpellyite assemblages (e.g. Bevins et al., 1985, p. 65 on the Lake District, and pp. 74–75, on southeastern Ireland), or simply by postulating the equivalence of the anchizone and the prehnite–pumpellyite facies after Kisch (e.g. Bevins et al., 1985, p. 69 on the Southern Uplands).

The data that have become available on the paratectonic Caledonides of Britain show that the prehnite–pumpellyite facies is associated with CAI values of not less than 4, which are equivalent to average reflectances of 2.5–3.0% R_0 (not less than 2% R_0) (see Figure 4.18, this volume). The broad range of reflectance values equivalent to the CAI values does not allow of very precise correlations, but the above conclusions are commensurate with the results on the Grès de Taveyanne, which correlate the onset of the prehnite–pumpellyite facies (without laumontite) with about 2.6 to 3.0% R_{max}, and would extend the reflectance range to at least $5\frac{1}{2}\%$ R_{max}. Associated illite 'crystallinities' are anchimetamorphic or even epimetamorphic.

7.7 Illite 'crystallinity' associated with the zeolite, prehnite–pumpellyite, and pumpellyite–actinolite facies

In dealing with the subject, it should be remembered that the zeolite facies is mainly developed in pyroclastic and volcaniclastic rocks, in which the disappearance of discrete smectite and the stage of I/S mixed-layering 'lags' behind that in other clastic rocks; major differences between the clay mineralogy of the zeolite-bearing rocks and the associated clastic rocks can be expected. For the clay mineralogy of zeolitic rocks, the persistence of smectite and I/S mixed-layers rich in expandable layers, and the almost ubiquitous association of laumontite with corrensite, the reader is referred to the preceding sections on these minerals.

Table 7.6 Coal ranks associated with prehnite–pumpellyite assemblages (without laumontite)

Terrain; age or formation; area; tectonic unit	Diagnostic minerals. (Reference if different from last column)	Illite crystallinity (from Table 7.7)	Coal rank R_m (%) unless otherwise indicated; (number of samples)
Helvetic zone of French and Swiss Alps, Grès de Taveyanne (U Eocene–L Oligocene)			
La Tièche, Valais	Pum, pr	high-diag, anchi	3.3 to 3.95 R_{max} (3) (Kisch, 1980b, Table 2)
Varneralp, Valais	Pum, pr	anchi	approx. 3.7* (Stalder, 1979, Table 1)
Reichenbachfall near Meiringen, Aare Valley (parautochthonous)	Pum, pr, epi (Martini and Vuagnat, 1965, Fig. 1)	anchi	3.8, 3.9 R_{max} (2) (Kisch, 1980b, Table 2)
Oberalp-Äsch, Schächen Valley (parautochthonous)	Pum, pr, epi (Martini and Vuagnat, 1968, Fig. 1)	anchi	3.3, 3.5 R_{max} (2) (Kisch, 1980b, Table 2)
Waldhüttli, Urner Boden Valley (Griesstock nappe)	Pum, pr, epi (Martini and Vuagnat, 1968, Fig. 1)	anchi	3.5 R_{max} (1) (Kisch, 1980b, Table 2)
Wespen, Schächen Valley (parautochthonous)	Pum, pr, epi	anchi	> 6 R_{max} * (Stalder, 1979, Table 1)
Chulm, Schächen Valley (parautochthonous)	Pum, pr, epi	anchi	5 R_{max} * (Stalder, 1979, Table 1)
Paratectonic Caledonides of the British Isles (Lower Palaeozoic)			
Southern Uplands accretionary complex	Pum, pr, (Oliver and Leggett, 1980; cf. Bevins and Rowbotham, 1983, p. 165)	anchi	CAI about 4.5 to 5 (Bergström, 1980)
Abington spilites and Elvan Fm, N belt of Southern Uplands	Pum, pr	anchi	3.91 to 8.22 R_{max} (graptolites) (Hepworth et al., 1982, p. 532)
Ordovician of Lake District (coal ranks from overlying U Ordovician and Silurian)	Pum, pr (Bevins et al., 1985, pp. 64–65)		CAI = 5 (Bergström, 1980) CAI 4 to 5 (Aldridge, in Bevins et al., 1985, p. 65)
Ordovician of Welsh Basin	Pum, pr; local pum, act (Bevins and Rowbotham, 1983)		CAI = 5 (Bergström, 1980) Up to 4.4 R_{max} (graptolites) (cf. Kemp et al., 1985, p. 336)
Silurian in centre of Welsh Basin	Pum, pr (Bevins et al., 1985, p. 62)		CAI about 4 (Aldridge, in Bevins et al., 1985, p. 62)

Internal W Carpathians of NE Hungary

'Anchizone'** of Bükkium	Pr–pum facies (Árkai, 1977, Fig. 6a–d, 7a–b; Árkai et al., 1981, p. 272 and 5a–d)	anchi, low-epi	3.5 to 6.0 R_{max} (Árkai, 1983, Fig. 1b)

Hercynides of South Cornwall (Devonian)

Roseland area	Pum-act facies (Barnes and Andrews, 1981)		CAI = 5 (Barnes and Andrews, 1981, p. 145) 7.2 R_{max} (Cook et al., 1972)

*Reflectance values calculated from V.M. yields in the reference.
**Postulated to be congruent with the prehnite–pumpellyite facies (Árkai, 1977, Fig. 6, 7; Árkai et al., 1983, p. 272 and Fig. 5)

7.7.1 *Illite 'crystallinity' in the zeolite facies*

In the zeolite facies sequence in the Murihiku Supergroup of southern New Zealand, Kisch (1981*b*) found that in many of the samples the 10–14 Å mixed-layer 'plateau' is too high to allow measurement of the 10 Å half-peak width without removal of the 'plateau' by EG solvation; the range of 0.27–0.43 °$\Delta 2\theta$ 10 Å peak widths found in virtually all the EG-solvated $-2\,\mu$m fractions is considered to reflect the persistence of degraded clastic illite at this low degree of burial alteration (ibid., p. 354).

In the higher-grade part of the zeolite facies, in which laumontite occurs without heulandite, analcime, and other zeolites, but locally with prehnite and/or pumpellyite, smectite and illite/smectite mixed layers with high contents of expandable layers appear to be absent, for instance in association with the heulandite-free laumontite- and pumpellyite-prehnite-bearing Grès de Taveyanne of Champsaur, Platé, Diablerets, Taveyanne, the lower Kien Valley, and middle Kander Valley (Stalder, 1979; Kisch, 1980*b*; Martini, 1972, p. 261 and Figure 2); the illite 'crystallinities' of the shales from these areas are uniformly 'diagenetic' (see also Aprahamian, 1974, p. 10). Hoffman and Hower (1979) mention the interbedding of laumontite-bearing volcanogenic rocks in the disturbed belt of Montana with shales and bentonites which contain an allevardite-type (IS) ordered illite/smectite. In the shelf areas marginal to the Welsh Basin zeolite facies seems to be associated with diagenetic clay mineralogy (Bevins *et al.*, 1985, p. 62; Fettes *et al.*, 1985, p. 49). Unequivocally high-diagenetic illite peak widths of 0.36 to 0.70 °$\Delta 2\theta$ characterize the laumontite-bearing Benmore Dam area in the Torlesse terrane of the upper Waitaki Valley on the northern margin of the Haast Schist terrane of Otago, southern New Zealand, phases rich in expandable layers being very subordinate or absent (Kisch, 1981*b*, p. 357).

The clay mineralogy of even the 'high-grade' part of the laumontite zone thus still appears to be almost uniformly diagenetic.

7.7.2 *'Anomalous' anchimetamorphic illite 'crystallinities' in the 'high-grade' part of the laumontite zone*

Very local association of anchimetamorphic illite 'crystallinities' with prehnite–pumpellyite–laumontite have been described. The occurrence of this association in the Grès de Taveyanne in an inverse illite-'crystallinity' gradient in the upper part of the Dérochoir section in the internal part of the Platé Massif, equivalent to the Morcles nappe in western Switzerland (Aprahamian *et al.*, 1975, pp. 104–105 and Figure 8) has been ascribed to the thrusting of the overlying Pre-Alpine nappes (ibid., pp. 116–117; Aprahamian and Pairis, 1981, p. 163). The development of the 'green facies' in the Grès de Taveyanne immediately underlying the thrust, in the most external part of the Morcles nappe at Arâches, is also thought to represent effect of this thrusting (Martini, 1968; Aprahamanian *et al.*, 1975, pp. 116–117). The usual association of laumontite is taken to be with advanced diagenetic illite 'crystallinites' and with the 1.75 or 1.55% R_m coals (cf. Kisch, 1974, p. 100 and Table III) at the Pernant mine.

The correlation of a prehnite–pumpellyite–laumontite assemblage at the front of the Diablerets nappe near Les Diablerets with a diagenetic to anchimetamorphic illite 'crystallinity' (Durney, 1974, p. 271) is questionable and contradicted by the uniformly diagenetic values found by Kisch (1974) in the Diablerets area. Subsequently, Kisch (1980b) has confirmed that the illite 'crystallinities' associated with laumontite-bearing Taveyanne greywacke in the Helvetic zone are uniformly 'diagenetic'.

7.7.3 Illite 'crystallinity' associated with the prehnite–pumpellyite facies

Earlier in this review, evidence has been presented that both the anchimetamorphic zone in regional terranes, and the prehnite–pumpellyite facies are associated with coal ranks of not less than about 2.5% R_m. These correlations strongly suggest that the onset of the anchizone and of the prehnite–pumpellyite facies should be directly correlated. Such correlations are made difficult particularly by the uncertainty regarding the comparability and correlation of the illite-'crystallinity' scales used (but see Kisch 1983, Table 5-IV).

In this section we will review the field evidence on the relationship and disregard, in the first instance, those authors that either (i) *presuppose* correlation and equivalence of Kubler's anchimetamorphic zone to the prehnite–pumpellyite facies, somewhat overworking the correlation suggested by Kisch (1974), e.g. Chen (1984) and Bevins *et al.* (1985); or (ii) re-define the anchizone using boundary values other than those of Kubler, for instance by *postulating* its assignment to Kossovskaya and Shutov's (1963) 'phase of initial metagenesis'—as suggested by Kubler (1967a, Figure 3) and Kisch (1974)—to be correlated with the prehnite–pumpellyite facies, e.g. Árkai (1973, Table II; 1977, Figure 6a–b, 7a–b; 1983, Figure 2–4; Árkai *et al.*, 1981, p. 272 and Figure 5a–d), or re-definition of its low-grade boundary in terms of coal rank (Teichmüller *et al.*, 1979; see comments on p. 258).

Illite 'crystallinities' associated with prehnite–pumpellyite facies assemblages are compiled in Table 7.7. The great majority of the prehnite–pumpellyite facies terranes on which illite-'crystallinity' data are available show anchimetamorphic values. In view of the many chemical controls on the appearance of the diagnostic metamorphic assemblages of the prehnite–pumpellyite facies it is rather surprising that there should be such a relatively consistent relationship with illite 'crystallinity'. However, there are discrepancies and divergences from the general correlation of the prehnite–

Table 7.7. Illite crystallinities associated with prehnite–pumpellyite and pumpellyite–actinolite assemblages

Terrain; age or formation*; area; tectonic unit	Diagnostic minerals**	Illite crystallinity ($-2\,\mu m$ fractions unless otherwise indicated). Values as given (number of samples); illite crystallinity zones
Helvetic zone of the Swiss Alps, Grès de Taveyanne (U Eocene–L Oligocene)		
La Tièche, Valais	pum, pr (near 'Varnerkumme' of Martini and Vuagnat, 1968, Fig. 1)	0.45–0.23 °$\Delta 2\theta$ (8): high-diag. and high-anchi (Kisch, 1980b, Table 1)
Varneralp, Valais (near La Tièche)	pum, pr; no corrensite	1.93 mm (1): high-anchi (Stalder, 1979, Table 1)
Reichenbachfall, Aare Valley / Oberalp-Äsch, Schächen Valley / Waldhüttli, Urner Boden Valley }	pum, pr, epi (Martini and Vuagnat, 1968, Fig. 1)	0.35–0.23 °$\Delta 2\theta$ (8): mid-high anchi (Kisch, 1980b, Table 1)
Wespen, Schächen Valley / Chulm, Schächen Valley }	pum, pr, epi; no corrensite	2.42–2.44 mm (2): low-anchi (Stalder, 1979, Table 1)
Leuk-Guttet, Valais	pum-act facies (Coombs et al., 1976)	0.13–0.21 °$\Delta 2\theta$ (2): anchi-epi, low-epi (Kubler, 1970; Kisch, 1980b, Table 1; see also Frey and Wieland 1975, p. 412)
Ophiolites in the Northern Apennines (Venturelli and Frey, 1977; Cortesogno and Venturelli, 1978, Fig. 4 and Fig. 1, caption)		
In Casanova Complex	pr-pum facies	9–5 mm (17): high-diag. and mid-anchi.
In Palombini shales and Calpionella limestone	pr-pum facies	8–6 $\frac{1}{2}$ mm (4): diag.-anchi and low anchi
In Lavagna shales	pr-pum facies	6 $\frac{1}{2}$–4 mm (14): mid- and high-anchi.
In Gariette shales and Gottero sandstones	pr-pum facies	7 $\frac{1}{2}$–5 $\frac{1}{2}$ mm (8): low-anchi
Northern Calcareous Alps of Austria		
Permian pillow lava, Wienern, Styria	pum (Kirchner, 1980)	anchi (Permoscythian inliers in general) (Schramm, 1982c, Fig. 1 and p. 78)

Alpine metamorphism of the Bükkium, NE Hungary (Árkai, 1983)

Darnó Hill (Triassic-Jurassic)	pr, pum	0.47–0.50 °Δ2θ (2): high-diag. ('diag.')[†]
SW Bükk Mountains (Triassic-Jurassic)	pr, pum, act, pyro	0.30–0.40 °Δ2θ (5): mid-anchi. ('diag. to anchi')[†]
E Bukk Mountains (Triassic)	pum, act, epi	0.20–0.33 °Δ2θ (5): low-epi. ('high-anchi and low-epi.')[†]

Permian metamorphism in Nambucca Slate Belt, NE New South Wales (Leitch, 1975)

Zone 1	pr (no stilp; absence of pum due to unsuitable bulk compos.)	0.165–0.14 in (0.31–0.27 °Δ2θ): anchi
Zone 2	pr-pum, pr-act, stilp (no pum-act)	0.155–0.09 in (0.29–0.17 °Δ2θ): anchi and low-epi
Zone 3	pum-act (no pr)	0.13–0.075 in (0.24–0.15 °Δ2θ): high-anchi and epi

Variscan metamorphism of diabase and slate in the 'Saxothuringian zone' (Palaeozoic) (Brand, 1980)

Zone 1 Bergaer Sattel	pr-pum	Hb_{rel} 118 (2–6 μm fraction): anchi-epi (Ludwig 1973, Fig. 8)
Zone 2 Berg-Joditz-Mödlareuth	pum-act, Fe-rich, epi (no pr)	Hb_{rel} 110 (2–6 μm fraction): epi
Zone 3 Hirschberg-Rudolphstein	Fe-poor epi-act (no pr, pum)	Hb_{rel} 103 (2–6 μm fraction): epi

Paratectonic Caledonides of Wales (Lower Palaeozoic)[‡]

Fishguard-St David's Head, N Pembrokeshire, S Wales	pum + pr; pum (Bevins, 1978; Bevins and Rowbotham, 1983)	0.31–0.17 °Δ2θ (17): high-anchi and low-epi (Robinson et al., 1980; Table I) epi (0.25–0.11 °Δ2θ) (Robinson and Bevins, 1986, Fig. 2)

K

Table 7.7 (*Contd.*)

Terrain; age or formation*; area; tectonic unit	Diagnostic minerals**	Illite crystallinity ($-2\,\mu m$ fractions unless otherwise indicated). Values as given (number of samples); illite crystallinity zones
Prescelly Hills, Dyfed, S Wales Cader Idris and SE Snowdonia, N Wales	greenschist (Bevins et al., 1981) act; pum + act (Bevins and Rowbotham, 1983)	epi (Bevins et al., 1981; Robinson and Bevins, 1986, Fig. 2)
Central Snowdonia, N Wales	pum + act (Bevins and Rowbotham, 1983)	0.21–$0.16\,°\Delta2\theta$: epi (Roberts and Merriman, 1985, Fig. 1)
Llŷn, N Wales	pr + pum; pum (Bevins et al., 1981; Bevins and Rowbotham, 1983)	0.23–$0.33\,°\Delta2\theta$: anchi (Roberts and Merriman, 1985, Fig. 1)
N Snowdonia, N Wales	pr + pum or pum (Bevins et al., 1981; Bevins and Rowbotham, 1983)	0.21–$0.27\,°\Delta2\theta$: high-anchi (Roberts and Merriman, 1985, Fig. 1)
Snowdonia and Llŷn, N Wales (schematic) (Merriman and Roberts, 1985, espec. Fig. 7)	Sub-pumpellyite zone (Roberts, 1981) Pumpellyite zone (Roberts, 1981) Clinozoisite zone (Roberts, 1981)	$< 0.43\,°\Delta2\theta$ diag.—Stage I (uncleaved or weakly cleaved) 0.43–$0.28\,°\Delta2\theta$ high-dia. and anchi—most of Stage II (common regul. parag/mus m–l; parag; survival of 1 Md mica; variable cleavage develop.) 0.28–$0.18\,°\Delta2\theta$ high-anchi and low-epi—highest-grade part of Stage II and low-grade part of Stage III (discrete K- and Na-rich 2M micas; strongly cleaved)
Berwyn Hills, eastern N Wales	pr + pum (Bevins and Rowbotham, 1983)	low-anchi to high-anchi (Robinson and Bevins, 1986, Fig. 2)
Breidden Hills, Shelve, and Builth Wells, eastern central Wales	pr + pum (Bevins and Rowbotham, 1983)	diag. (Robinson and Bevins 1986, Fig. 2; Bevins, 1985, p. 452)
Paratectonic Caledonides of the Lake District		
L and M Ordovician volcanics (Eycott Group and Borrowdale Volcanic Group)	pr, pum	Hb_{rel} 112–166 (75), av. 144, in underlying Ordovician and Hb_{rel} 99–150 (40), av. 130, in overlying U Ordovician and Silurian: low- and mid-anchi (Bevins et al., 1985, pp. 63–65;

Paratectonic Caledonides of Scotland and Ireland (Lower Palaeozoic)

N and central belts of Southern Uplands accretionary complex	pr-pum facies (Oliver and Leggett, 1980)	Hb_{rel} 90–160 (317), av. 110: predominantly high-anchi (Bevins et al., 1985, p. 69; Fettes et al., 1985, p. 47)
Longford-Down Massif, Ireland	pr-pum facies (Oliver, 1978)	Hb_{rel} 100–260 (145), av. 140: predominantly mid-anchi (Bevins et al., 1985, p. 69; Fettes et al., 1985, p. 47)
Ordovician of Bail-Hill-Abington area, N belt of the Southern Uplands complex	pr-pum	exclusively anchi (Hepworth et al., 1982, p. 532)

Haast Schist and adjoining Caples-Pelorus terrain of South Island, New Zealand (Triassic) (Kisch, 1981b)

Caples-Pelorus terrain of South Otago (textural zone 1)	pr-pum facies	mainly 0.30–0.45 °Δ2θ (9): high-diag. to mid-anchi
Watson's Beach, South Otago (textural zone 1, or 1 to 2a)	pum (no pr or act)	0.18–0.23 °Δ2θ (3): anchi-epi
Akatore and Taieri Mouth, South Otago (textural zone 1 to 2a)	pum-act facies; stilp; or pr-pum to pum-act (Nelson, 1982, p. 627)	0.13–0.19 °Δ2θ (4): epi
Rambling Creek, Dansey's Pass area, N Otago (text zone 1 to 2a)	pum-act facies (Bishop, 1972)	0.20–0.27 °Δ2θ (10): high-anchi. and anchi.-epi. (Kisch, 1981b and unpublished data)
Routeburn, Humboldt Mountains, W Otago (textural zone 1 to 2) and Maitai terrane (Key Summit)	laws-pum; no act (metamorphic zone II of Kawachi, 1974, 1975)	0.15–0.18 °Δ2θ (4): epi 0.24 °Δ2θ (1): high-anchi

Table 7.7 (*Contd.*)

Terrain; age or formation*; area; tectonic unit	Diagnostic minerals**	Illite crystallinity ($-2\,\mu m$ fractions unless otherwise indicated). Values as given (number of samples); illite crystallinity zones
Low-pressure terrains		
Hercynides of Brittany		
Devonian of Bodennec area, Finistère	pr-pum zone	predominantly diag. but pyro, chld, and parag. (Bril and Thiry, 1976)
Carboniferous metamorphism in Lachlan Fold Belt of New South Wales, Australia (Offler and Prendergast, 1985)		
Lower Paleozoic of North Hill End Synclinorium	actinolite zone (Z4) (no pum)	
	lowest-grade strip	predom. 0.30–0.21 °$\Delta 2\theta$: anchi.
	bulk of actinolite zone	predom. 0.20–0.12 °$\Delta 2\theta$: anchi.
	biotite zone (Z5)	predom. 0.24–0.16 °$\Delta 2\theta$: anchi-epi and low epi

† Illite-crystallinity zone designations of Kubler according to correlation of Árkai (1983, Fig. 4; Árkai and Tóth, 1983, Fig. 2–5; Árkai's zone denominations for NE Hungary (limits of the anchizone for $-2\,\mu m$ fractions: 0.34 and 0.25 ° $\Delta 2\theta$) between quotation marks.

‡ The illite-crystallinity values in this section (Robinson *et al.*, 1980; Merriman and Roberts, 1985; Robinson and Bevins, 1986) were measured on smear mounts which according to Brime (1980) and Kisch (unpublished) give somewhat broader peaks than thin sedimented slides (see p. 302).

Notes

*   The age or formation given is that of the rocks containing the basic metamorphic mineral assemblages

**   Reference to source of mineral data is given only if different from or additional to reference in heading or in last column

pumpellyite facies with the anchizone. Such divergences are particularly evident from detailed studies which show the relation between mineral assemblages in metabasites and 'crystallinity' of illite, such as the study of Merriman and Roberts (1985; see also Roberts and Merriman, 1985) in Snowdonia and Llŷn, North Wales. The main correlation is shown in Table 7.7, but locally the 'isograds' in the metabasites and the 'isocrysts' in metapelites intersect at high angles (Merriman and Roberts, 1985, Figure 1B). In the following discussion we will concentrate on the discrepancies which may be ascribed to various factors.

(i) *Non-diagnostic nature of the basic assemblages.* For instance, Robinson and Bevins (1986, p. 111) suggest that the prehnite–pumpellyite assemblages of the Builth, Breidden, and Shelve inliers of eastern central Wales (Bevins and Rowbotham, 1983) represent non-buffered (and therefore non-diagnostic) assemblages, and that the diagenetic grade indicated by the clay assemblages and illite 'crystallinity' values (generally $> 0.4°\Delta 2\theta$, with a mean of $0.47°\Delta 2\theta$) is a more reliable interpretation. Bevins (1985) has suggested that the alteration of the Builth Volcanic Series may in part be due to a submarine hydrothermal circulation system.

(ii) *Inhibition of metamorphic grain growth by induration of metapelites during prior metamorphism.* Induration of pelites by incipient recrystallization prior to regional metamorphism may inhibit or even prevent the otherwise initial stages of cleavage formation and the formation of P and Q domains (Knipe, 1981), and therefore the subsequent metamorphic growth processes which are usually concentrated at their interfaces. The net effect would be the survival of anomalously poor illite 'crystallinities'. Such induration could be due to two causes: (i) intrusive or hydrothermal activity preceding regional metamorphism and (ii) prior burial metamorphism.

Considering the mineral paragenesis of the schists (pyrophyllite, chloritoid, paragonite), the predominantly diagenetic illite 'crystallinities' associated with the prehnite–pumpellyite zone in the Hercynian low-pressure metamorphic area of Bodennec, Brittany, are anomalously poor, and may be related to the intrusion of the Lower-Devonian dolerites. Merriman and Roberts (1985 pp. 312–313) found that 90% of the occurrences of metapelites with pyrophyllite, rectorite, and corrensite in Snowdonia and Llŷn, N Wales, are located close to igneous intrusions and are of contact-hydrothermal origin; they are usually characterized by poor 'crystallinity', domination of 1 Md mica, and often failed to develop penetrative cleavage. Merriman and Roberts (ibid., p. 318) conclude that the pelites, indurated hydrothermally prior to regional metamorphism, then resisted the effects of regional metamorphism by failing to develop penetrative cleavage.

Breaks in grade of metamorphism across unconformities commonly reflect higher grade of the lower unit due to a burial or deformation metamorphic episode before deposition of the overlying rocks. However, the opposite effect is also observed.

In a detailed study of 'isocryst' patterns in Snowdonia and Llŷn, N Wales, Roberts and Merriman (1985, pp. 618–619) found marked breaks in 'crystallinity' patterns across the sub-Arenig unconformity. Pronounced inversions in 'crystallinity' and degree of deformation occur—for instance at the northern closure of the Ynyscynhaiarn Anticline around Dolbenmaen, where weakly cleaved Upper-Cambrian metapelites of Stage II are overlain by strongly cleaved Lower-Ordovician metapelites of Stage III—which they explain by a mild burial metamorphism of the Cambrian rocks before the uplift and erosion which preceded the Arenig marine transgression. Induration by incipient recrystallization inhibited or even prevented the effects of subsequent Caledonian regional metamorphism, and preserved their pre-Arenig metamorphic grade.

(iii) *Relation between strain and illite 'crystallinity'.* In their study, Roberts and Merriman (1985) found marked coincidence of areas of high strain and high 'crystallinity', and areas of low strain and low 'crystallinity,' demonstrating a causative relationship between strain and metamorphic grade as apparent from illite 'crystallinity' (see Figure 2.5). They suggest that the cause of the Caledonian metamorphism was the deformation itself, invoking strong heating associated with development of cleavage.

The three above-mentioned effects pertain to illite 'crystallinity' in pelitic rocks. It is quite conceivable that they do not affect metamorphic mineral assemblages in basic rocks in the same way, or at all: this could then cause discrepancies in the correlation of grade.

7.7.4 *Relation of the pumpellyite–actinolite facies to illite 'crystallinity'*

Kisch's (1981b) studies of illite 'crystallinity' from areas bordering on the Haast Schist terrane of Otago, New Zealand (see Table 7.7), show that the pumpellyite–actinolite facies is associated with high anchimetamorphic and transitional anchi-epimetamorphic values in north Otago, but with epimetamorphic values in south Otago; the lawsonite-bearing pumpellyite zone of western Otago and of the Maitai terrain to the SW show predominantly epimetamorphic values. The pumpellyite–actinolite assemblage in the Grès de Taveyanne near Leuk (e.g. Coombs *et al.*, 1976), at the W extremity of the Aar Massif, is also associated with epimetamorphic illite 'crystallinities' (see Table 7.7); whereas the onset of the actinolite zone in the low-pressure terrane of the Lachlan Fold Belt of N.S.W. is associated with anchimetamorphic values, the bulk of the zone still being associated with transitional anchi-epimetamorphic to low-grade epimetamorphic values.

The comparatively scanty data available thus far do not provide a consistent picture of the relationship between the pumpellyite–actinolite facies and illite 'crystallinity': the relationship appears to vary, the facies being associated with 'crystallinities' varying from area to area between anchimetamorphic and epimetamorphic. There seems to exist a tendency for the prehnite-free pumpellyite facies in HP metamorphic terrains (as indicated by the appearance of lawsonite) to be associated with advanced illite 'crystallinities', and for pumpellyite-free actinolite zones in LP metamorphic terrains to be associated with comparatively poor illite 'crystallinities'; such effects may reflect an effect of pressure in enhancing the 'crystallinity' of mica.

7.7.5 Summary

The vast majority of prehnite–pumpellyite facies assemblages are associated with predominantly or entirely anchimetamorphic illite 'crystallinities'.

In some isolated cases the onset of the prehnite–pumpellyite facies is associated with predominantly diagenetic 'crystallinities', while local association of laumontite assemblages with anchimetamorphic 'crystallinities' appears to be related to vicinity of major nappe thrusts. The greenschist facies is associated with epimetamorphic and only locally with anchimetamorphic illite 'crystallinities'. The pumpellyite–actinolite facies occupies an intermediate position: the transition between the prehnite–pumpellyite and the pumpellyite–actinolite facies may be associated with anchimetamorphic or with already epizonal illite 'crystallinities' (particularly in high-pressure terrains).

The general congruence between the anchizone and the prehnite–pumpellyite facies does not justify the assignment of an area with anchimetamorphic illite 'crystallinities' to the prehnite–pumpellyite facies merely on the grounds of prevalence of anchimetamorphic illite crystallinities any more than a terrain can be referred to the amphibolite facies merely on the basis of the occurrence of garnet-biotite schists: statements must be qualified, such as 'the terrain may provisionally be taken to have undergone metamorphism under conditions close to the prehnite–pumpellyite facies'. Kisch (1983) proposed the term 'incipient metamorphism' (better: incipient regional metamorphism) as a general term encompassing the prehnite–pumpellyite and pumpellyite–actinolite facies, anchimetamorphic illite 'crystallinities', and anthracitic coal ranks (cf. section 1.2).

7.8 Illite 'crystallinity' and coal rank associated with the glaucophane–lawsonite schist facies and other high-pressure terrains

Data on illite 'crystallinity' and coal rank from high-pressure terrains are still scanty. However, the data available (Table 7.8) show a wide range of values. In association with the lawsonite metagreywackes of the Diablo Range illite 'crystallinities' interpreted by the writer as diagenetic (there is a calibration problem with the data of Cloos) are associated with reflectances of $1–2\%\ R_0$; a similar association might be found in the higher-grade part of the zeolite facies,

Table 7.8 Illite-crystallinity and coal-rank data from glaucophane–lawsonite–schist and lawsonite–albite–chlorite facies terrains

Terrain; age or formation; area; tectonic unit. Diagnostic minerals; temperatures from minerals	Illite crystallinity and diagnostic phyllosilicate minerals. Values as given (number of samples)	Coal rank R_0 (%) unless otherwise indicated; calculated temperatures
Shales interbedded with lawsonite-bearing metagraywacke, Franciscan of Diablo Range, California (Bostick, 1971; 1974, pp. 12–14)		
Mount Hamilton Road anticline		
Upper Unit, SW flank		1.1 to 1.3% R_0 (4); 115 to 130 °C*
Lower unit (core)		1.7 to 2.1% R_0 (3); 140 to 150 °C*
Upper unit, NE flank; with jadeitic px		1.15% R_0 (1); 115 °C*
Garzas Creek area (SE of Mt Hamilton)		
Chaotically disturbed strata		1.7 to 1.8% R_0 (2); 140 to 145 °C*
Coherent metagreywacke		1.2 to 1.3% R_0 (3); 120 to 130 °C*
Jadeitic px-bearing metagreywacke near E boundary of Franciscan		1.1% R_0 (1); 115 °C*
Metashales from Pacheco Pass and Panoche Pass, Diablo Range, and melange matrix, Franciscan of northern California (Cloos, 1983)**		
High-pressure shale (ass. with jadeite)	10.2 to 5.8 mm (9) (1.0 to 0.6° $\Delta 2\theta$)	
Shale matrix of melange (many with pumpellyite and/or lawsonite)	11.2 to 4.8 (24) (1.1 to 0.5 ° $\Delta 2\theta$)	
Tertiary high-pressure belt of Ouegoa area, northern New Caledonia (Diessel *et al.*, 1978; unpubl. illite data by Kisch)		
Prehnite–pumpellyite zone, low-grade	0.29 to 0.39° $\Delta 2\theta$ (3)	1.39 to 2.16% R_{max} (3), mean 1.88% R_{max}
high-grade	0.34 ° $\Delta 2\theta$ (1); parag.	2.80 to 3.34% R_{max} (3)
Lawsonite zone (255 °C and up)	0.24 ° $\Delta 2\theta$ (1)	2.86 to 4.0% R_{max}; (6) $d_{002} = 3.50–3.55$ Å
Ferroglaucophane zone (300 °C and up)	0.20 to 0.25 ° $\Delta 2\theta$ (5); 3 with parag.	4.77 to 6.53% R_{max} (5); $d_{002} = 3.36–3.44$ Å
Epidote zone (no lawsonite) (390 °C and up)	0.18 to 0.22 ° $\Delta 2\theta$ (3); 1 with parag.	Graphite with ordered layer structure, and $d_{002} = 3.355$ Å
Vanoise, Briançonnais zone of the French Alps (Goffé, 1982; Goffé and Velde, 1984)		
Green 'chloritic marbles' (U Cretaceous to Paleogene)	2.4 to 2.7 mm (3)—near to maximum; paragonite (Dunoyer, 1969, pp. 179–180)	

Black Fe–Mg carpholite schists (Dogger), Aiguilles de Chanrossa, W Vanoise; with chloritoid, lawsonite, pyrophyll. (Goffé, 1982, pp. 110–114); 300 °C, 6 kbar (from mineral equilibria)

Mica crystallinity near 2.5 mm (p. 33) Regular I/S with c. 30% sm and allevardite-type order (p. 33)

Kerogen corr. approx. with R_0 of 2% (p. 133); bitumoids corr. with R_0 between 1.4–1.5 and 2% (p. 134) LOM 11 to 12 (eq. to 1.1 to 1.5% R_0) (Goffé and Velde, 1984)

Lawsonite zone in Caples terrain of Humboldt Mountains and Maitai terrain of Livingstone Mountains, western Otago, New Zealand (largely Permian)

Routeburn, Caples terrain. Laws-pump, no actin. (metam. zone II of Kawachi, 1974, 1975)

0.15 to 0.18 °$\Delta 2\theta$ (4); 3 with parag. (Kisch, 1981b, p. 357)

Key Summit, Maitai terrain. Laws-pump; no actin.

0.24 °$\Delta 2\theta$ (1); parag. (Kisch, 1981b, p. 357)

Penninic domain of western and central Alps (Frey, 1986, Plate 1 and p. 20)

Pumpellyite ± glaucophane + chlorite + albite ± quartz

Epimetamorphic

*Calculated using Bostick's (1974, Fig. 8) 'well-standard' curve for samples buried roughly 30 to 80 Ma.)

**The scale to the abscissa of Cloos (1983) Fig. 3 is in error; the values given are based on the corrected values (Cloos, personal communication).

which not only seems to confirm the view that coal rank is not enhanced by high lithostatic pressures, but suggests the same for illite 'crystallinity'. In the prehnite–pumpellyite facies of New Caledonia similar and slightly higher ranks (but still rather lower than are common in the prehnite–pumpellyite facies of 'medium-pressure' terrains) are associated with diagenetic-anchimetamorphic and low-anchimetamorphic illite 'crystallinities', suggesting that better illite 'crystallinities' are reached in association with given ranks than in medium-pressure terrains.

More striking is the difference in the progress of both coal rank and illite 'crystallinity' between the lawsonite zones of different terrains: in the Franciscan of the Diablo Range both illite 'crystallinity' and coal rank have advanced very little, whereas in the lawsonite zones of New Caledonia and western Otago illite 'crystallinities' are high-anchimetamorphic and epimetamorphic, coal ranks (New Caledonia only) being anthracitic. The 'crystallinities' in the Vanoise appear to be epimetamorphic, but the presence of allevardite-type order indicates anchimetamorphism, in line with the coal rank, which appears to be somewhere around $1–1.5\%$ R_0.

The temperatures derived for the lawsonite zone in various areas from mineral assemblages and oxygen-isotope thermometry are rather similar: 250 °C at 3 kbar, 300 °C at 4.7 kbar and 400 °C at 6.5 kbar respectively for the lawsonite-in, ferroglaucophane-in and lawsonite-out isograds in New Caledonia (Diessel *et al.*, 1978—see Figure 4.33), 265—310 °C at 3.3–4.0 kbar for the lawsonite zone in western Otago, and 270–330 °C in the stability field of aragonite in the Franciscan (compilations of field gradients by Turner, 1981, p. 427 ff.), and 300 °C at 6 kbar calculated for the metamorphic peak in the western Vanoise. These temperatures are not sufficiently different to explain the wide difference between the poor illite 'crystallinities' and coal ranks of the Franciscan and the Vanoise on one hand, and of New Caledonia and western Otago on the other.

One variable can be taken to explain the differences, namely duration of burial, or of exposure to heating. For instance, even a very low temperature estimate of 260° for the metamorphism of the Franciscan would give the measured rank range of 1.1 to 2.1% R_0 after effective heating times (t_{eff}) of less than 0.1 Ma to 0.4 Ma according to the curves of Hood *et al.* (1975). Higher temperatures would allow of even shorter t_{eff} duration to reach these ranks; lower temperature estimates, such as that of 170 °C for the first appearance of clinopyroxene in Franciscan greywackes (e.g. Liou *et al.*, this volume, Figure 3.27) would require a longer effective heating time of some 7 Ma to give the rank of 1.1% R_0 measured in the jadeitic clinopyroxene-bearing grey-wackes in the Franciscan, and less time for higher temperatures (curves of Hood *et al.*, 1975, Figure 3). Metamorphic temperatures above 270 °C would therefore require very brief burial durations for the Franciscan.

Goffé and Velde (1984, p. 356) similarly have noted that according to the curves of Hood *et al.* (1975) organic matter of LOM 11 to 12 (1.1 to 1.5% R_0) could not have been subjected to temperatures of 260 °C for more than 0.1 Ma. On the other hand, a rank of about 2.86% R_{max} for the lawsonite-in isograd in northern New Caledonia would be in accordance with a much longer effective heating time of some 4 Ma at a temperature of 255 °C. It is therefore tentatively concluded that the low coal ranks and poor illite 'crystallinities' associated with some blueschist-facies terrains (Franciscan, Vanoise) must be related to very short-lived burial.

7.9 Order in graphite and degree of incipient metamorphism

Powder XRD characteristics of carbonaceous materials have been used as an indicator of the more advanced degree of very low-grade metamorphism in the meta-anthracite to graphite range. The method is very approximate, since electron diffraction studies by Landis (1971), and recent HRTEM studies by Buseck and Huang (1985) have shown that even at these 'high' grades carbonaceous matter within a given rock sample may show a range of structural order, which reflects the chemical and structural character of the carbonaceous precursors (cf. section 4.4.1.3).

Landis (1971, Table 2) has shown that the zeolite facies (including the prehnite–pumpellyite–laumontite subfacies) and the 'lawsonite–prehnite–pumpellyite subfacies' of the lawsonite–albite–chlorite facies of the Wakatipu belt of New Zealand exclusively shows the nearly amorphous graphite-d_3 type, which is in agreement with the bituminous coal ranks found by Kisch (1981*b*; see also Table 7.5); the lawsonite–epidote–pumpellyite subfacies of the lawsonite–albite–chlorite facies and the higher-grade pumpellyite–actinolite facies in this belt, and the pumpellyite–actinolite–glaucophane and epidote–actinolite–glaucophane subfacies of the blueschist facies of the Sanbagawa belt in Shikoku show various mixtures of the graphite-d_1 and d_{1A} types (with d_{002} at 1/3 peak height at 3.38–3.41 Å), with local graphite-d_2 present (Landis, 1971).

Carbonaceous matter in northern New Caledonia has been studied by Diessel *et al.* (1978). In the prehnite–pumpellyite facies of the 'melange zone' all carbonaceous material is X-ray amorphous. At the lawsonite isograd it corresponds to Landis's type d_3, with a broad low peak centred on 3.50–3.55 Å, the strong decrease in the d_{002} values taking place through the lower-grade part of the lawsonite zone (as in southern New Zealand): in the middle of the lawsonite zone (beyond the ferroglaucophane-in isograd) the carbonaceous material has the character of graphite-d_2 or d_{1A}, with d_{002} spacings of 3.38–3.44 Å, migrating towards the 3.36 Å at the epidote isograd. Three-dimensionally ordered graphite with $d_{002} = 3.355$ Å and sharp 100, 112, and 114 lines only appears at the epidote isograd.

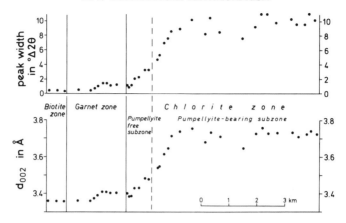

Figure 7.6   X-ray data of carbonaceous material in pelitic schists along the Asemi River traverse through the Sanbagawa metamorphic belt in central Shikoku, Japan. Peak width was measured at half height. (Modified from Itaya, 1981, Figure 6).

Itaya (1981, Figure 6—see Figure 7.6) has shown that a marked change in the crystal parameters takes place in the highest-grade part of the chlorite zone, in which pumpellyite disappears: from $B_{002} = 8-11°2\theta$ and apparent $d_{002} = 3.65-3.75$ Å (corresponding to Landis' graphite-d_3) to $B_{002} = 1-2°2\theta$ and $d_{002} = 3.38-3.41$ Å (corresponding to Landis' graphite-d_{1A}), with a further drop to $d_{002} = 3.363$ Å and B_{002} c. $0.5°2\theta$ taking place within the garnet zone (i.e., before the biotite zone). In the Narragansett Basin the first change also takes place in the chlorite zone, the second in the high-grade chlorite and biotite zone.

The relationship between the grade of metamorphism and the d_{002} of carbonaceous matter in various metamorphic terranes is shown in Figure 7.7. The lower grade at which well-ordered graphite is attained in high-pressure than in medium-low pressure regional metamorphism and contact metamorphism suggests an effect of deformation, duration of heat treatment, hydrostatic pressure, and an ambient fluid phase under pressure (Grew, 1974., p. 69). In this connection it is noteworthy that both in southern New Zealand and in northern New Caledonia, on which some illite-'crystallinity' data are available (Kisch, 1981b and unpublished data; see Tables 7.7 and 7.8), the zones with predominance of carbonaceous material with d_{002} spacings at 3.38–3.41 Å (graphite-d_2, d_{1A}, and d_1) are associated with anchi-epimetamorphic and epimetamorphic values. This might indicate that both indicators of very low-grade metamorphism are affected by the factors listed by Grew, by deformation in particular.

Well-ordered graphite, with $d_{002} = 3.35-3.36$ Å, is approached at different grades in different terrains: in high-pressure terrains in the garnet-oligoclase subfacies of the amphibolite facies in the Wakatipu metamorphic belt of southern New Zealand (Landis, 1971), in the garnet-epidote zone (glaucophanitic greenschist facies) of northern New Caledonia (Diessel et al., 1978),

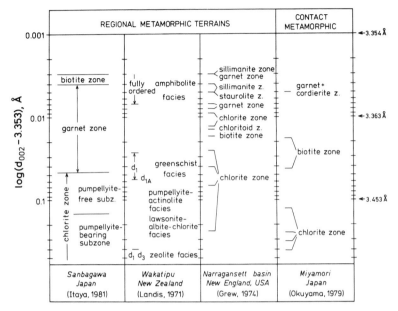

Figure 7.7 Relationship between the metamorphic grade and d_{002} of metamorphosed carbonaceous material in some regional and contact metamorphic terrains. (Modified after Itaya, 1981, Figure 10).

and in the biotite-zone (i.e., beyond the garnet zone) in the Sanbagawa belt of central Shikoku (Itaya, 1981); in medium-low pressure terrains in the garnet zone of the Narragansett Basin of Rhode Island (Grew, 1974), and in the andalusite–staurolite zone of the Mt Lofty ranges, South Australia (Diessel and Offler, 1975).

7.10   Relationship of temperatures derived for the anchizone and pumpellyitic facies

The approximate correlation of the onset of the prehnite–pumpellyite facies and the anchizone, at least for regional terrains, has consequences for the evaluation of the temperatures that have been derived independently for these degrees of very low-grade metamorphism using oxygen-isotope and fluid-inclusion thermometry, experimentally by establishing temperatures of reversible mineral equilibria and by direct measurement in geothermal fields. These temperatures will be briefly reviewed in the following sections.

7.10.1   Oxygen-isotope and fluid-inclusion temperatures of the anchizone

The beginning of the anchizone is reasonably well documented with oxygen-isotope temperatures; some recent work tends to place the onset of anchimetamorphism at temperatures somewhat below 200 °C.

Brauckmann (1984, p. 150) found the following quartz–illite oxygen isotope temperatures:

'Low diagenesis'—$Hb_{rel} = 400–500$ 90–110 °C
    (associated with 0.7–1% R_m)
'Advanced diagenesis'—$Hb_{rel} = 300$ 125–130 °C
    (associated with 1.5–1.9% R_m)
'High diagenesis'—anchimetamorphism—
$Hb_{rel} = 150–200$ 145–155 °C
    (associated coal ranks R_m 4% near
    Vlotho, and 2.5–4% R_m in the
    southern Bramsche Massif).

These coal ranks indicate much higher temperatures, respectively > 240 °C and 180–240 °C; this contradiction is ascribed to slow establishment of isotopic equilibrium, particularly by quartz, under contact metamorphism.

Primmer (1985) found that the diagenetic zone and the low-grade anchizone on the north coast of Cornwall are associated with oxygen-isotope quartz–illite temperatures of less than 200 °C (although the well-defined transition from diagenesis in the north to anchizone in the south is not recorded by the isotopic temperatures); in the low-grade epizone one sample gave a quartz–illite temperature of 321–377 °C and quartz–chlorite temperatures approaching 300 °C. However, temperatures obtained throughout the study were lower than expected, which Primmer suggests to be due to the fact that the geothermometers used are derived from essentially open systems.

The appearance of biotite in low-pressure metamorphic terrains takes place at a carbon-isotope calcite–graphite temperature of about 330 °C (e.g. Morikiyo, 1984), and at 300 °C in the Cerro Prieto geothermal system (Schiffman et al., 1985).

One might possibly be inclined to ascribe these oxygen-isotope temperatures to disequilibrium, were it not for the fact that they are roughly matched by current fluid-inclusion temperatures. Some earlier fluid-inclusion studies placed the onset of the anchizone at slightly higher temperatures. For instance, Durney (1974) found a *Th* of 245 °C in laumontite–prehnite–pumpellyite greywackes at Diablerets, and Saliot et al. (1982) found *P–T* conditions of 170–310 °C at 200–650 bar for the laumontite–prehnite–pumpellyite metamorphism in the Grès du Champsaur (French Alps); both areas have high-grade diagenetic illite 'crystallinities' (see Kisch, 1980b, and Aprahamian, 1974, Figure 1). Barlier (1974; also Barlier et al., 1974; Touray and Barlier, 1975) correlated the disappearance of kaolinite and of expandable layers in the French Alps with a 'ligne de démarcation' at c. 200 °C, based on reflectance and fluid-inclusion geothermometry, so that the onset of the anchizone would be at slightly higher temperatures.

Robinson et al. (1980) reported somewhat lower *Th* values between 150° and 250 °C in quartz veins in the Pembroke coal field, in association with illite of the diagenesis-anchimetamorphism transition, and anthracites with 5–10% V.M.; Kisch (unpublished) found similar minimum homogenization temperatures of 185–230 °C from presumably in part methane-saturated aqueous fluid inclusions in quartz veins in somewhat higher-grade low- and middle-anchimetamorphic Silurian and Devonian rocks of the Valley-and-Ridge province of the Appalachians.

Duba and Williams-Jones (1983b) found Th temperatures of 300 °C in association with $0.24°\Delta 2\theta$ and 4.6% R_m, i.e. close to the anchi-epiboundary.

Extensive fluid-inclusion data from the Helvetic Alps by Mullis (1979) and Frey et al. (1980) locate the transition from the HHC zone (> 1 mole %) to the CH_4 zone of fluid inclusions—which is associated with illite 'crystallinities' of 7 to 9 mm (highest-grade diagenesis and diagenesis-anchimetamorphism boundary) and $2.5–3\%$ R_m—at minimum temperature and pressure of 200 °C and 1200 bars, whereas the minimum temperatures and pressures for the transition from the CH_4 zone to the H_2O zone—at illite crystallinities of 4.5–5 mm and $4.5–5\%$ R_m (i.e., the high-grade part of the anchizone)—are 270 °C and 1700 bars, respectively. The low- and medium-grade anchizone are thus dominated by methane-bearing fluids with minimum formation temperatures between 200° and 270 °C (similar temperatures in Breitschmid, 1982). Niedermayr et al. (1984) also found that cleft quartz in the Scythian of the N. Drauzug associated with transitional anchi-epimetamorphic 'crystallinities' contained aqueous (with < 1 mol % CH_4, CO_2, etc.) fluids, and adopted $\sim 180–200$ °C and $\sim 300–350$ °C as 'the generally accepted temperature boundaries of the anchizone' (Figure 3). The homogenization temperatures have been taken as minimum temperatures, since no pressure correction has been made, on the assumption of saturation with HHC or CH_4; this writer would therefore submit that there is no real evidence for temperatures of less than 200 °C for the onset of the anchizone.

7.10.2   Temperature–pressure conditions for the prehnite–pumpellyite facies

Zeolite-facies condition are summarized by Liou (1979, p. 18) to be in the range of 100–300 °C and total pressures below 3 kbar. On the basis of estimates of minimum formation temperatures of about 50°, 90° and 190 °C for laumontite, prehnite and pumpellyite in the Hokonui section (Boles and Coombs, 1977, p. 1007), the same author estimated the temperatures of metamorphism in laumontite- and Fe-pumpellyite- bearing extrusive rocks in the East Taiwan ophiolite at about 150° to 250°C at p_{fluid} of less than 0.5 kbar (Liou, 1979, p. 12; see also Ishizuka, 1985).

For a discussion of petrogenetic grids relating to the zeolite, prehnite–pumpellyite, and pumpellyite–actinolite facies the reader is referred to Liou et al. (1985, and this volume) where the $P–T$ diagram showing the continuous reactions around the invariant points lm + pm + pr + ep + chl (XII) and pm + pr + ep + tr + chl + qz (VI) in the Fe-free model system, and displacement of the invariant points along discontinuous reactions upon introduction of Fe_2O_3 into the model system, is shown as Figure 3.8.

In the Fe-free chlorite-excess model system the assemblage lm + pr disappears at a temperature of c. 225–230 °C at pressures < 1.4 kbar—below that of the invariant point XII—representing the transition of the zeolite facies to the prehnite–actinolite facies at low pressures. For pressures above the invariant point (> 1.4 kbar) the temperature of the continuous reaction pm + lm = ep + chl—marking the transition between the zeolite and prehnite–pumpellyite facies at intermediate pressures—increases with pressure from 225–230 °C to approximately 260 °C at 2.5 kbar.

The equilibrium curve for the continuous reaction pr + chl = pm + tr—representing the limit between the prehnite–pumpellyite and the pumpellyite–actinolite facies at pressures above the invariant point VI at 340 °C and 2.2 kbar—has a strongly negative slope in P–T space, so that the prehnite–pumpellyite facies appears to be restricted to pressures of less than c. 3.3 kbar.

However, the positions of these invariant points and of the continuous reactions emanating from them are lowered by up to 45 °C by the presence of Fe^{3+}, so that all the above temperatures must be considered maximum temperatures for the boundaries of the prehnite–pumpellyite facies. This effect must be taken into account particularly in view of the consistent reports of high Fe-contents of pumpellyites in the prehnite–pumpellyite facies (and in the zeolite facies) compared to those in the pumpellyite–actinolite facies (Surdam, 1969; Kawachi, 1975; Tzeng and Lidiak, 1976; Kuniyoshi and Liou, 1976b; Bevins, 1978; Smith et al., 1982; Hepworth et al., 1982; Árkai, 1983; Nyström, 1983; Bauer, 1983; Schiffman and Liou, 1983; Primmer, 1985).

In this light the temperature ranges for the prehnite–pumpellyite facies given in earlier papers, e.g. 250° to 350 °C at lithostatic pressures of 1 to 2.5 kbar (Kuniyoshi and Liou, 1976a,b) for the (initially wairakite-bearing) prehnite–pumpellyite facies and the prehnite–actinolite facies in the Karmutsen Volcanics of Vancouver Island at a high geothermal gradient of 50–60 °C/km (Liou et al., this volume, Table 3.1 and Figure 3.22), and 250–300 °C at 2–3 kbar for the prehnite–pumpellyite facies at a lower geothermal gradient of 30–25 °C/km in western New Caledonia (Campbell, 1984)—though in agreement with estimated temperatures of up to 250 °C of the assemblage laumontite + Fe-rich pumpellyite (Liou, 1979; see also Ishizuka, 1985)—appear to correspond to the Fe-free system and thus may be somewhat on the high side. This is certainly true for the 315–370 °C at 1–2.5 in New Brunswick (Strong et al., 1979).

At high thermal gradients of 80–100 °C/km in the Proterozoic Hamersley Basin, Western Australia (Smith et al., 1982), the prehnite–pumpellyite–epidote zone ZI is indicated by the appearance of epidote to extend from approximately 200 °C (Liou, 1979) at $P_{fl} = P_{load}$ of 0.5 to 1 kbar, whereas the prehnite–pumpellyite–epidote–actinolite zone ZIII (a low-pressure subfacies of the pumpellyite–actinolite facies) extends from 300 °C to about 360 °C at $P_{fl} = P_{load}$ of 1.2 to 1.7 kbar.

The earlier-mentioned fluid inclusion data on the laumontite–prehnite–pumpellyite greywackes of the Alps by Durney (1974) and Saliot et al. (1982) do not allow of much lower temperatures for the onset of the prehnite–pumpellyite facies in medium-pressure terrains.

7.10.3 Pressure-temperature conditions of the pumpellyite–actinolite facies

Nakajima et al. (1977) have established a temperature range of 90 °C for the stability of the assemblage haematite–pumpellyite–epidote with excess chlorite, actinolite, albite and quartz in ordinary metabasites, the upper temperature limit of the pumpellyite–actinolite facies being defined by the disappearance of pumpellyite in the presence of epidote with $X_{Fe} = 0.10$–0.15, the minimum value in most metabasites, at about 370 °C.

Most current temperature-range estimates for the pumpellyite–actinolite facies, such as 270–360 °C at a pressure of about 5 kbar (Katagas and Panagos, 1979), minimum values of 290–380 °C at total P 3.5 kbar, and possibly as high as 5 kbar (with occurrence of lawsonite; Baltatzis and Katagas, 1984) are based on the experiments of Nitsch (1971) and the model of Nakajima et al. (1977). However, Schiffman and Liou (1980; see also Liou et al., 1985a) have criticized Nitsch's disposition of the univariant curves about the invariant point GR (equivalent to VI) in the Mg-rich system, and concluded that the critical assemblage pumpellyite–tremolite–chlorite–albite–quartz is stable at temperatures below 350 °C at P_{fl} less than 5 kbar, and at temperatures of 350 to 380 °C and pressures below 5 and 8 kbar P_{fl} (as in the petrogenetic grid of Liou et al., 1985, Figure 2; this volume, Figure 3.6). Compatible quartz-magnetite oxygen-isotope temperatures of 320° and 340 °C are indicated by Brown and O'Neill (1982) for the jadeitic clinopyroxene-, lawsonite-, and pumpellyite-bearing Shuksan suite, Washington state, at some 7 kbar.

The prehnite-bearing pumpellyite–actinolite subfacies, which appears at high geothermal gradients, and can be regarded as a lower-pressure subfacies of the pumpellyite–actinolite facies, may approach the conditions of the invariant point VI of Liou et al. (1985a, and this volume); taking into account the Fe content of the associated epidotes, it extends from 300° to 360 °C at c. 1.5 kbar (geothermal gradient of c. 40 °C/km) according to Smith et al. (1982), and from 315 °C at 1.85 kbar according to Robinson and Bevins (1986).

At the even lower pressures of geothermal fields, epidote, prehnite, and actinolite appear at somewhat lower temperatures, i.e. at respectively 230°, 270° and 280 °C at P_{fl} below 0.3 in Cerro Prieto (Schiffman et al., 1985) and at similar temperatures in Iceland (Kristmansdóttir, 1979) at thermal gradients of 80–90 °C/km (Robinson et al., 1982); at an even higher thermal gradient of 100–200 °C, Schiffman et al. (1984) place the boundary between the pumpellyite zone and the epidote zone (both with minor prehnite and laumontite) in the Del Puerto ophiolite at *c.* 225 °C. Temperatures in the same range are indicated by *Th* values of 225–285 °C in a propylitic epidote–prehnite–K-feldspar–chlorite zone, interpreted to represent a temperature range of *c.* 250–280 °C (Shikazono, 1985).

7.10.4 *Temperature range for the anchizone*

If the approximate correlation of the onset of the prehnite–pumpellyite facies and the anchizone for regional incipient metamorphism are accepted, the temperatures for these stages have to correspond. At this stage, this writer would accept the arguments that the oxygen-isotope temperatures for the onset of the anchizone may be somewhat too low, and the experimental equilibrium temperatures for the onset of the prehnite–pumpellyite facies somewhat too high: the temperatures should lie in the range 200–250 °C.

The rather scanty oxygen-isotope and fluid-inclusion data available constrain the high-grade boundary of the anchizone to a temperature around 300 °C. A similar temperature for this boundary would be suggested by the oxygen-isotope temperatures of 255 °C and 300 °C for the lawsonite-in and ferroglaucophane-in isograds in northern New Caledonia (Diessel *et al.*, 1978), associated with highest-grade anchimetamorphic and anchi-epimetamorphic illite 'crystallinities' (Kisch, unpubl. data; see Table 7.8).

7.10.5 *Coal rank associated with very low-grade metamorphism and the effect of duration of heating on coalification*

Earlier in this review we have presented the differences in current views on the effect of duration of heating on continuation of coal-rank increase with time. Without presuming to be able to solve this question, the correlation of indicators of very-low-grade metamorphism can contribute towards a solution. The correlation of a stage of mineral alteration with coal rank implies that both have formed under the same temperature conditions. The fact that some indicators of metamorphism, whose evolution does not appear to be time-dependent, exhibit a reasonably uniform correlation with coal rank in a wide variety of terrains by itself suggests that the effect of duration of burial cannot constitute a major factor in the establishment of that rank.

We may take as an example the widespread association of the onset of anchimetamorphism in 'regional' environments with coal ranks of 2.5–3.1% R_m. This relative uniformity can only be explained with respect to the effect of the duration of residence at maximum temperature of heating on the coal rank attained in either of two ways: (1) the effect is negligible, or (2) the effect exists, but the duration of heating between the different terrains varies in very narrow limits, because of the similarity in uplift and cooling rates.

We can test these two possibilities against the temperature estimated from the mineral data.

(1) Price's (1983, Figure 19) linear extrapolation of his regression line of log vitrinite reflectance v. burial temperature for burial times ranging from 200 000 to 240 million years beyond a rank of 1.3% R_0 gives temperatures of 280° to 330 °C for ranks of 2.5–3.1% R_0; Barker's (1983, Figure 2) exponential regression curve for six geothermal systems gives 255–285 °C, and Barker and Pawlewicz' (1986) least-squares regression equation $\ln (R_m) = 0.0078 T_{max} - 1.2$ gives 271–295 °C. These temperatures are rather higher than the temperature range of 200–250 °C estimated for the beginning of the anchizone.

(2) Using Hood et al.'s (1975, Figure 3) relation of LOM to maximum temperature and effective heating time (t_{eff} = time within 15 °C of T_{max}), a temperature of 225 °C would give rise to a coal rank of 2.5% R_m (LOM = 16) as a result of t_{eff} of 15 Ma, or after a burial time of 25 Ma using the curve of Karweil (1956) as modified by Bostick (1973, Figure 5). These times are excessively long, particularly for the Grès de Taveyanne — whose burial was almost entirely due to the overlying nappe pile, Alpine uplift rates of 1 mm/y giving a t_{eff} of less than 1 Ma for thermal gradients of 15–30 °C/km — and probably also for most other areas considered. A temperature of 250 °C would require a somewhat more reasonable t_{eff} of 2.5–3 Ma, but at a t_{eff} of 1 Ma, temperatures of 225° and 250 °C would produce coal ranks of only 1.5 and 2.2% R_m respectively, lower than found in association with the onset of the anchizone anywhere.

The two coalification models thus generate results that are not only irreconcilable, but seem to be both in conflict with the temperatures of 200–250 °C for the onset of the anchizone, albeit for opposite reasons: (1) the Price–Barker model gives excessively high temperatures; (2) the Karweil model gives excessive duration of heating for the estimated temperatures, or excessive temperatures for realistic times of heating.

Price's (1983) assumption that burial of more than 2 million years does not appreciably affect organic metamorphism could account for the uniformity of the correlation between mineral alteration stages and coal rank; however, his regression curve of log vitrinite reflectance v. burial temperature should be non-linear, and its extrapolation beyond ranks of 1.3% R_0 modified towards higher ranks and lower temperatures in order to match the mineral-alteration temperatures. Bostick (1984) has separated Barker's (1983, Figure 2) data on borehole temperatures and vitrinite reflectance for geothermal systems into 'long' and 'short' duration of heating (respectively about 10^6 and 10^4 years); the 'long-duration' regression curve (Bostick 1984, Figure 1) gives temperatures of 225–250 °C, and therefore would satisfactorily explain the correlation data. However, we have shown above (p. 288) that the low coal ranks associated with some lawsonite–albite and glaucophane–lawsonite–schist facies terrains (Franciscan, Vanoise) can only be explained by assuming very brief durations of heating. This makes it difficult to accept Price and Barker's (1983) claims that full organic maturation may be achieved only after about 1000 to 10000 years.

Provisionally, this writer would submit that the correlation data for the onset of anchimetamorphism necessitate a significant effect of duration of heating only for brief periods, until 'stabilization' at a coal rank which is essentially a function of temperature. After attainment of this 'stabilization', continued heating at constant temperature would have only negligible effects on coalification. This concept would be analogous to laboratory organic-maturation experiments in the presence of water (Price 1983).

The heating times necessary for the attainment of 'stabilization' of the rank are likely to be longer for low than for high temperatures, by analogy with Ishiwatari et al.'s (1972; see Price, 1983, Figure 18) plot of reaction extent v. heating time for kerogen at various temperatures. This author would suggest that for the temperatures corresponding with the onset of the anchizone, such 'stabilization' should be attained within 1 Ma.

For the commonly used plots of burial time and temperature versus coal rank this concept would imply that in Karweil's (1956) coal rank v. temperature plot the equal log burial-time curves should be wider spaced for brief, and closer spaced for extended burial times, and that the equal-rank (equal-LOM) lines on Hood et al.'s (1975) and Bostick et al.'s (1978; see Figure 4.23 in this volume) time–temperature plots should be strongly non-linear, flattening out with increasing t_{eff}, t_{eff} being more important at low than at high temperatures, as shown schematically on Figure 7.8.

It should become possible to further test the models of the effect of duration of heating on coalification not only in sequences presently at maximum temperature, but also where correlation between coal rank and a mineral alteration-stage thermometer, and the burial and uplift history have been well established.

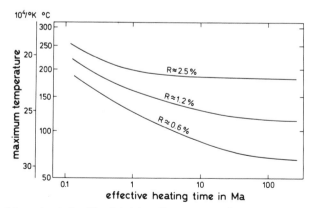

Figure 7.8 Schematic relationship of vitrinite reflectance to maximum temperature and effective heating time. Note stronger curvature of the equal-rank curves compared to the analogous plot by Bostick et al. (1978; cf. Figure 4.23 in this volume), to show 'stabilization' of the coal-rank increase, which is taken to be reached after shorter heating times at high than at low temperatures.

7.11 Conclusions

We will not here repeat the conclusions as to the relations between the various indicators of very low-grade metamorphism that have been drawn in the various sections of this review. Considering the chemical, mineralogical and lithological controls on the evolution of the various indicators, the correlation is much better than might be expected, particularly at the higher diagenetic and anchimetamorphic grades. At the lower grades these compositional controls are apparent in the range of coal ranks in association with which the clay minerals smectite and kaolinite disappear upon burial, and laumontite does appear. The poorest correlation appears to be that between the lower grades of the zeolite facies and coal rank, probably reflecting the fact that the appearance of the diagnostic minerals of the zeolite facies is very strongly subject to chemical effects.

Discrete smectite will disappear from clastic sediments, and laumontite may form at the expense of plagioclase in association with coal ranks of about 0.5% R_0. However, disappearance of discrete smectite, and the appearance of I/S mixed-layers with low contents of expandable layers and stacking order are delayed to much higher coal ranks in tuffaceous and in calcareous rocks; the extended range of grades in which heulandite is replaced by laumontite may be due to the metastable formation of heulandite–clinoptilolite from siliceous volcanic glass.

Kinetic differences in the evolution of mineralogic and organic parameters are apparent in the 'lag' in illitization of smectite in areas of post-kinematic magmatic heating.

The correlation between illite 'crystallinity', pumpellyitic facies and coal rank in the anchimetamorphic range is rather consistent: the approximate equivalence of the prehnite–pumpellyite facies, anchimetamorphism, and low- and medium-anthracite ranks (ASTM classification) is well established for regional terrains. The major divergence appears to be in terrains that underwent post-kinematic magmatic heating, due to the different kinetics of the evolution of the various parameters, and high-pressure/low-temperature metamorphism. In 'regional' terrains the correlation may be assumed as a 'prima-facie' probability, but not accepted as established fact unless documented by observation of the relevant minerals, mineral assemblages, or coal ranks. In the divergent terrains the correlation is not justified even as a prima-facie assumption. In any case it is confusing, misleading, and—chiefly— unnecessary to refer to anchimetamorphic terrains without prehnite– pumpellyite assemblages as being in the prehnite–pumpellyite facies, when it is quite sufficient and unequivocal to refer to them as 'anchimetamorphic'. It is important to keep the terminology in order (cf. Chapter 1 of this volume).

The temperatures of the onset of anchimetamorphism can be narrowed down by its correlation with the beginning of the prehnite–pumpellyite facies to a range between 200 and 250 °C on the basis of oxygen-isotope, fluid- inclusion, and experimental reversed equilibrium temperatures. However, the calibration with temperatures based on coal rank remains somewhat un-certain due to the lack of agreement as to the role of duration of heating upon coalification: there are strongly opposed schools of thought on this point. It seems to this writer that the effect of time on coalification has been overstressed in the past. It is to be hoped that this uncertainty will be eliminated by further studies on the dependence of the evolution of coal rank on temperature and duration of heating during the next few years.

In summary, the study of the correlation of indicators of very low-grade metamorphism has come of age, its main features having been established. In the coming years the divergences for these main features may be expected to be applied increasingly to the evaluation of chemical, mineralogical, lithological, and kinetic features of very low-grade metamorphism.

Acknowledgements

The writer would like to thank Doug Robinson for inviting him to the conference "Diagenesis and Low-Temperature Metamorphism" at the University of Bristol in April 1984, where a preliminary version of this paper was presented. He thanks the editor of this volume, Martin Frey, for helpful suggestions and for his meticulous editing of the manuscript.

Appendix: Effect of sample preparation on the measured 10Å peak width of illite (illite 'crystallinity')

HANAN J. KISCH and MARTIN FREY

Depending on instrumental (diffractometer) settings, different peak widths are obtained. Kisch (1983, Table 5-IV) has compared the various scales in use, and has converted them to angular values ($°2\theta$), the use of which he recommended, and which is increasingly being adopted.

However, there is a wide variation in peak-width values as a result of differences in preparation of the X-rayed samples as discussed below.

A1   Disaggregation methods

Subsequent to grinding of the samples, some authors (Foscolos and Kodama, 1974; Merriman and Roberts, 1985, p. 307; Kemp et al., 1985, p. 341; Robinson and Bevins, 1986, p. 103) routinely use ultrasonic disaggregation for various periods; most laboratories, including Neuchâtel (Kübler), Strasbourg (Dunoyer), Bochum–Göttingen (Weber) and Beer-Sheva (the first writer), avoid it.

Weber (1972) recommends disaggregation after gentle grinding (to about 1 mm), followed by treatment with H_2O_2 and washing with ammonia solution, avoiding the use of an ultrasonic cleaner (see also Teichmüller et al., 1979, pp. 210–211). Toselli and Weber (1982, p. 194) have shown that ultrasonic disaggregation causes an increase in 'crystallinity' (decrease in peak width) relative to dispersion with ammonia solution—presumably due to reduction of more crystalline coarser clastic mica grains. Frey (unpublished data), on the other hand, observed no variation in 'crystallinity' after ultrasonic treatment up to 30 minutes for an illite 0.1–2 μm fraction from the low-grade anchizone.

A2   Effect of acid treatment

An acid treatment, which may be indispensable for the enrichment of illite in carbonate-rich sediments, leads to an improved 'crystallinity index'. This is especially true for diagenetic illites containing expandable layers (Kübler, 1984, p. 577; see also Krumm, 1984, p. 230).

A3   Methods of slide preparation

The most common method of slide preparation is pipetting on glass slides (Dunoyer, 1969; Kisch, 1980a, b).

Weber (1972) has studied the 'crystallinity' of illite as measured in 'thin' and 'thick' sedimentation slides (respectively 1.5 and 7.5 mg/cm^2 of clay mineral). The thick slides show half-height peak widths up to 50% greater (on the average, by 20–25%). In 'diagenetic' samples of the wells Münsterland 1 and Alfhausen Z1, the 10 Å peaks are up to 100% broader in thick than in thin slides (Teichmüller *et al.*, 1979). Weber (1972, pp. 271–272) ascribes this peak broadening in the thick sedimentation samples to more extensive grain-size fractionation during sedimentation, the finest fraction being concentrated in the top layer as a result of longer settling and drying periods. Such differential settling during sedimentation, or centrifuging on to glass slides is also apparent from the marked increase in the quartz/illite ratio in the X-ray diffraction trace of the bottom surface as compared with the top surface of preparations (Hosterman and Whitlow, 1983); use of the suction method after Kinter and Diamond (1956) minimizes or eliminates this differential settling.

Brime (1980) compared peak width measured by three methods of slide preparation. She found that smear mounts usually show somewhat broader peaks than sedimented slides. As a result of their greater thickness, smear mounts also show much higher I_{002}/I_{001} intensity ratios than pipetted slides (Brime, 1980; Robinson and Bevins, 1986; Kisch, unpublished results), confirming the doubtful value of this parameter (cf. Kisch, 1983, pp. 352–354).

A4   Differences between grain-size fractions

Illite 'crystallinity' is also dependent on the grain size of a sample (e.g. Weber, 1972; Teichmüller *et al.*, 1979; Brauckmann, 1984; Weaver, 1984, p. 77). In most studies the so-called clay size fraction $< 2 \mu$m is used. Because very small grain sizes ($< \sim 0.01 \mu$m) cause line broadening (Brindley, 1980, p.128), Weber (1972) proposed working with the 2–6 μm fraction. A disadvantage of using this coarser grain-size fraction is, however, that it may contain a higher percentage of detrital illite with inherited better 'crystallinity'. To master these difficulties the second author concentrates the 0.1–2 μm size fraction whereby the fraction $< 0.1 \mu$m is removed with the aid of a millipore filter.

Weber (1972) has shown appreciable differences between the peak widths of the $< 2 \mu$m fractions (thin slides) and the 2–6 μm fractions. For Hb_{rel} values of less than 130 there is no significant difference between the 10 Å peak widths as measured or rock slabs, 2–6 μm fractions, and $< 2 \mu$m fractions (thin slides) (Weber, 1972, Figure 5). However, the $< 2 \mu$m fractions (thick slides) continue to show up to 50% (average 20%) broader peaks. Thus, Teichmüller *et al.* (1979) show that even in the high-grade anchizone to epizone (the Soest-Erwitte 1/1a well) the 10 Å peak widths of the $< 2 \mu$m fractions (thick slides) remain in the order of 10 to 40% broader than in the corresponding 2–6 μm fractions; considering that the thick slides have peaks wider by an average of 20–25% than the thin slides, the peak widths of the latter at this grade would be very similar to those of the 2–6 μm fractions.

Teichmüller *et al.* (1979) show that peak widths in the diagenetic zone (e.g. in the upper part of the Münsterland 1 borehole) are between 40% to over 200% broader in $< 2\,\mu m$ fractions (thin slides) than in the 2–$6\,\mu m$ fractions, and up to 60% (generally not more than 20%) broader at the diagenesis-anchimetamorphism transition (e.g. in the deepest part of the Münsterland 1 borehole and the Alfhausen Z1 borehole).

Kemp *et al.* (1985) report Hb_{rel} values up to 20% higher in the $< 2\,\mu m$ than in the 2–$6\,\mu m$ fractions for the transition diagenesis–anchimetamorphism, but essentially the same values in the mid-anchizone.

Árkai (1983, Figures 2–4; Árkai *et al.*, 1981, Figure 5/c–d) also places a given degree of metamorphism (e.g. his limit between the laumontite-prehnite–quartz facies and the pumpellyite–prehnite–quartz facies) at a somewhat higher peak width ($0.38°\ \Delta 2\theta$ or $Hb_{rel} = 160$) in whole rock and solution-residue samples than in the $< 2\,\mu m$ sedimented fractions ($0.34°\ \Delta 2\theta$ or $Hb_{rel} = 150$–155); at higher degrees of metamorphism, e.g. for the anchi-epizone boundary, there is virtually no difference.

A5   Effect of cation saturation

Several authors have compared the same samples after saturation with K- and Mg-ions.

Van Biljon and Bensch (1970) and Kisch (1980a, 1980b) found narrower peaks upon K- than upon Mg-saturation; in samples from the diagenetic zone and the low-grade anchizone ($B_{001} > 0.30°\ \Delta 2\theta$) of Jämtland differences of 0.05–$0.10°\ \Delta 2\theta$ between the fractions were found (Kisch, 1980a); in samples from Jämtland and the Helvetides the differences in B in most samples were less than 20% (Kisch, 1980a, Table 2 and Figures 3 and 4; 1980b, Figures 2 and 3). However, in the diagenetic range peaks measured by authors that routinely use K-saturation rather than saturation with Mg or Ca (e.g. Foscolos and Kodama, 1974; Foscolos and Stott, 1975, p. 7) peaks may be expected to be somewhat narrower than those of authors using saturation with Mg or Ca. The first writer has for several years routinely used both K- and Mg-saturation, but has lately discontinued routine K-saturation.

A6   Effect of ethylene glycol (EG)

The illites of the diagenetic and the low-grade anchimetamorphic zone commonly show marked asymmetry of the 10Å diffraction peak, with a low-angle (high-spacing) tail (Kisch, 1980a, 1980b; Ivanova *et al.*, 1980; Árkai and Tóth, 1983). The narrowing of these peaks upon EG solvation has been reported by several authors, and appears to be almost ubiquitous; this narrowing is usually accompanied by an increase in symmetry of the peaks (Triplehorn, 1970; Kisch, 1980a, 1980b, 1981).

Determinations of illite 'crystallinity' are usually carried out on air-dried

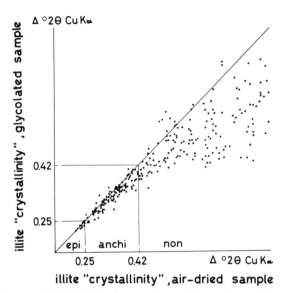

Figure A.1 Illite 'crystallinity' (Kübler index) of air-dried samples *v.* illite 'crystallinity' of glycolated samples. Data from Breitschmid (1980) and Frey *et al.* (1980).

samples. In the presence of expandable layers the humidity will affect 'crystallinity' (Kübler, 1968, p. 390; Kübler, 1984, p. 577). For this reason some authors (e.g. Kisch, 1980*b*; Frey *et al.*, 1980; Brauckmann, 1984) give peak widths both air-dried and after EG solvation.

Percentages of narrowing upon EG solvation range up to 30%, even in the higher-grade part of the diagenetic range (Brime and Perez-Estaun, 1980; Rohde, 1980), or even somewhat more (e.g. in the upper part of the Münsterland 1 well, Teichmüller *et al.*, 1979, Figure 15; Kisch, 1980*a*, 1980*b*), but by the diagenesis–anchimetamorphism transition rarely more than 20% (Kisch, 1980*b*; Teichmüller *et al.*, Figure 15—Münsterland 1 borehole below 4655 m depth), and in the anchizone usually less than 10% (Kisch, 1980*a*, Table 3; Brime and Perez-Estaun, 1980; Frey *et al.*, 1980; Dandois, 1981; Brauckmann, 1984, Table A6; Ahrendt *et al.*, 1977; Toselli and Toselli, 1982, see Figure A.1); the latter two papers (see also Toselli and Weber, 1982) show occasional peak broadening upon glycolation in the anchizone, particularly near the anchi-epizone transition.

The 2–6 μm fraction shows negligible changes upon glycolation: exceptionally narrowing by 10% occurs, and in some samples some broadening (Brauckmann, 1984, Table A6).

References

Abbas, M. (1974) Métamorphisme des argiles dans le Rhétien des Alpes sud-occidentales—étude minéralogique et géochimique. Thèse Doct. Spec. (3ème Cycle), Univ. Louis Pasteur Strasbourg.

Adams, C.J., Bishop, D.G., and Gabites, J.E. (1985) Potassium–argon studies of low-grade, progressively metamorphosed greywacke sequence, Dansey Pass, South Island, New Zealand. *J. geol. Soc. London* **142**(2), 339–349.

Ahn, J.H. and Peacor, D.R. (1985) Transmission electron microsopic study of diagenetic chlorite in Gulf Coast argillaceous sediments. *Clays and Clay Minerals* **33**(3), 228–236.

Ahn, J.H. and Peacor, D.R. (1985) Transmission electron microscopic study of diagenetic chlorite implications for the origin and structure of "fundamental particles". *Clays and Clay Minerals* **34**(2), 180–186.

Ahrendt, H., Behr, H.J., Clauer, N., Hunziker, J.C., Porada, H. and Weber, K. (1983b) The Northern Branch: Depositional development and timing of the structural and metamorphic evolution within the framework of the Damara Orogen. In *Intracontinental Fold Belts*, eds. Martin, H. and Eder, F.W., Springer, Berlin, 723–744.

Ahrendt, H., Clauer, N., Hunziker, J.C. and Weber, K. (1983a) Migration of folding and metamorphism in the Rheinisches Schiefergebirge deduced from K–Ar and Rb–Sr age determinations. In *Intracontinental Fold Belts*, eds. Martin, H. and Eder, F.W., Springer, Berlin, 323–338.

Ahrendt, H., Hunziker, J.C., and Weber, K. (1977) Age and degree of metamorphism and time of nappe emplacement along the southern margin of the Damara orogen/Namibia (SW-Africa). *Geol. Rdsch.* **67**, 719–742.

Ahrendt, H., Hunziker, J.C., and Weber, K. (1978) K/Ar-Altersbestimmungen an schwach-metamorphen Gesteinen des Rheinischen Schiefergebirges. *Z. dtsch. geol. Ges.* **129**, 229–247.

Aldahan, A.A. and Morad, S. (1986) Mineralogy and chemistry of diagenetic clay minerals in Proterozoic sandstones from Sweden. *Amer. J. Sci.* **286**(1), 29–80.

Aldridge, R.J. (1984) Thermal metamorphism of the Silurian strata of the Oslo region, assessed by conodont colour. *Geol. Mag.* **121**, 347–349.

Aldridge, R.J. (1986) Conodont palaeobiogeography and thermal maturation in the Caledonides. *J. Geol. Soc. London* **143**, 177–184.

Allamagny, P. (1976) *Encyclopédie des Gaz: L'air liquide.* Elsevier, Amsterdam.

Alpern, B. (1970) Classification petrographique des constituants organiques fossiles des roches sédimentaires. *Rev. Inst. franç. Pétrole* **25**, 1233–1266.

Alpern, B. (1980) Pétrographie du kérogène. In *Kerogen*, ed. Durand, B., Technip, Paris, 339–371.

Alpern, B. and Lemos de Sousa, M.J. (1970) Sur le pouvoir réflecteur de la vitrinité et de la fusinité des houilles. *C.R. Acad. Sci. Paris* **271**, 956–959.

Alt, J.C. (1985) Alteration of the upper oceanic crust: mineralogy and processes in deep sea drilling project hole 504B, Leg 83. In *Initial Report Deep Sea Drilling Project*, **83**, eds. Anderson, R.N., Honnorez, J. et al., 217–247.

Alt, J.C. and Honnorez, J. (1984) Alteration of the upper oceanic crust, DSDP site 417: mineralogy and chemistry. *Contrib. Miner. Petrol.* **87**, 149–169.

Althaus, E. and Johannes, W. (1969) Experimental metamorphism of NaCl-bearing aqueous solutions by reaction with silicates. *Amer. J. Sci.* **267**, 87–98.

Ammosov, I.I. (1968) Coal organic matter as a parameter of the degree of sedimentary rock lithification. *Int. Geol. Congr., Rep. 23rd Sess. Czechoslovakia 1968, Sect. 11 'Origin of Coal Deposits'*, 23–30.

Ammosov, I.I., Babashkin, B.G., and Sharkova, L.S. (1975) Bituminite of Lower Cambrian deposits in the Irkutsk oil and gas region (in Russian). In *Paleotemperatury zon Nefteobrazovanya*, ed. Yeremin, I.V., Nauka Press, Moscow, 25–59 (quoted by Price, 1983).

Ammosov, I.I., Gorshkov, V.J., Greshnikov, N.P., and Kalmykov, G.S. (1977) Paleo-geothermic criteria of the location of petroleum deposits (in Russian). Nedra Press, Moscow (quoted by Price, 1983).

Aparicio, A. and Galán, E. (1980) Las caracteristicas del metamorfismo Hercinico de bajo y muy bajo grado en el sector oriental del Sistema Central. *Estudios geol.* **36**, 75–84.

Aprahamian, J. (1974) La cristallinité de l'illite et les minéraux argileux en bordure des massifs cristallins externes de Belledonne et du Pelvoux (Variations et relations possibles avec des événements tectoniques et métamorphiques alpins). *Géol. alp.* **50**, 3–15.

Aprahamian, J. and Pairis, J.-L. (1981) Very low grade metamorphism with a reverse gradient induced by an overthrust in Haute-Savoie (France). In *Thrust and Nappe Tectonics*, eds. McClay, K.R. and Price, N.J., Geol. Soc. London/Blackwell, Oxford, 159–165.

Aprahamian, J., Pairis, B., and Pairis, J.L. (1975) Nature des minéraux argileux et cristallinité des illites dans le massif de Platé et le revers occidental des Aiguilles Rouges: implications possibles d'un point de vue sédimentaire, structural, et métamorphique. *Ann. Centre. Univ. Savoie* **2**, 95–119.

Apted, M.J. and Liou, J.G. (1983) Phase relations among greenschist, epidote amphibolite and amphibolite in a basaltic system. *Amer. J. Sci.*, **283-A**, 328–354.

Árkai, P. (1973) Pumpellyite-prehnite-quartz facies Alpine metamorphism in the Middle Triassic volcanogenic–sedimentary sequence of the Bükk Mountains, northeast Hungary. *Acta. Geol. Acad. Sci. Hung.* **17**, 67–83.

Árkai, P. (1977) Low-grade metamorphism of Paleozoic sedimentary formations of the Szendrö Mountains (NE-Hungary). *Acta. Geol. Acad. Sci. Hung.* **21**, 53–80.

Árkai, P. (1983) Very low- and low-grade Alpine regional metamorphism of the Paleozoic and Mesozoic formations of the Bükkium, NE-Hungary. *Acta Geol. Hung.* **26**(1–2), 83–101.

Árkai, P., Horváth, Z.A., and Tóth, M. (1981) Transitional very low- and low-grade regional metamorphism of the Paleozoic formations, Uppony Mountains, NE-Hungary: mineral assemblages, illite–crystallinity, b_0 and vitrinite reflectance data. *Acta Geol. Acad. Sci. Hung.* **24**, 265–294.

Árkai, P. and Tóth, M.N. (1983) Illite crystallinity: combined effects of domain size and lattice distortion. *Acta Geol. Hung.* **26**(3–4), 341–358.

Árkai, P. and Vizcián, I. (1975) Transformation of clay minerals in the sedimentary rocks (in Hungarian; Eng. summ.) *Osztályának Közleményei* **8**(3–4), 373–381.

Arnold, A. and Jäger, E. (1965) Rb-Sr Altersbestimmungen an Glimmern im Grenzbereich zwischen voralpinen Alterswerten und alpiner Verjüngung der Biotite. *Eclogae geol. Helv.* **58**, 369–390.

Artru, Ph., Dunoyer de Segonzac, G., Combaz, A., and Giraud, A. (1969) Variations d'origine sédimentaire et évolution diagénétique des caractères palynologiques et géochimiques des Terres Noires jurassiques en direction de l'arc alpin. *Bull. Centre Rech. Pau—SNPA* **3**, 357–376.

Ashworth, J.R. and Evirgen, M.M. (1984) Mineral chemistry of regional chloritoid assemblages in the chlorite zone, Lycian Nappes, South-West Turkey. *Miner. Mag.* **48**, 159–165.

Bailey, S.W. (1980) Structures of layer silicates. In *Crystal Structures of Clay Minerals and their X-ray Identification*, eds. Brindley, G.W. and Brown, G., Min. Soc. London, 2–123.

Bailey, S.W. (1982) Nomenclature for regular interstratifications. *Amer. Miner.* **67**, 394–398.

Bailey, S.W. (1984) Classification and structures of the micas. In *Micas*, ed. Bailey, S.W., Rev. Mineral. **13**, Mineralogical Society of America, 1–12.

Bailey, S.W. and Brown, B.E. (1962) Chlorite polytypism: I. Regular and semirandom one-layer structures. *Amer. Miner.* **47**, 819–850.

Baltatzis, E.G. and Katagas, C.G. (1984) The pumpellyite–actinolite and contiguous facies in part of the phyllite–quartzite series, central northern Peloponnesus, Greece. *J. metamorphic Geol.* **2**(4), 349–363.

Banno, S. (1986) The high pressure metamorphic belts in Japan: a review. *Geol. Soc. Amer. Memoir* **164**, 365–374.

Barker, C. and Smith, M.P. (1986) Mass spectrometric determination of gases in individual fluid inclusions in natural minerals. *Anal. Chem.* **58**, 1330–1333.

Barker, C.E. (1979) Vitrinite reflectance geothermometry in the Cerro Prieto geothermal system, Baja California, Mexico. Unpubl. M.Sc. Thesis, Univ. California, Riverside.

Barker, C.E. (1983) Influence of time on metamorphism of sedimentary organic matter in liquid-dominated geothermal systems, western North America. *Geology* **11**, 384–388.

Barker, C.E. and Elders, W.A. (1981) Vitrinite reflectance geothermometry and apparent heating duration in the Cerro Prieto geothermal field. *Geothermics* **10**, 207–223.

Barker, C.E. and Pawlewicz, M.J. (1986) The correlation of vitrinite reflectance with maximum temperature in humic organic matter. In *Paleogeothermics*, eds. Buntebarth, G. and Stegena, L., Springer Verlag, Berlin, 79–93.

Barker, C.G. (1966) Volatile content of rocks and minerals with special reference to fluid inclusions (abstr.). *Geol. Soc. Amer. Spec. Paper* **101**, 10.

Barlier, J. (1974) Recherches paléothermométriques dans le domaine des Terres Noires subalpines méridionales. Thèse Doct. 3e Cycle, Univ. Paris-Sud (Centre d'Orsay).

Barlier, J., Ragot, J.-P., and Touray, J.-C. (1974) L'évolution des Terres Noires subalpines méridionales d'après l'analyse minéralogique des argiles et la réflectométrie des particules carbonées. *Bull. B.R.G.M. (2me sér.), sect. II* (6), 533–548.

Barlier, J., Touray, J.-C. and Guilhaumou, N. (1973) Des témoins d'une genèse d'hydrocarbures liquides et gazeux inclus dans des quartz en gisement dans la nappe de l'Autapie (Alpes-de-Haute-Provence). *C.R. Acad. Sci. Paris* **277**, Série D, 2297–2300.

Barnes, R.P. and Andrews, J.R. (1981) Pumpellyite-actinolite grade regional metamorphism in south Cornwall. *Proc. Ussher Soc.* **5**(2), 139–146.

Bartenstein, H. and Teichmüller, R. (1974) Inkohlungsuntersuchungen, ein Schlüssel zur Prospektierung von paläozoischen Kohlenwasserstoff–Lagerstätten? *Fortschr. Geol. Rheinld. Westf.* **24**, 129–160.

Bartenstein, H., Teichmüller, M., and Teichmüller, R. (1971) Die Umwandlung der organischen Substanz im Dach des Bramscher Massivs. *Fortschr. Geol. Rheinld. Westf.* **18**, 501–538.

Bates, R.L. and Jackson, J.A., eds. (1980) *Glossary of Geology*. 2nd edn., Am. Geol. Inst., Falls Church, Virginia.

Bauer, R.L. (1983) Low-grade metamorphism in the Heritage Range of the Ellsworth Mountains, West Antarctica. In *Antarctic Earth Science, International Symposium Adelaide 1982*, Cambridge Univ., Cambridge, 256–260.

Bayliss, P., Loughnan, F.C. and Standard, J.C. (1965) Dickite in the Hawkesbury sandstone of the Sydney Basin, Australia. *Amer. Miner.* **50**, 418–425.

Bennett, J.N. and Grant, J.N. (1980) Analysis of fluid inclusions using a pulsed laser microprobe. *Miner. Mag.* **43**, 945–947.

Bergström, S.M. (1980) Conodonts as paleotemperature tools in Ordovician rocks of the Caledonides and adjacent areas in Scandinavia and the British Isles. *Geol. Fören. Stockholm Förh.* **102**, 377–392.

Berman, R.G., Engi, M. and Brown, T.H. (1985) Optimization of standard state properties and activity models for minerals: methodology and application to an 11 component system. *Codata Symposium on Chemical Thermodynamics and Thermophysical Properties Databases*, Paris, August 1985, 166–173.

Bertrand, R., Humbert, L., Achab, A., Calise, G., Chagnon, A., Héroux, Y., and Globensky, Y. (1983) Recristallisation des calcaires micritiques en fonction de la maturation thermique dans les Basses-Terres du Saint-Laurent du Québec. *Can. J. Earth Sci.* **20**(1), 66–83.

Beuf, S., Biju-Duval, B., Stevaux, J. and Kulbicki, G. (1966) Ampleur des glaciations 'siluriennes' au Sahara: leurs influences et leurs conséquences sur la sédimentation. *Rev. Inst. Franç. Petrol.* **XXI/3**, 363–381.

Bevins, R.E. (1978) Pumpellyite-bearing basic igneous rocks from the Lower Ordovician of North Pembrokeshire, Wales. *Miner. Mag.* **42**, 81–83.

Bevins, R.E. (1985) Pumpellyite-dominated metadomain alteration at Builth Wells, Wales— evidence for a fossil submarine hydrothermal system? *Miner. Mag.* **49**(3), 451–456.

Bevins, R.E. and Rowbotham, G. (1983) Low-grade metamorphism within the Welsh sector of the paratectonic Caledonides. *Geol. J.* **18**, 141–167.

Bevins, R.E., Oliver, G.J.H., and Thomas, L.J. (1985) Low-grade metamorphism in the paratectonic Caledonides of the British Isles. In *The Tectonic Evolution of the Caledonide-Appalachian Orogen*, ed. Gayer, R.A., Friedr. Vieweg & Sohn, Braunschweig, 57–79.

Beutner, E.C. (1978) Slaty cleavage and related strain in Martinsburg slate, Delaware Water Gap, New Jersey. *Amer. J. Sci.* **278**, 1–23.

Beutner, E.C. and Diegel, F.A. (1985) Determination of fold kinematics from syntectonic fibers in pressure shadows, Martinsburg slate, New Jersey. *Amer. J. Sci.* **285**(1), 16–50.

Beutner, E.C., Jancin, M.D., and Simon, R.W. (1977) Dewatering origin of cleavage in light of deformed calcite veins and clastic dikes in Martinsburg slate, Delaware Water Gap, New Jersey. *Geology* **5**, 118–122.

Bickle, M. and Powell, R. (1977) Calcite-dolomite geothermometry for iron-bearing carbonates. *Contrib. Miner. Petrol.* **59**, 281–292.

Biljon, W.J. van and Bensch, J.J. (1970) The "crystallinity" of illite as a measure of contract metamorphism in mudstone of the Karroo System, South Africa. In *Second Gondwana Symposium, Proceedings and Papers.* CSIR, Pretoria, South Africa, 451–453.

Birch, F. and LeComte, P. (1960) Temperature-pressure plane for albite composition. *Amer. J. Sci.* **258**, 209–217.

Bird, D.K., and Helgeson, H.C. (1981) Chemical interaction of aqueous solutions with epidote-feldspar mineral assemblages in geologic systems. II. Equilibrium constraints in metamorphic/geothermal processes. *Amer. J. Sci.* **281**, 576–614

Bird, D.K., Schiffman, P., Elders, W.A. Williams, A.E., and McDowell, D.D. (1984). Calc-silicate mineralization in active geothermal systems. *Econ. Geol.* **769**, 671–695.

Biscaye, P.E. (1965) Mineralogy and sedimentation of recent deep-sea clay in the Atlantic Ocean and adjacent seas and oceans. *Bull. Geol. Soc. Amer.* **76**, 803–832.

Bishop, D.G. (1972) Progressive metamorphism from prehnite-pumpellyite to greenschist facies in the Dansey Pass area, Otago, New Zealand. *Bull. Geol. Soc. Amer.* **83**, 3177–3198.

Black, L.P., Bell, T.H., Rubenach, M.J. and Withnall, J.W. (1979) Geochronology of discrete structural-metamorphic events in a multiply deformed Precambrian terrain. *Tectonophysics* **54**, 103–137.

Black, P.M. (1973) Mineralogy of New Caledonian metamorphic rocks. I. Garnets from the Ouégoa District. *Contrib. Miner. Petrol.* **38**, 221–235.

Black, P.M. (1975) Mineralogy of New Caledonian metamorphic rocks. IV. Sheet silicates from the Ouégoa District. *Contrib. Miner. Petrol.* **49**, 269–284.

Black, P.M. (1977) Regional high-pressure metamorphism in New Caledonia: phase equilibria in the Ouégoa District. *Tectonophysics* **43**, 89–107.

Blake, M.C., Jr., and Jones, D.L. (1981) The Franciscan assemblage and related rocks in northern California, a reinterpretation. In *The Geotectonic Development of California*, ed. Ernst, W.G., Prentice Hall, New Jersey, 306–328.

Blake, M.C., Irwin, W.P., Coleman, R.G. (1967) Upside-down metamorphic zonation, blueschist facies, along a regional thrust in California and Oregon. *US Geol. Survey Prof. Paper* 575–C, 1–9.

Blank, P. and Seifert, W. (1976) Zur Untersuchung diagenetischer Tonmineralbildungen und deren experimentelle Modellierung. *Z. angew. Geol.* **22**, 560–564.

Blount, Ch. W., Wenger, L.M., Tarullo, M. and Price, L.C. (1980) Methane solubility in aqueous NaCl solutions at elevated temperatures and pressures. *Geol. Soc. Amer. Abstr. Programs* **12**, 276.

Bocquet, J., Delaloye, M., Hunziker, J.C. and Krummenacher, D. (1974) K–Ar and Rb–Sr dating of blue amphiboles, micas, and associated minerals from the western Alps. *Contrib. Miner. Petrol.* **47**, 7–26.

Bocquet-Desmons, J. (1974) Etudes minéralogiques et pétrologiques sur les métamorphismes d'âge alpin dans les alpes françaises. *Thèse, Univ. Grenoble.*

Bodnar, R.J. and Bethke, P.M. (1980) Systematics of 'stretching' of fluid inclusions as a result of overheating (abstr.). *Trans. Amer. Geophys. Union* **61**, 393.

Bodnar, R.J. and Bethke, P.M. (1984) Systematics of stretching of fluid inclusions I: Fluorite and sphalerite at 1 atmosphere confining pressure. *Econ. Geol.* **79**, 141–161.

Boles, J.R. and Coombs, D.S. (1975) Mineral reactions in zeolitic Triassic tuff, Hokonui Hills, New Zealand. *Bull. Geol. Soc. Amer.* **86**, 163–173.

Boles, J.R. and Coombs, D.S. (1977) Zeolite facies alteration of sandstones in the Southland Syncline, New Zealand. *Amer. J. Sci.* **277**, 982–1012.

Boles, J.R. and Franks, S.G. (1979) Clay diagenesis in Wilcox sandstones of southwest Texas: implications of smectite diagenesis on sandstone cementation. *J. sediment. Petrol.* **49**, 55–70.

Boles, J.R. and Landis, C.A. (1984) Jurassic sedimentary mélange and associated facies, Baja California, Mexico. *Bull. Geol. Soc. Amer.* **95**, 513–521.

Bombicci, L. (1898) Sulla probabilità che talune anomalie di forma nei cristalli dipendano da durevoli movimenti negli spazi naturalmente cristalligeni. *Mem. Accad. Sci. Ist. Bologna* V/VII, 761–780.

Bonhomme, M.G., Baubron, J.C. and Jebrak, M. (1987) Mineralogie, géochimie, terres rares et âge K–Ar des argiles associées aux minéralisations filoniennes. *Isotope Geosciences* (in press).

Bonijoly Roussel, M. (1980) Evolution des substances carbonées naturelles. Formation de graphite. Thèse, Univ. Orléans.

Bonijoly, M., Oberlin, M. and Oberlin, A. (1982) A possible mechanism for natural graphite formation. *Int. J. Coal Geol.* **1**, 283–312.

Borisenko, A.S. (1977) Study of the salt composition of solutions of gas-liquid inclusions in minerals by the cryometric method. *Soviet Geol. and Geophys.* **18**, 11–19.

Bostick, N.H. (1971) Thermal alteration of clastic organic particles as an indicator of contact and burial metamorphism in sedimentary rocks. *Geosci. and Man* **3**, 83–92.

Bostick, N.H. (1973) Time as a factor in thermal metamorphism of phytoclasts (coaly particles). *Congr. int. Stratigr. Geol. Carbonifère, 7me, Krefeld 1971, C.R.* **2**, 183–193.

Bostick, N.H. (1974) Phytoclasts as indicators of thermal metamorphism, Franciscan assemblage and Great Valley sequence (Upper Mesozoic), California. In *Carbonaceous Materials as Indicators of Metamorphism*, ed. Dutcher, R.R., Hacquebard, P.A., Schopf, J.M., and Simon, J.A., *Geol. Soc. Amer. Spec. Paper* **153**, 1–17.

Bostick, N.H. (1979) Microscopic measurement of the level of catagenesis of solid organic matter in sedimentary rocks to aid exploration for petroleum and to determine former burial temperatures—a review. In *Aspects of Diagenesis*, eds. Scholle, P.A. and Schluger, P.R., Soc. Econ. Paleont. Miner., Spec. Publ. **26**, 17–43.

Bostick, N.H. (1984) Comment on 'Influence of time on metamorphism of sedimentary organic matter in liquid-dominated geothermal systems, western North America'. *Geology* **12**(11), 689–691.

Bostick, N.H., Cashman, S.M., McCulloh, T.H., and Waddell, C.T. (1978) Gradients of vitrinite reflectance and present temperature in the Los Angeles and Ventura basins, California. In *Low Temperature Metamorphism of Kerogen and Clay Minerals*, ed. Oltz, D.F., Pacific Section, Soc. Econ. Paleontol. Mineral., Los Angeles, CA, 65–69.

Botscharnikowa, A.I. (1965) Über die Metamorphose der Kohle und der Nebengesteine im Petschora-Kohlenrevier (in Russian), *Materialy po Geol. i poleznym Iskopajemym ssevero-wost. jewrop. Tschasti SSSR* **4**, 39–43 (abstract in *Zentralbl. Geol. Paläontol. I*, **1966**, 2259).

Boudier, F. and Nicolas, A. (1968) Découverte de chloritoide dans les schistes ardoisiers d'Angers. *Bull. Soc. fr. Minéral. Cristallogr.* **91**, 92–94.

Bouquillon, A., Chamley, H., Debrabant, P., and Piqué, A. (1985) Étude minéralogique et géochimique des forages de Jeumont et Épinoy (Paléozoïque du Nord de la France). *Ann. Soc. géol. Nord* **54**, 167–178.

Bowers, T.S. and Helgeson, H.C. (1983) Calculation of the thermodynamic and geochemical consequences of nonideal mixing in the system H_2O-CO_2-NaCl on phase relations in geologic systems: metamorphic equilibria at high pressures and temperatures. *Amer. Miner.* **68**, 1059–1075.

Bozzo, A.T., Chen, H.S., Kass, J.R. and Barduhn, A.J. (1973) The properties of the hydrates of chlorine and carbon dioxide. *Int. Symp. Fresh Water from the Sea* **3**, 437–451.

Brady, G.A. and Griffiths, H.H. (1968) Properties of anthracite from the Bottom Ross bed. *US Bur. Mines Rept. Invest.* **7086**.

Brand, R. (1980) Die niedriggradige Metamorphose einer Diabas-Assoziation im Gebiet von Berg/Frankenwald. *Neues. Jb. Miner. Abh.* **139**, 82–101.

Brandt, S.B. and Voronovsky, S.N. (1964) Dehydration and diffusion of radiogenic argon from micas. *Akad. Nauk. USSR* **11**.

Brauckmann, F.J. (1984) Hochdiagenese im Muschelkalk der Massive von Bramsche und Vlotho. *Bochumer geol. geotech. Arb.* **14**.

Brauckmann, F.J. and Füchtbauer, H. (1983) Alterations of Cretaceous siltstones and sandstones near basalt contacts (Nûgssuaq, Greenland). *Sediment. Geol.* **35**, 193–213.

Brazier, S., Robinson, D. and Matthews, S.C. (1979) Studies of illite crystallinity in southwest England. Some preliminary results and their geological setting. *Neues Jb. Geol. Paläont. Mh.*, 641–662.

Breitschmid, A. (1982) Diagenese und schwache Metamorphose in den sedimentären Abfolgen der Zentralschweizer Alpen (Vierwaldstätter See, Urirotstock), *Eclogae geol. Helv.* **75**, 331–380.

Bril, H. and Thiry, M. (1976) Le métamorphisme de basse pression anchi à mèsozonal de la région de Bodennec (Finistère): essai methodologique. *C. R. Acad. Sci. Paris. Sér. D*, **283**, 227–230.

Brime, C. (1980) Influencia del modo de preparación de las muestras en la relación I(002)/I(001) de las ilitas. *Breviora Geol. Astúrica* **24** (3–4), 24–28.

Brime, C. (1985) A diagenesis to metamorphism transition in the Hercynian of north-west Spain. *Miner. Mag.* **49**, 481–484.

Brime, C. and Perez-Estaun, A. (1980) La transicion diagenesis-metamorfismo en la region del Cabo Peñas. *Cuadernos do Lab. Xeolóxico de Laxe* **1**, 85–97.

Brindley, G.W. (1980) Order-disorder in clay mineral structures. In *Crystal Structures of Clay Minerals and their X-ray Identification*, eds. Brindley, G.W. and Brown, G., Min. Soc. London, 126–195.

Brindley, G.W. and Wardle, R. (1970) Monoclinic and triclinic forms of pyrophyllite and pyrophyllite anhydride. *Amer. Miner.* **55**, 1259–1272.

Brooks, J.D. and Smith, J.W. (1967) the diagenesis of plant lipids during the formation of coal, petroleum and natural gas. I. Changes in the *n*-paraffin hydrocarbons. *Geochim. Cosmochim. Acta* **31**, 2389–2397.

Brothers, R.N. (1970) Lawsonite-albite schists from northernmost New Caledonia. *Contrib. Miner. Petrol.* **25**, 185–202.

Brothers, R.N. and Yokoyama, K. (1982) Comparison of the high-pressure schist belts of New Caledonia and Sanbagawa, Japan. *Contrib. Miner. Petrol.* **79**, 219–229.

Brown, E.H. (1974) Comparison of the mineralogy and phase relations of blueschists from the North Cascades, Washington and greenschists from Otago, New Zealand, *Bull. Geol. Soc. Amer.* **85**, 333–344.

Brown, E.H. (1975) A petrogenetic grid for reactions producing biotite and other Al–Fe–Mg silicates in the greenschist facies. *J. Petrol.* **16**, 258–271.

Brown, E.H. (1977a) Phase equilibria among pumpellyite, lawsonite, epidote and associated minerals in low-grade metamorphic rocks. *Contrib. Miner. Petrol.* **64**, 123–136.

Brown, E.H. (1977b) The crossite content of Ca-amphibole as a guide to pressure of metamorphism. *J. Petrol.* **18**, 53–72.

Brown, E.H. and O'Neill, J.R. (1982) Oxygen isotope geothermometry and stability of lawsonite and pumpellyite in the Shuksan Suite, North Cascades, Washington. *Contrib. Miner. Petrol.* **80**, 240–244.

Brown, G. and Brindley, G.W. (1980) X-ray diffraction procedures for clay mineral identification. In *Crystal Structures of Clay Minerals and their X-ray Identification*, eds. Brindley, G.W. and Brown, G., Min. Soc. London, 305–359.

Brown, P.E., Essene, E.J., and Kelly, W.C. (1978) Sphalerite geobarometry in the Balmat-Edwards district, New York. *Amer. Miner.* **63**, 250–257.

Browne, P.R. (1978) Hydrothermal alteration in active geothermal fields. *Ann. Rev. Earth Planetary Sci.*, **6**, 229–250.

Bruce, C.H. (1984) Smectite dehydration–its relation to structural development and hydrocarbon accumulation in northern Gulf of Mexico basin. *Bull. Amer. Assoc. Petrol. Geol.* **68**(8), 673–683.

Buntebarth, G. (1978/79) The degree of metamorphism of organic matter in sedimentary rocks as paleo-geothermometer, applied to the Upper Rhine Graben. *Paleophys.* **117**, 83–91.

Buntebarth, G. and Stegena, L. (1986) *Palaeogeothermics.* Lecture Notes in Earth Sciences 5, Springer Verlag, Heidelberg etc.

Buntebarth, G., Koppe, I. and Teichmüller, M. (1982) Palaeogeothermics in the Ruhr Basin. In *Geothermics and Geothermal Energy*, eds. Cermak, V. and Haenel, R., Schweizerbart, Stuttgart, 44–55.

Burne, R.V. and Kantsler, A.J. (1977) Geothermal constraints on the hydrocarbon potential of the Canning Basin, western Australia. *J. Austral. Geol. Geophys.* **2**, 271–288.

Burruss, R.C. (1977) Analysis of fluid inclusions in graphitic metamorphic rocks from Bryant Pond, Maine, and Khtada Lake, British Columbia: thermodynamic basis and geologic interpretation of observed fluid compositions and molar volumes. Ph.D. thesis, Princeton University.

Burruss, R.C. (1981a) Hydrocarbon fluid inclusions in studies of sedimentary diagenesis. In *Short course in Fluid Inclusions: Applications to Petrology*, eds. Hollister, L.S. and Crawford, M.L., Miner. Assoc. Canada, *Short Course Handbook* **6**, 138–156.

Burruss, R.C. (1981b) Analysis of phase equilibria in C–O–H–S fluid inclusions. In *Short Course in Fluid Inclusions: Applications to Petrology*, eds. Hollister, L.S. and Crawford, M.L., Miner. Assoc. Canada, *Short Course Handbook* **6**, 39–74.

Buryanova, E.Z. and Bogdanov, V.V. (1967) Distribution of the authigenic zeolites laumontite and heulandite in the sedimentary rocks of the Tarbagatai coal deposits (in Russian). *Litol Polezn. Iskop.* **1967**(2), 59–68 (transl. in *Lithol. Miner. Resourc.* **1967**(2), 195–202).

Buseck, P.R. and Huang Bo-Jun (1985) Conversion of carbonaceous material to graphite during metamorphism. *Geochim. Cosmochim. Acta* **49**(10), 2003–2016.

Bustin, R.M. (1983) Heating during thrust faulting in the Rocky Mountains: friction or fiction? *Tectonophysics* **95**, 309–328.

Bustin, R.M. (1984) Coalification levels and their significance in the Groundhog coalfield, north-central British Columbia. *Int. J. Coal. Geol.* **4**, 21–44.

Campbell, H.J. (1984) Petrography and metamorphism of the Téremba Group (Permian—Lower Triassic) and the Baie de St.—Vincent Group (Upper Triassic–Lower Jurassic), New Caledonia. *J. Roy. Soc. N. Zeal.* **14**(4), 335–348.

Cann, J.R. (1979) Metamorphism in the ocean crust. In *Deep Sea Drilling Results in the Atlantic Ocean: Ocean Crust*, eds. Talwani, M. *et al.*, Amer. Geophys. Union, Maurice Ewing Series 2, 230–238.

Carlson, W.D. (1983) The polymorphs of $CaCO_3$ and the aragonite-calcite transformation. Reviews in Mineralogy **11**, Mineralogical Society of America, 191–225.

Caron, J.-M. (1977) Lithostratigraphie et tectonique des schistes lustrés dans les Alpes cottiennes septentrionales et en Corse orientale. *Sci. Géol. Mémoire* **48**, 326 pp.

Caron, J.M., Kienast, J.R. and Triboulet, C. (1981) High-pressure low-temperature metamorphism and polyphase alpine deformation at Sant-Andrea di Cotone (eastern Corsica, France). *Tectonophysics* **78**, 419–451.

Cassinis, G., Mattavelli, L. and Morelli, G.L. (1978) Studio petrografico e mineralogico della formazione di Collio nel permiano inferiore dell'alta Val Trompia (prealpi Bresciane). *Mem. 1st. Geol. Miner. Univ. Padova* **32**, 13.

Castano, J.R. and Sparks, D.M. (1974) Interpretation of vitrinite reflectance measurements in sedimentary rocks and determination of burial history using vitrinite reflectance and authigenic minerals. In *Carbonaceous Materials as Indicators of Metamorphism*, ed. Dutcher, R.R., Hacquebard, P.A., Schopf, J.M., and Simon, J.A., *Geol. Soc. Amer. Spec. Paper.* **153**, 31–52.

Cavarretta, G., Gianelli, G., and Puxeddu, M. (1982) Formation of authigenic minerals and their use as indicators of the chemicophysical parameters of the fluid in the Larderello-Travale geothermal field. *Econ. Geol.* **77**, 1071–1084.

Chateauneuf, J.-J., Debelmas, J., Feys, R., Lemoine, M., and Ragot, J.-P. (1973) Premiers résultats d'une étude des charbons jurassiques de la zone briançonnaise. *C.R. Acad. Sci. Paris, Sér. D.* **276**, 1649–1652.

Chatterjee, N.D. (1973) Low-temperature compatibility relations of the assemblage quartz-paragonite and the thermodynamic status of the phase rectorite. *Contrib. Miner. Petrol.* **42**, 259–271.

Chatterjee, N.D., Johannes, W. and Leistner, H. (1984) The system $CaO–Al_2O_3–SiO_2–H_2O$: new phase equilibria data, some calculated phase relations, and their petrological applications. *Contrib. Miner. Petrol.* **88**, 1–13.

Cheilletz, A., Dubessy, J., Kosztolanyi, C., Masson-Peretz, N., Ramboz, C. and Zimmermann, J.-L. (1984) Les fluides moléculaires d'un filon de quartz hydrothermal: comparaison de techniques analytiques ponctuelles et globales, contamination des fluides occlus par des composés carbonés. *Bull. Minér.* **107**, 169–180.

Chen, Chao-Hsia (1984) Determination of lower greenschist facies boundary by K-mica–chlorite crystallinity in the Central Range of Taiwan. *Proc. Geol. Soc. China* **27**, 41–53.

Chennaux, G. and Dunoyer de Segonzac, G. (1967) Etude pétrographique de la pyrophyllite du Silurien et du Dévonien au Sahara. Répartition et origine. *Bull. Serv. Carte géol. Alsace Lorraine* **20**, 195–210.

Chennaux, G., Dunoyer de Segonzac, G. and Petracco, F. (1970) Genèse de la pyrophyllite dans le Paléozoique du Sahara occidental. *C. R. Acad. Sci. Paris, Sér. D*, **270**, 2405–2408.

Cho, M. and Fawcett, J.J. (1986) A kinetic study of clinochlore and its high-temperature equivalent forsterite-cordierite-spinel at 2 kbar water pressure. *Amer. Miner.* **71**, 68–77.

Chopin, C. and Schreyer, W. (1983) Magnesiocarpholite and magnesiochloritoid: two index minerals of pelitic blueschists and their preliminary phase relations in the model system $MgO–Al_2O_3–SiO_2–H_2O$. *Amer. J. Sci.* **283**–A, 72–96.

Cho, M. and Liou, J.G. (1987) Prehnite-pumpellyite to greenschist facies transition in the Karmutsen metabasites, Vancouver Island, British Columbia. *J. Petrol.* **37**.

Cho, M., Liou, J.G. and Maruyama, S. (1987) An experimental investigation of heulandite-laumontite equilibrium at 1000 to 2000 bar P_{fluid} *Contrib. Miner. Petrol.* (in press).

L

Cho, M., Maruyama, S. and Liou, J.G. (1986) Transition from the zeolite to prehnite-pumpellyite facies in the Karmutsen metabasites, Vancouver Island, British Columbia. *J. Petrol.* **27**, 467–494.

Chudaev, O.V. (1978) Occurrence of clay minerals in flyschoid sediments of eastern Kamchatka (in Russian). *Litol. Polezn. Iskop.* **1978** (1), 105–115 (transl. in Lithol. Miner. Resour. **13** (1), 89–97).

Christensen, O.D. (1975) Metamorphism of the Manning Canyon and Chainman formations (abstract). *Geol. Soc. Amer. Abstr. Programs* **7**, 303–304.

Clauer, N. (1976) Géochimie isotopique du strontium des milieux sédimentaires. *Mém. Sci. Géol.* **45**, 256 pp.

Clauer, N. (1979) A new approach to Rb–Sr dating of sedimentary rocks. In *Lectures in Isotope Geology*, eds. Jäger, E. and Hunziker, J.C., Springer, Berlin etc., 30–51.

Clauer, N., Chaudhuri, S. and Massey, K.W. (1986) Relationship between clay minerals and environment by the Sr isotopic composition of their leachates. *Geol. Soc. Amer. Abstr. Programs* **18**, 565–566.

Clauer, N., Ey, F., Gautier-Lafaye, F. (1985) K–Ar dating of different rock types from the Cluff Lake Uranium-ore deposits (Saskatchewan-Canada). In *The Carswell Structure Uranium Deposits, Saskatchewan*, eds. Lainé, R., Alonso, D. and Svab, M., Geol. Assoc. Canada Spec. Paper **29**, 47–53.

Clauer, N. and Lucas, J. (1970) Minéralogie de la fraction fine des schistes de Steige–Vosges septentrionales. *Bull. Groupe fr. Argiles* **22**, 223–235.

Clausen, C.D. and Teichmüller, M. (1982) Die Bedeutung der Graptolithen-Fragmente im Paläozoikum von Soest-Erwitte für Stratigraphie und Inkohlung. *Fortschr. Geol. Rheinld. Westf.* **30**, 145–167.

Cloos, M. (1983) Comparative study of melange matrix and metashales from the Franciscan subduction complex with the basal Great Valley sequence, California. *J. Geol.* **91**, 291–306.

Coleman, R.G. (1977) *Ophiolites: Ancient Oceanic Lithosphere?* Springer Verlag, New York.

Coleman, R.G. and Lee, D.E. (1962) Metamorphic aragonite in the glaucophane schists of Cazadero, California. *Amer. J. Sci.* **260**, 577–595.

Coleman, R.G. and Lee, D.E. (1963) Glaucophane-bearing metamorphic rock types of the Cazadero area, California. *J. Petrol.* **4**, 260–301.

Collins, P.L.F. (1979) Gas-hydrates in CO_2-bearing fluid inclusions and the use of freezing data for estimation of salinity. *Econ. Geol.* **74**, 1435–1444.

Connan, J. (1974) Time-temperature relation in oil genesis. *Bull. Amer. Assoc. Petrol. Geol.* **58**, 2516–2521.

Cook, A.C., Murchison, D.C., and Scott, E. (1972) A British meta-anthracitic coal of Devonian age. *Geol. J.* **8**, 83–94.

Cook, A.C., Murchison, D.C. and Scott, E. (1972) Optically biaxial anthracitic vitrinites. *Fuel* **51**, 180–184.

Coombs, D.S. (1954) The nature and alteration of some Triassic sediments from Southland, New Zealand. *Roy Soc. New Zealand Trans.* **82**, 65–109.

Coombs, D.S. (1960) Lower grade mineral facies in New Zealand. *21st Int. Geol. Congr.*, Copenhagen, 1960, **13**, 339–351.

Coombs, D.S. (1961) Some recent work on the lower grades of metamorphism. *Austr. J. Sci.* **24**, 203–215.

Coombs, D.S., Ellis, A.J., Fyfe, W.S. and Taylor, A.M. (1959) The zeolite facies, with comments on the interpretation of hydrothermal synthesis. *Geochim. Cosmochim. Acta* **17**, 53–107.

Coombs, D.S., Horodyski, R.J., and Naylor, R.S. (1970) Occurrence of prehnite–pumpellyite facies metamorphism in northern Maine. *Amer. J. Sci.* **268**, 142–156.

Coombs, D.S., Kawachi, Y., Houghton, B.F., Hyden, G., Pringle, I.J., and Williams, J.G. (1977) Andradite and andradite-grossular solid solution in very low-grade regionally metamorphosed rocks in southern New Zealand. *Contrib. Miner. Petrol.* **63**, 229–246.

Coombs, D.S., Nakamura, Y. and Vuagnat, M. (1976) Pumpellyite–actinolite facies schists of the Taveyanne Formation near Loeche, Valais, Switzerland. *J. Petrol.* **17**, 440–471.

Cooper, H.M., Snyder, N.H., Abernethy, R.F., Tarpley, E.C., and Swingle, R.J. (1944) Analysis of mine, breaker, and delivered samples. In *Analyses of Pennsylvania Anthracitic Coals, U.S. Dept. Mines, Tech. Paper.* **659**, 40–173.

Cortesogno, L. and Verturelli, G. (1978) Metamorphic evolution of the ophiolite sequences and associated sediments in the Northern Apennines-Voltri Group, Italy. In *Alps, Apennines, Hellenides*, ed. Closs, H., Roeder, D., and Schmidt, K., Schweizerbart, Stuttgart, 253–260.

Crawford, M.L. (1981a) Fluid inclusions in metmorphic rocks—low and medium grade. In *Short Course in Fluid Inclusions: Applications to Petrology*, eds. Hollister, L.S. and Crawford, M.L., Miner. Assoc. Canada, *Short Course Handbook* **6**, 157–181.

Crawford, M.L. (1981b) Phase equilibria in aqeous fluid inclusions. In *Short Course in Fluid Inclusions: Applications to Petrology*, eds. Hollister, L.S. and Crawford, M.L., Miner. Assoc. Canada, *Short Course Handbook* **6**, 75–100.

Crawford, M.L., and Hollister, L.S. (1986) Metamorphic fluids: The evidence from fluid inclusions. In *Fluid–Rock Interactions During Metamorphism*, eds. Walter, J.V. and Wood, B.J., Advances in Physical Geochemistry 5, Springer, New York etc.

Crawford, M.L., Kraus, D.W. and Hollister, L.S. (1979) Petrologic and fluid-inclusion study of calc-silicate rocks, Prince Rupert, British Columbia. *Amer. J. Sci.* **9**, 1135–1159.

Curtis, C.D., Hughes, C.R., Whiteman, J.A. and Whittle, C.K. (1985) Compositional variations within some sedimentary chlorites and some comments on their origin. *Miner. Mag.* **49**, 375–386.

Czolbe, P. (1975) Beitrag zur Bestimmung der Löslichkeit von Gasen und ihrer Sättigungsdrücke als Voraussetzung für die Lösung erdölgeologischer Erkundungsaufgaben. *Freiberger Forschungshefte* C **319**.

Dahme, A. and Mackowsky, M.-Th. (1951) Mikroskopische, chemische und röntgenographische Untersuchungen an Anthraziten. *Brennstoff-Chemie* **32**, 73–77.

Dalrymple, G.B., Lanphere, M.A., Clague, D.A. (1980) Conventional and $^{40}Ar/^{39}Ar$ ages of volcanic rocks from Ojin (Site 430), Nintoku (Site 432) and Suiko (Site 433) seamounts and the chronology of volcanic propagation along the Hawaiian Emperor chain. *Init. Repts Deep Sea Drilling Project* **55**, 659–676.

Damberger, H.H. (1974) Coalification patterns of Pennsylvanian coal basins of the eastern United States. In *Carbonaceous Materials as Indicators of Metamorphism*, ed. Dutcher, R.R., Hacquebard, P.A., Schopf, J.M., and Simon, J.A., *Geol. Soc. Amer. Spec. Paper* **153**, 53–74.

Dandois, Ph. (1981) Diagenèse et métamorphisme des domaines calédonien et hercynien de la vallée de la Meuse entre Charleville-Mézières et Namur (Ardennes franco-belges). *Bull. Soc. belge Géol.* **90**, 299–316.

Dapples, E.C. (1967) The diagenesis of sandstones. In *Diagenesis in Sediments*, eds. Larsen, G. and Chilingar, G.V., Elsevier, Amsterdam, 91–125.

Dapples, E.C. (1979) Diagenesis of sandstones. In *Diagenesis in Sediments and Sedimentary Rocks*, Developments in Sedimentology Vol. 1, eds. Larsen, G. and Chilingar, G.V., Elsevier, Amsterdam, 31–97.

Davis, A. (1978) The reflectance of coal. In *Analytical Methods for Coal and Coal Products*, ed. Karr, C., Academic Press, New York, 27–81.

Davis, A. and Spackman, W. (1964) The role of the cellulosic and lignitic components of wood in artificial coalification. *Fuel* **43**, 215–224.

Davies, V.M. (1983) Alpine structure and metamorphism in a traverse from the Grandes Rousses Massif to the internal Briançonnais. Unpubl. Ph.D. thesis, Univ. Liverpool.

Day, H.W. (1976) A working model of some equilibria in the system alumina-silica-water. *Amer. J. Sci.* **276**, 1254–1284.

Deer, W.A., Howie, R.A. and Zussmann, J. (1962) *Rock Forming Minerals II: Sheet Silicates*. Longman, London.

Deicha, G. (1950) Essais par écrasement de fragments minéraux pour la mise en evidence d'inclusions de gaz sous pression. *Bull. Soc. franç. Miner. Crist.* **73**, 439–445.

Délaloye, M.F. (1966) Contribution a l'étude des silicates de fer sédimentaires, Le gisement de Chamoson (Valais). *Beitr. Geol. Schweiz. Geotechn. Série* **13/9**.

Delhaye, M. and Dhamelincourt, P. (1975) Raman microprobe and microscope with laser excitation. *J. Raman Spectrosc.* **3**, 33–43.

Deloule, E. and Eloy, J.F. (1982) Improvements of laser probe mass spectrometry for the chemical analysis of fluid inclusions in ores. *Chem. Geol.* **37**, 191–202.

Dempster, T.J. (1986) Isotope systematics in minerals: biotite rejuvenation and exchange during Alpine metamorphism. *Earth Planet. Sci. Lett.* **78**, 355–367.

De Swardt, A.M.J. and Roswell, D.M. (1975) Note on the relationship between diagenesis and deformation in the Cape fold-belt. *Trans. Geol. Soc. S. Afr.* **77**, 239–245.

Deutloff, O., Teichmüller, M., Teichmüller, R., and Wolf, M. (1980) Inkohlungs-untersuchungen im Mesozoikum des Massivs von Vlotho (Niedersächsisches Tektogen). *Neues Jb. Geol. Paläont. Mh.* **1980**, 321–341.

Dhamelincourt, P., Beny, J.M., Dubessy, J. and Poty, B. (1979) Analyse d'inclusions fluides à la microsonde MOLE à effet Raman. *Bull. Minér.* **102**, 600–610.

Diamond, L.W. and Wiedenbeck, M. (1986) K–Ar radiometric age of the gold quartz veins at Brusson, Val d'Ayas, N. Italy: Evidence of mid-Oligocene hydrothermal activity in the Northwestern Alps. *Schweiz. Miner. Petrogr. Mitt.* **66**, 385–393.

Dickson, B.L., Gulson, B.L. and Snelling, A.A. (1985) Evaluation of lead isotopic methods for uranium exploration, Koongarra area, Northern territories, Australia. *J. Geochem. Explor.* **24**, 81–102.

Diessel, C.F.K., Brothers, R.N., and Black, P.M. (1978) Coalification and graphitization in high-pressure schists in New Caledonia. *Contrib. Miner. Petrol.* **68**, 63–78.

Diessel, C.F.K. and Offler, R. (1975) Change in physical properties of coalified and graphitised phytoclasts with grade of metamorphism. *Neues Jb. Miner. Mh.* **1975**, 11–26.

Dietrich, H. (1983) Zur Petrologie und Metamorphose des Brennermesozoikums (Stubaier Alpen, Tirol). *Tschermaks Min. Petr. Mitt.* **31**, 235–257.

Dobretsov, N.L., Khlestov, V.V. and Sobolev, V.S. (1973) The facies of regional metamorphism at moderate pressures. Dept. Geol. Publ. 236, Univ. Canberra (transl. by D.A. Brown).

Dobretsov, N.L. and Sobolov, N.V. (1984) Glaucophane schists and eclogites in the folded systems of northern Asia. *Ofioliti* **9**, 401–424.

Dobretsov, N.L., Sobolev, V.S., Sobolev, N.V. and Khlestov, V.V. (1975) The facies of regional metamorphism at high pressures. Dept. Geol. Publ. 266, Univ. Canberra (transl. by D.A. Brown).

Doe, B.R. (1970) *Lead Isotopes.* Springer, Heidelberg.

Dodson, M.H. (1979) Theory of cooling ages. In *Lectures in Isotope Geology*, eds. Jäger, E. and Hunziker, J.C., Springer, Heidelberg, 194–202.

Doll, C.G., Cady, W.M., Thompson, J.B. and Billings, M.P. (1961) Centennial geologic map of Vermont. Vermont Geol. Survey, scale 1:250,000.

Doluda, M.E., Litvin, S.V., and Kharchenko, S.D. (1968) Regional epigenesis of Carboniferous deposits of the Dnepr-Donets depression and its effect on reservoir properties. *Litol. Polezn. Iskop.* **1968** (4), 144–147 (transl. in *Lithol. Miner. Resourc.* **1968** (4), 516–520).

Dow, W.G. (1977) Kerogen studies and geological interpretations. *J. Geochem. Explor.* **7**, 79–99.

Duba, D. and Williams-Jones, A.E. (1983*a*) Studies of burial metamorphism in the post-Taconic stage of the Appalachian orogen, southwestern Gaspé. *Can. J. Earth Sci.* **20**, 1152–1158.

Duba, D. and Williams-Jones, A.E. (1983*b*) The application of illite crystallinity, organic matter reflectance, and isotopic techniques to mineral exploration: a case study in southwestern Gaspé, Quebec. *Econ. Geol.* **78**, 1350–1363.

Dubessy, J. (1984) Simulation des équilibres chimiques dans le système C–O–H., Conséquences méthodologiques pour les inclusions fluides. *Bull. Miné.* **107**, 155–168.

Dubessy, J. (1985) Contribution à l'étude des interactions entre paléo-fluids et minéraux à partir de l'étude des inclusions fluides par microspectrometry Raman. Conséquences métallogéniques. Ph. D. thesis, Univ. Nancy.

Dubessy, J., Burneau, A. and Dhamelincourt, P. (1987) Control parameters of gas analysis in fluid inclusions by micro Raman spectroscopy. *Terra cognita*, in press.

Dubessy, J., Guilhaumou, N., Mullis, J. and Pagel, M. (1984) Reconnaissance par microspectrométrie Raman, dans les inclusions fluides, de H_2S et CO_2 solides à domaine de fusion comparable. *Bull. Minér.* **107**, 189–192.

Dubessy, J., Pagel, M., Poty, B., Kosztolanyi, C. and Beny, J.-M. (1980) Evidence by Raman spectroscopy of free hydrogen and free oxygen in fluid inclusions from two uranium deposits. *I.M.A., 12th general meeting*, Orléans.

Dunoyer de Segonzac, G. (1968) The birth and development of the concept of diagenesis. *Earth Sci. Rev.* **4**, 153–201.

Dunoyer de Segonzac, G. (1969) Les minéraux argileux dans la diagenèse—passage au métamorphisme. *Mém. Serv. Carte géol. Alsace Lorraine* **29**.

Dunoyer de Segonzac, G. (1970) The transformation of clay minerals during diagenesis and low-grade metamorphism: a review. *Sedimentology* **15**, 281–346.

Dunoyer de Segonzac, G. and Abbas, M. (1976) Métamorphisme des argiles dans le Rhétien des Alpes sud-occidentales. *Sci. Géol. Bull.* **29**, 3–20.

Dunoyer de Segonzac, G., Artru, P. and Ferrero, J. (1966) Sur une transformation des minéraux argileux dans les 'terres noires' du bassin de la Durance: influence de l'orogénie alpine. *C.R. Acad. Sci. Paris, Sér. D*, **262**, 2401–2404.

Dunoyer de Segonzac, G. and Bernoulli, D. (1976) Diagenèse et métamorphisme des argiles dans le Rhétien Sud-alpin et Austro-alpin (Lombardie et Grisons). *Bull. Soc. géol. France* **18**, 1283–1293.

Dunoyer de Segonzac G. and Chamley, H. (1968) Sur le rôle joué par la pyrophyllite comme marqueur dans les cycles sédimentaires. *C.R. Acad. Sci. Paris* **267**, 274–277.

Dunoyer de Segonzac, G. and Heddebaut, C. (1971) Paléozoique anchi-métamorphique à illite, chlorite, pyrophyllite, allevardite et paragonite dans les Pyrénées Basques. *Bull. Serv. Carte Géol. Alsace Lorraine* **24**, 277–290.

Dunoyer de Segonzac, G., Ferrero, J., and Kubler, B. (1968) Sur la cristallinité de l'illite dans la diagenèse et l'anchimétamorphisme. *Sedimentology* **10**, 137–143.

Dunoyer de Segonzac, G. and Millot, G. (1962) Pyrophyllite de diagenèse dans le Dévonien inférieur du synclinal de Laval (Massif Armoricain). *C.R. Acad. Sci. Paris* **255**, 3438–3440.

Durand, B. (1980) *Kerogen. Insoluble Organic Matter from Sedimentary Rocks*. Technip, Paris.

Durand, B. (1980) Sedimentary organic matter and kerogen. Definition and quantitative importance of kerogen. In *Kerogen*, ed. Durand, B., Technip, Paris, 13–33.

Durand, B. and Espitalié, J. (1976) Geochemical studies on the organic matter from the Douala basin (Cameroon)—II. Evolution of kerogen. *Geochim. Cosmochim. Acta* **40**, 801–808.

Durney, D. (1974) Relations entre les températures d'homogénéisation d'inclusions fluides et les minéraux métamorphiques dans les nappes helvétiques du Valais. *Bull. Soc. géol. Fr.*, (7) **16**, 269–272.

Durney, D.W. and Ramsay, J.G. (1973) Incremental strains measured by syntectonic crystal growths. In *Gravity and Tectonics*, eds. De Jong, K.A. and Scholten, R., Wiley, New York etc., 67–96.

Dypvik, H. (1983) Clay mineral transformations in Tertiary and Mesozoic sediments from North Sea. *Bull. Amer. Assoc. Petrol. Geol.* **67**, 160–165.

Eberl, D. (1978) Reaction series for dioctahedral smectites. *Clays and Clay Minerals* **26**, 327–340.

Eberl, D. (1979) Synthesis of pyrophyllite polytypes and mixed layers. *Amer. Miner.* **64**, 1091–1096.

Eberl, D. (1984) Clay mineral formation and transformation in rocks and soils. *Phil. Trans. Roy Soc. London* **A311**, 241–257.

Eberl, D. and Hower, J. (1977) The hydrothermal transformation of sodium and potassium smectite into mixed-layer clay. *Clays and Clay Minerals* **25**, 215–227.

Echle, W., Plüger, W.L., Zielinski, J., Frank, B. and Scheps, V. (1985) Petrography, mineralogy, and geochemistry of the Salmian rocks from research borehole Konzen, Hohes Venn (West Germany). *Neues Jb. Geol. Paläont. Abh.* **171**, 31–50.

Eckhardt, F.-J. (1965) Ueber den Einfluss der Temperatur auf den kristallographischen Ordnungsgrad von Kaolinit. *Proc. Int. Clay Conf.*, Stockholm, 1963, Vol. 2, 137–145.

Eckhardt, F.-J. and Von Gaertner, H.R. (1962) Zur Entstehung und Umbildung der Kaolin-Kohlentonsteine. *Fortschr. Geol. Rheinl. Westf.* **3/2**, 623–640.

Ehlmann, A.J. and Sand, L.B. (1959) Occurrences of shales partially altered to pyrophyllite. *Clays and Clay Minerals, 6th Nat. Conf.*, 386–391.

Ellenberger, F. (1960) Sur une paragénèse éphémère à lawsonite et glaucophane dans le métamorphisme alpin en Haute-Maurienne (Savoie). *Bull. Soc. géol. France* **7**, 190–194.

England, P.C. and Richardson, S.W. (1977) The influence of erosion upon the mineral facies of rocks from different metamorphic environments. *J. geol. Soc. London* **134**, 201–213.

England, P.C. and Thompson, A.B. (1984): Pressure-temperature-time paths of regional metamorphism. I. Heat transfer during the evolution of regions of thickened continental crust. *J. Petrol.* **25**, 894–928.

England, R.N. (1972) Lamellar intergrowths of pyrophyllite and muscovite and the assemblage kyanite-pyrophyllite-quartz in quartzites from the Petermann Ranges, Northern Territory. *Bur. Miner. Res. Austral., Bull.* **125**, 67–73.

Epprecht, W. (1946) Die Eisen- und Manganerze des Gonzen. *Beitr. Geol. Schweiz, Geotechn. Serie* **24**.

Epstein, J.B. (1974) Metamorphic origin of slaty cleavage in eastern Pennsylvania (abstr.). *Geol. Soc. Amer. Ann. Meet. 1974, Miami Beach, Abstr. w. Progr.* **6**(7), 724.

Epstein, J.B. and Epstein, A.G. (1969) Geology of the Valley and Ridge province between Delaware Water Gap and Lehigh Gap, Pennsylvania. In *Geology of Selected Areas in New Jersey and eastern Pennsylvania and Guidebook of Excursions*, ed. Subitzky, S., Rutgers Univ. Press, New Brunswick, N.J., 132–205.

Epstein, A.G., Epstein, J.B., and Harris, L.D. (1977) Conodont color alteration—an index to organic metamorphism. *US Geol. Surv. Prof. Paper* **955**.

Ergun, S. (1967) Determination of longitudinal and transverse optical constants of absorbing uniaxial crystals. Optical anisotropy of graphite. *Nature* **14**, 135–136.

Ernst, W.G. (1975) *Subduction Zone Metamorphism*. Benchmark papers in Geology, 19, Halsted Press, New York.

Ernst, W.G. (1984) California blueschist, subduction, and the significance of tectonostratigraphic terranes. *Geology* **12**, 436–440.

Ernst, W.G., Seki, Y., Onuki, H., and Gilbert, M.C. (1970) Comparative study of low-grade metamorphism in the California Coast Ranges and the outer metamorphic belt of Japan. *Geol. Soc. Amer. Memoir* **124**.

Eskola, P. (1915) On the relations between the chemical and mineralogical composition in the metamorphic rocks of the Orijarvi region. *Comm. geol. Finlande Bull.*, No. 44.

Eslinger, E.V. and Savin, S.M. (1973) Mineralogy and oxygen isotope geochemistry of the hydrothermally altered rocks of the Ohaki-Broadlands, New Zealand geothermal area. *Amer. J. Sci.* **273**, 240–267.

Eslinger, E. and Sellars, B. (1981) Evidence for the formation of illite from smectite during burial metamorphism in the Belt Supergroup, Clark Fork, Idaho. *J. sediment. Petrol.* **51**, 203–216.

Eslinger, E., Highsmith, P., Albers, D., and deMayo, B. (1979) Role of iron reduction in the conversion of smectite to illite in bentonites in the Disturbed Belt, Montana. *Clays and Clay Minerals* **27**, 327–338.

Espitalié, J. (1979) Charakterisierung der organischen Substanz und ihres Reifegrades in vier Bohrungen des mittleren Oberrheingrabens sowie Abschätzung der paläogeothermischen Gradienten. *Fortschr. Geol. Rheinld. Westf.* **27**, 87–96.

Esquevin, J. (1969) Influence de la composition chimique des illites sur leur cristallinité. *Bull. Centre Rech. Pau-SNPA* **3**, 147–153.

Esquevin, J. and Kulbicki, G. (1963) Les minéraux argileux de l'Aptien supérieur du bassin d'Arzacq (Aquitaine). *Bull. Serv. Carte géol. Alsace Lorraine* **16/4**, 197–203.

Essene, E.J. (1982) Geologic thermometry and barometry. In *Characterization of Metamorphism through Mineral Equilibria*, ed. Ferry, J.M. Reviews in Mineralogy **10**, Mineralogical Society of America, 153–206.

Essene, E.J. and Fyfe, W.S. (1967) Omphacite in California metamorphic rocks. *Contrib. Miner. Petrol.* **15**, 1–23.

Eugster, H.P. and Skippen, G.B. (1967) Igneous and metamorphic reactions involving gas equilibria. In *Researches in Geochemistry*, ed. Abelson, Ph. H., Wiley, New York, 492–520.

Eugster, H.P. and Wones, D.R. (1962) Stability relations of the ferruginous biotite, annite. *J. Petrol.* **3**, 82–125.

Evarts, R.C. and Schiffman, P. (1983) Submarine hydrothermal metamorphism of the Del Puerto Ophiolite, California. *Amer. J. Sci.* **283**, 289–341.

Exley, R.A. (1982) Electron microprobe studies of Iceland Research Drilling Project: High-temperature hydrothermal mineral geochemistry. *J. Geophys. Res.* **87**, 6547–6557.

Fairchild, I.J. (1985) Petrography and carbonate chemistry of some Dalradian dolomitic metasediments: preservation of diagenetic textures. *J. geol. Soc. London* **142**, 167–185.

Fechtig, H. and Kalbitzer, S. (1966) The diffusion of argon in potassic bearing solids. In *Potassium Argon Dating*, eds. Schaeffer, O.A. and Zähringer, J., Springer, Heidelberg, 68–107.

Ferla, P. and Lucido, G. (1972) Pyrophyllite nelle filladi presso Gioiosa Vecchia (M. Peloritani-Sicilia). *Periodico Miner. Roma* **41**, 241–252.

Ferrero, J. and Kubler, B. (1964) Présence de dickite et de kaolinite dans les grès Cambriens d'Hassi Messaoud. *Bull. Serv. Carte géol. Alsace Lorraine.* **17/4**, 247–261.

Fersman, A.E. (1922) *The Geochemistry of Russia* (in Russian). Goskhimizdat, Leningrad.

Fettes, D.J., Long, C.B., Bevins, R.E., Max, M.D., Oliver, G.J.H., Primmer, T.J., Thomas, L.J., and Yardley, B.W.D. (1985) Grade and time of metamorphism in the Caledonide Orogen of Britain and Ireland. In *The Nature and Timing of Orogenic Activity in the Caledonian Rocks of the British Isles*, ed. Harris, A.L., *Mem. geol. Soc. London* **9**, 41–53.

Fieremans, M. and Bosmans, H. (1982) Colour zones and the transition from diagenesis to low-grade metamorphism of the Gedinnian shales around the Stavelot Massif (Ardennes, Belgium). *Schweiz. Mineral. Petrogr. Mitt.* **62**, 99–112.

Flehmig, W. (1973) Kristallinität und Infrarotspektroskopie natürlicher dioktaedrischer Illite. *Neues Jb. Miner. Mh.* 351–361.

Tonschiefern und ihrer Paragenese mit Paragonit und Pyrophyllit (abstract). *Fortschr. Miner.* **61**, Beih. 1, 61.

Flehmig, W. and Langheinrich, G. (1974) Beziehung zwischen tektonischer Deformation und Illit-Kristallinität, *Neues Jb. Geol. Paläontol. Abh.* **146**, 325–326.

Fleischer, R.L., Price, P.B. and Walker, R.M. (1965) Effects of temperature, pressure and ionization on the formation and stability of fission tracks in minerals and glasses. *J. Geophys. Res.* **70**, 1497–1502.

Fleischer, R.L., Price, P.B. and Walker, R.M. (1975) *Nuclear Tracks in Solids: Principles and Applications.* Univ. California Press, Berkeley.

Forbes, R.B., Evans, B.W., and Thurston, S.P. (1984) Regional progressive high-pressure metamorphism, Seward Peninsula, Alaska. *J. Metamorphic Geol.* **2**, 43–54.

Foscolos, A.E. and Kodama, H. (1974) Diagenesis of clay minerals from Lower Cretaceous shales of north-eastern British Columbia. *Clays and Clay Minerals* **22**, 319–335.

Foscolos, A.E. and Powell, T.G. (1979a) Mineralogical and geochemical transformation of clays during burial-diagenesis (catagenesis): relation to oil generation. In *International Clay Conference 1978*, ed. Mortland, M.M. and Farmer, V.C., Elsevier Sci. Publ. Co., Amsterdam, 261–270.

Foscolos, A.E. and Powell, T.G. (1979b) Catagenesis in shales and occurrence of authigenic clays in sandstones, North Sabine H–49 well, Canadian Arctic Islands. *Can. J. Earth Sci.* **16**, 1309–1314.

Foscolos, A.E., Powell, T.G., and Gunther, P.R. (1976) The use of clay minerals and organic geochemical indicators for evaluating the degree of diagenesis and oil generating potential of shales. *Geochim. Cosmochim. Acta* **40**, 953–966.

Foscolos, A.E. and Stott, D.F. (1975) Degree of diagenesis, stratigraphic correlations and potential sediment sources of Lower Cretaceous shale of northeastern British Columbia. *Geol. Surv. Can. Bull.* **250**,

Franceschelli, M., Leoni, L., Memmi, I. and Puxeddu, M. (1986) Regional distribution of Al–silicates and metamorphic zonation in the low grade 'Verrucano' metasediments from the northern Apennines (Italy). *J. metamorphic Geol.* **4**, 309–332.

Franceschelli, M., Pandeli, E. and Puxeddu, M. (1984) Kyanite bearing early Alpine metapsammite in the Larderello Geothermal Region (Italy) and its implications to Alpine metamorphism and Triassic paleogeography. *Schweiz. Miner. Petrogr. Mitt.* **64**, 405–422.

Frank, E. and Stettler, A. (1979) K–Ar and $^{39}Ar^{-40}$Ar systematics of white K-mica from an Alpine metamorphic profile in the Swiss Alps. *Schweiz. Miner. Petrogr. Mitt.* **59**, 375–394.

Frank, W., Alber, H. and Thöni, M. (1977) Jungalpine K/Ar-Alter von Hellgimmern aus dem Permotriaszug von Mauls-Penser Joch (Südtirol). *Anz. Oesterr. Akad. Wiss., math. -naturwiss. Kl.* **7**, 102–107.

French, B.M. (1966) Some geological implications of equilibrium between graphite and a C–H–O gas phase at high temperatures and pressures. *Rev. Geophys.* **4**, 223–253.

French, B.M. and Eugster, H.P. (1965) Experimental control of oxygen fugacities by graphite-gas equilibriums. *J. geophys. Res.* **70**, 1529–1539.

Frey, M. (1969a) Die Metamorphose des Keupers vom Tafeljura bis zum Lukmanier-Gebiet. *Beitr. geol. Karte Schweiz, N.F.* **137**.

Frey, M. (1969b) A mixed-layer paragonite/phengite of low-grade metamorphic origin. *Contrib. Miner. Petrol.* **24**, 63–65.

Frey, M. (1970) The step from diagenesis to metamorphism in pelitic rocks during Alpine orogenesis. *Sedimentology* **15**, 261–279.

Frey, M. (1978) Progressive low-grade metamorphism of a black shale formation, Central Swiss Alps, with special reference to pyrophyllite and margarite bearing assemblages. *J. Petrol.* **19**, 95–135.

Frey, M. (1986) Very low-grade metamorphism of the Alps—an introduction. *Schweiz. Miner. Petrogr., Mitt.* **66**, 13–27.

Frey, M. (1987) The reaction-isograd kaolinite + quartz = pyrophyllite + H_2O, Helvetic Alps, Switzerland. *Schweiz. Miner. Petrogr. Mitt.* **67**.

Frey, M., Bucher, K., Frank, E. and Mullis, J. (1980b) Alpine metamorphism along the Geotraverse Basel-Chiasso—a review. *Eclogae geol. Helv.* **73**, 527–546.

Frey, M., Bucher, K., Frank, E. and Schwander, H. (1982) Margarite in the Central Alps. *Schweiz. Miner. Petrogr. Mitt.* **62**, 21–45.

Frey, M., Hunziker, J.C., Jäger, E. and Stern, W.B. (1983) Regional distribution of white K-mica polymorphs and their phengite content in the Central Alps. *Contrib. Miner. Petrol.* **83**, 185–197.

Frey, M., Hunziker, J.C. O'Neil, J.R. and Schwander, H.W. (1976) Equilibrium-disequilibrium relations in the Monte Rosa granite, western Alps: Petrological, Rb–Sr and stable isotope data. *Contrib. Miner. Petrol.* **55**, 147–179.

Frey, M., Hunziker, J.C., Roggwiller, P. and Schindler, C. (1973) Progressive neidriggradige Metamorphose glaukonitführender Horizonte in den helvetischen Alpen der Ostschweiz. *Contrib. Miner. Petrol.* **39**, 185–218.

Frey, M., Schwander, H. and Saunders, J. (1987) The mineralogy and metamorphic geology of low-grade metasediments, Northern Range, Trinidad. *J. geol. Soc.* (London) (submitted)

Frey, M., Teichmüller, M., Teichmüller, R., Mullis, J., Künzi, B., Breitschmid, A., Gruner, U., and Schwizer, B. (1980) Very low-grade metamorphism in external parts of the Central Alps: Illite crystallinity, coal rank and fluid inclusion data. *Eclogae geol. Helv.* **73**, 173–203.

Frey, M. and Wieland, B. (1975) Chloritoid in autochthon-parautochthonen Sedimenten des Aarmassivs. *Schweiz. Miner. Petrogr. Mitt.* **55**, 407–418.

Frost, B.R. (1979) Mineral equilibria involving mixed-volatiles in a C–O–H fluid phase: The stabilities of graphite and siderite. *Amer. J. Sci.* **279**, 1033–1059.

Füchtbauer, H. and Goldschmidt, H. (1959) Die Tonminerale der Zecksteinformation. *Beitr. Miner. Petrologie* **6**, 320–345.

Gambari, L. (1868) Descrizione dei quarzi di Porretta. *Ann. Soc. nat. Modena* **III**, 1–19.

Gavish, E. and Reynolds, R.C. (1970) Structural changes and isomorphic substitution in illites from limestones of variable degrees of metamorphism. *Israel J. Chem.* **8**, 477–485.

Gavrilov, A.A. and Aleksandrova, V.A. (1968) Post sedimentation argillization in Paleozoic clastic deposits of the southern Urals and northern Mugodzhars. *Doklady Acad. Sci. USSR. Earth. Sci. Sect.* **182**, 178–180.

Gebauer, D. and Grünenfelder, M. (1976) U–Pb zircon and Rb–Sr whole rock dating of low-grade metasediments. Example: Montagne Noire (Southern France). *Contrib. Miner. Petrol.* **59**, 13–32.

Gerling, E.K., Levskii, L.K. and Morozova, I.M. (1963) On the diffusion of radiogenic argon from minerals. *Geochemistry* **6**, 551–592.

Ghent, E.D. (1979) Problems in zeolite facies geothermometry, geobarometry and fluid compositions. In *Aspects of Diagenesis*, ed. Scholle, P.A. and Schluger, P.R., Soc. Econ. Paleontol. Mineral. Spec. Publ. **26**, 81–87.

Gibbons, W. and Gyopari, M. (1986) A greenschist protolith for blueschist in Anglesey, U.K. *Geol. Soc. Amer. Memoir* **164**, 217–228.

Gibson, R.G. and Gray, D.R. (1985) Ductile-to-brittle transition in shear during thrust sheet emplacement, Southern Appalachian thrust belt. *J. struct. Geol.* **7**(5), 513–525.

Giletti, B.J. (1974) Diffusion related to geochronology. In *Geochemical Transport and Kinetics*, eds. Hofmann, A.W., Giletti, B.J., Yoder, Jr. H.S. and Yund, R.A., Carnegie Inst. Publ. 34, 61–76.

Gill, W.D., Khalaf, F.I., and Massoud, M.S. (1977) Clay minerals as an index of the degree of metamorphism of the carbonate and terrigenous rocks in the South Wales coalfield. *Sedimentology* **24**, 675–691.

Gleadow, A.J.W. and Brooks, C.K. (1979) Fission track dating, thermal histories and tectonics of igneous intrusions in East Greenland. *Contrib. Miner. Petrol.* **71**, 45–60.

Gleadow, A.J.W. and Duddy, I.R. (1981) A natural long-term annealing experiment for apatite. *Nucl. Tracks* **5**, 169–174.

Gleadow, A.J.W. and Duddy, I.R. (1984) Fission track dating and thermal history analysis of apatite from wells in the northwest Canning Basin. In *The Canning Basin, Western* Australia, ed. Purcell, P.G., Geol. Soc. Australia and Australian Petrol. Expl. Soc., 377–386.

Gleadow, A.J.W., Duddy, I.R. and Lovering, J.F. (1983) Fission track analysis: a new tool for the evaluation of thermal histories and hydrocarbon potential. *Australian Petrol. Expl. Assoc. J.* **23**, 93–102.

Goffé B. (1979) La lawsonite et les associations à pyrophyllite calcite dans les métasédiments alumineux di Briançonnais. Premières occurrences. *C.R. Acad. Sci. Paris* **289**, Sér. D, 813–816.

Goffé, B. (1982) Définition du faciès à FeMg carpholite-chloritoide, un marqueur de métamorphisme de HP–BT dans les métasédiments alumineux. Thèse d'Etat, Univ. P. et M. Curie, Paris.

Goffé, B. and Saliot, P. (1977) Les associations minéralogiques des roches hyperalumineuses du Dogger de Vanoise. Leur signification dans le métamorphisme régional. *Bull. Soc. franç. Minéral. Crist.* **100**, 302–309.

Goffé, B. and Velde, B. (1984) Contrasted metamorphic evolutions in thrusted cover units of the

Briançonnais zone (French Alps): a model for the conservation of HP–LT metamorphic mineral assemblages. *Earth. Planet. Sci. Lett.* **68**(2), 351–360.

Goffé, B., Goffé-Urbano, G. and Saliot, P. (1973) Sur la présence d'une variété magnésienne de ferrocarpholite en Vanoise (Alpes françaises). Sa signification probable dans le métamorphisme alpin. *C.R. Acad. Sci. Paris* **277**, Sér D, 1965–1968.

Gomez-Pugnaire, M.T., Sassi, F.P. and Visona, D. (1978) Sobre la presencia de paragonita y pirofilita en las filitas del Complejo Nevado-Filabride en la Sierra de Baza (Cordilleras Béticas, España). *Bol. Geol. Miner. LXXXIX-V*, 468–474.

Gonzales Martinez, J., Fenoll Hach-Ali, P. e Martin Vivaldi, J.L. (1970) Estudio minéralogico de niveles arcillosos del trîas alpujárride. *Bol. Geol. Miner.* **81**, 620–629.

Goodarzi, F. (1984) Organic petrography of graptolite fragments from Turkey. *Marine Petrol. Geol.* **1**, 202–210.

Goodarzi, F. (1985*a*) Optical properties of vitrinite carbonized at different pressures. *Fuel* **84**, 156–162.

Goodarzi, F. (1985*b*) Reflected light microscopy of chitinozoan fragments. *Marine Petrol. Geol.* **2**, 72–78.

Goodarzi, F. and Norford, B.S. (1985) Graptolites as indicators of the temperature histories of rocks. *J. geol. Soc. London* **142**, 1089–1099.

Grayson, J.F. (1975) Relationship of palynomorph translucency to carbon and hydrocarbons in clastic rocks. In *Pétrographie organique et potentiel pétrolier*, ed. Alpern, B., Centre National de la Recherche Scientifique, Paris, 261–273.

Gratier, J.-P. (1982) Approche expérimentale et naturelle de la déformation des roches par dissolution—cristallisation, avec transfert de matière. *Bull. Minér.* **105**, 291–300.

Gratier, J.-P. (1984) La déformation des roches par dissolution—cristallisation. Ph.D. thesis, Univ. Grenoble.

Green, P.F. (1981) 'Track-in-track' length measurements in annealed apatites. *Nucl. Tracks* **5**, 121–128.

Greenwood, H.J. (1975) Thermodynamically valid projections of extensive phase relationships. *Amer. Miner.* **60**, 1–8.

Gretener, P.E. and Curtis, C.D. (1982) Role of temperature and time on organic metamorphism. *Bull. Amer. Assoc. Petrol. Geol.* **66**, 1124–1149.

Grew, E.S. (1974) Carbonaceous material in some metamorphic rocks of New England and other areas. *J. Geol.* **82**, 50–73.

Groshong, R.H. (1976) Strain and pressure solution in the Martinsburg slate. Delaware Water Gap, New Jersey. *Amer. J. Sci.* **276**, 1131–1146.

Groshong, R.H., Pfiffner, O.A., and Pringle, L.R. (1984) Strain partitioning in the Helvetic thrust belt of eastern Switzerland from the leading edge to the internal zone. *J. struct. Geol.* **6**(1/2), 5–18.

Grubenmann, U. (1904) *Die kristallinen Schiefer*. Bornträger, Berlin.

Gruner, U. (1976) Geologie des Falknis-Glegghorn-Gebietes (W–Rhätikon). Unpubl. Liz.–Arbeit Univ. Bern.

Gruner, U. (1981) Die jurassischen Breccien der Falknis-Decke und altersäquivalente Einheiten in Graubünden. *Beitr. geol. Karte Schweiz*, N.F. 154.

Guidotti, C.V. (1984) Micas in metamorphic rocks. In *Micas*, ed. Bailey, S.W., Reviews in Mineralogy 13, Mineralogical Society of America, 357–467.

Guidotti, C.V. and Sassi, F.P. (1976) Muscovite as a petrogenetic indicator mineral in pelitic schists. *Neues Jb. Miner. Abh.* **127**, 97–142.

Guidotti, C.V. and Sassi, F.P. (1986) Classification and correlation of metamorphic facies series by means of muscovite b_0 data from low-grade metapelites. *Neues Jb. Miner. Abh.* **153**, 363–380.

Guilhaumou, N. and Beny, C. (1985) Caracterisation des hydrocarbures inclus dans les quartz des Terres Noires subalpines meridionales (France). *European Current Res. in Fluid Inclusions, 8th Symposium*, 10–12 April 1985, Göttingen, 62.

Guilhaumou, N., Dhamelincourt, P., Touray, J.-C. and Touret, J. (1981) Etude des inclusions fluides du système N_2–CO_2 de dolomites et de quartz de Tunesie septentrionale. Données de la microcryoscopie et de l'analyse à la microsonde à effet Raman. *Geochim. Cosmochim. Acta* **45**, 657–673.

Guilhaumou, N., Velde, B. and Beny, C. (1984) Raman microprobe analysis of gaseous inclusion in diagenetically recrystallized calcites. *Bull. Minér.* **107**, 193–202.

Gulson, B.L. and Mizon, K.J. (1980) Lead isotopic studies at Jabiluka. *Proc. Int. Uranium Symp. on the Pine Creek Syncline*, Int. Atomic Energy Agency, Vienna.

Gürler, B. (1982) Geologie der Val Tasna und Umgebung (Unterengadin). Unpubl. Ph.D. Thesis, Univ. Basel.

Guthrie, J.M., Houseknecht, D.W., and Johns, W.D. (1986) Relationships among vitrinite reflectance, illite crystallinity, and organic geochemistry in Carboniferous strata, Ouachita Mountains, Oklahoma and Arkansas. *Bull. Amer. Assoc. Petrol. Geol.* **70**(1), 26–33.

Gutjahr, C.C.M. (1966) Carbonization measurements of pollen grains and spores and their application. *Leidse geol. Meded.* **38**.

Haas, J.L. (1978) An empirical equation with tables of smoothed solubilities of methane in water and aqueous sodium chloride solutions up to 25 weight percent, 360 °C, and 138 M Pa. *US Geol. Surv. Open-file Rep.* 78–100.

Hahn, C. (1969) Mineralogisch-sedimentpetrographische Untersuchungen an den Flussbettsanden im Einzugsbereich des Alpenrheins. *Eclogae geol. Helv.* **62**, 227–278.

Hall, M.G. and Lloyd, G.E. (1981) The SEM examination of geological samples with a semiconductor back-scattered electron detector. *Amer. Miner.* **66**, 362–368.

Hamilton, L.H., Ramsden, A.R. and Stephens, J.F. (1970) Fossiliferous graphite from Undercliff, New South Wales. *J. Geol. Soc. Austral.* **17**, 31–37.

Hamilton, W. (1978) Mesozoic tectonics of the western United States. In *Mesozoic Paleogeography of the Western United States* eds. Howell, D.G. and McDougall, K.A., SEPM, Pacific Coast Paleogeography Symposium **2**, 33–70.

Hammerschmidt, K. (1982) K/Ar and $^{40}Ar/^{39}Ar$ age resolution from illites of the Trias of Mauls; Mesozoic cover of the Austroalpine basement, eastern Alps (South Tyrol). *Schweiz. Miner. Petrogr. Mitt.* **62**, 113–133.

Hammerschmidt, K. and Wagner, M. (1983) K/Ar Bestimmungen an Biotiten aus den kristallinen Gesteinen der Forschungsbohrung Urach III. *Neues Jb. Miner. Mh.* **1**, 35–48.

Hanor, J.S. (1980) Dissolved methane in sedimentary brines: Potential effect on the PVT properties of fluid inclusions. *Econ. Geol.* **75**, 603–617.

Harland, W.B., Francis, E.H. and Evans, P. (1971) *The Phanerozoic Time-scale, a Supplement.* Spec. Publ. 5, Geol. Soc., London.

Harland, W.B., Smith, A.G. and Wilcock, B. (1964) *The Phanerozoic Time-scale.* Geol. Soc., London.

Harper, C.T. (1967) On the interpretation of potassium-argon ages from Precambrian shields and Phanerozoic orogens. *Earth. Planet. Sci. Lett.* **5**, 128–132.

Harper, C.T. (1970) Graphical solutions to the problem of radiogenic Argon-40 loss from metamorphic minerals. *Eclogae geol. Helv.* **63**, 119–140.

Harris, A.G. (1979) Conodont color alteration, an organo-mineral metamorphic index, and its application to Appalachian basin geology. In *Aspects of Diagenesis*, eds. Scholle, P.A. and Schluger, P.R., Soc. Econ. Paleont. Miner., Spec. Publ. **26**, 3–16.

Harris, A.G., Harris, L.D., and Epstein, J.B. (1978) Oil and gas data from Paleozoic rocks in the Appalachian basin: maps for assessing hydrocarbon potential and thermal maturity (conodont color alteration isograds and over-burden isopachs). US Geol. Surv., Misc. Invest. Ser. Map I–917–E (4 sheets).

Harris, A.G., Wardlaw, B.R., Rust, C.C. and Merrill, G.K. (1980) Maps for assessing thermal maturity (conodont color alteration index maps) in Ordovician through Triassic rocks in Nevada and Utah and adjacent parts of Idaho and California. US Geol. Surv., Misc. Invest. Ser., Map I—1249, scale 1:2 500 000.

Hashimoto, M. (1966) On the prehnite-pumpellyite metagraywacke facies. *J. Geol. Soc. Japan* **72**, 253–265.

Hashimoto, M. and Kanehira, K. (1975) Some petrological aspects of stilpnomelane in glaucophanitic metamorphic rocks. *J. Japan. Assoc. Miner. Petr. Econ. Geol.* **70**, 377–387.

Hayakawa, N., Suzuki, S., Oda, Y., Hamaji, A., and Nambu, M. (1979) Vitrinite reflectivity and authigenic minerals in the Neogene sediments of the Green Tuff region, Japan (in Japanese; Eng. abstr.). *Mining Geol.* **29**, 103–111.

Hayashi, H. (1980) Pyrophyllite shales from South Africa. *Earth Resource Inst. Akita Univ. Rep.* **45**, 110–123.

Haynes, F.M. (1985) Determination of fluid inclusion compositions by sequential freezing *Econ. Geol.* **80**, 1436–1439.

Hayes, J.B. (1970) Polytypism of chlorite in sedimentary rocks. *Clays and Clay Minerals* **18**, 285–306.

Heek, K.H. van Jüntgen, H., Luft, K.F. and Teichmüller, M. (1971) Aussagen zur Gasbildung in frühen Inkohlungsstadien auf Grund von Pyrolyseversuchen. *Erdöl und Kohle* **24**, 566–572.

Helgeson, H.C., Delany, J.M., Nesbitt, H.W., and Bird, D.K. (1978) Summary and critique of the thermodynamic properties of rock-forming minerals. *Amer. J. Sci.* **278**-A, 1–229.

Heling, D. (1974) Diagenetic alteration of smectite in argillaceous sediments of the Rheingraben (SW Germany). *Sedimentology* **21**, 463–472.

Heling, D. and Teichmüller, M. (1974) Die Grenze Montmorillonite/Mixed Layer-Minerale und ihre Beziehung zur Inkohlung in der Grauen Schichtenfolge des Oligozäns im Oberrheingraben. *Fortschr. Geol. Rheinld. Westf.* **24**, 113–128.

Helmold, K.P., Fontana, D. and Loucks, R.G. (1982) Diagenetic provinces of the Verrucano Lombardo and Val Gardena sandstone (Permian), southern Alps, Italy. *Rend. Soc. ital. Miner. Petrol.* **38**, 1361–1386.

Helmold, K.P. and van de Kamp, P.C. (1984) Diagenetic mineralogy and controls on albitization and laumontite formation in Paleogene arkoses, Santa Ynez Mountains, California. In *Clastic Diagenesis*, ed. McDonald, D.A. and Surdam, R.C., *Amer. Assoc. Petrol. Geol. Mem.* **37**, 239–276.

Helmold, K.P. and Snider, E.H. (1981) Geothermal gradients, organic metamorphism, and mineral diagenesis in Tertiary rocks, Santa Ynez Mountains, California. *Bull. Amer. Assoc. Petrol. Geol.* **65**, 937.

Hemley, J.J. and Jones, W.R. (1964) Chemical aspects of hydrothermal alteration with emphasis on hydrogen metasomatism. *Econ. Geol.* **59**, 538–569.

Henderson, G.V. (1970) The origin of pyrophyllite and rectorite in shales of north central Utah. *Clays and Clay Minerals* **18**, 239–246.

Henderson, G.V. (1971) The origin of pyrophyllite-rectorite in shales of north central Utah. *Utah. Geol. Miner. Survey. Spec. Studies* **34**.

Hepworth, B.C., Oliver, G.J.H., and McMurthy, M.J. (1982) Sedimentology, volcanism, structure and metamorphism of the northern margin of a Lower Palaeozoic accretionary complex; the Bail Hill-Abington area of the Southern Uplands of Scotland. In *Trench-Forearc Geology: Sedimentation and Tectonics on Modern and Ancient Active Plate Margins*, ed. Leggett, J.K., Geol Soc. London/Blackwell, Oxford, 521–534.

Herold, R. (1970) Sedimentpetrographische und mineralogische Untersuchungen an pelitischen Gesteinen der Molasse Niederbayerns. Diss. Ludwig-Maximilians-Univ., München, 132p.

Héroux, Y., Bertrand, R., Chagnon, A., Connan, J., Pittion, J.-L., and Kübler, B. (1981) Évolution thermique et potential pétroligène par l'étude des kérogènes, des extraits organiques, des gaz adsorobés, des argiles, du sondage Karlsefni H–13 (offshore Labrador, Canada). *Can. J. Earth Sci.* **18**, 1856–1877.

Héroux, Y., Chagnon, A., and Bertrand, R. (1979) Compilation and correlation of major thermal maturation indicators. *Bull. Amer. Assoc. Petrol. Geol.* **63**, 2128–2144.

Hevia Rodriguez, V. (1977) Le concept de la reflectivité moyenne statistique des substances anisotropes. Relations avec le rang des anthracites et de la matière organique dispersée dans les sédiments. In *Advances in Organic Geochemistry 1975*, eds. Campos, R. and Goni, J., Enadimsa, Madrid, 655–673.

Heyen, G., Ramboz, C. and Dubessy, J. (1982) Simulation des équilibres de phases dans le système CO_2–CH_4 en dessous de 50 °C et de 100 bar. Application aux inclusions fluides. *C.R. Acad. Sci. Paris Sér. II*, **294**, 203–206.

Hinckley, D.N. (1963) Variability in 'crystallinity' values among the kaolin deposits of the coastal plain of Georgia and South Carolina. *Clays and Clay Minerals* **11**, 229–235.

Hirsch, P.B. (1954) X-ray scattering from coals. *Proc. Roy. Soc. London* **A266**, 143–169.

Hoefs, J. and Morteani, G. (1979) The carbon isotopic composition of fluid inclusions in Alpine fissure quartzes from the western Tauern Window (Tyrol, Austria). *Neues Jb. Miner. Mh.*, 123–134.

Hoefs, J. and Stalder, H.A. (1977) Die C-Isotopenzusammensetzung von CO_2–haltigen Flüssigkeitseinschlüssen in Kluftquarzen der Zentralalpen. *Schweiz. Miner. Petrogr. Mitt.* **57**, 329–347.

Hoffman, J. and Hower, J. (1979) Clay mineral assemblages as low grade metamorphic geothermometers: application to the thrust faulted disturbed belt of Montana, USA. In *Aspects of Diagenesis*, ed. Scholle, P.A. and Schluger, P.R., Soc. Econ. Paleontol. Mineral., Spec. Publ. **26**, 55–79.

Holland, T.J.B. (1980) The reaction albite = jadeite + quartz determined experimentally in the range 600–1200 °C. *Amer. Min.* **65**, 129–134.

Holland, T.J.B. (1983) The experimental determination of activities in disordered and short-range ordered jadeitic pyroxenes. *Contrib. Miner. Petrol.* **82**, 214–220.

Holland, T.J.B. and Powell, R. (1985) An internally consistent thermodynamic dataset with uncertainties and correlations: 2. Data and results. *J. metamorphic Geol.* **3**, 343–370.

Hollerbach, A., Hagemann, H.W., and Kasig, W. (1982) Organisches Material in paläozoischen Sedimenten des Stavelot-Venn-Massivs. *Compend. Ergänzungsb. Erdöl Kohle* **1982**, 13–14.

Hollister, L.S. and Burruss, R.C. (1976) Phase equilibria in fluid inclusions from the Khtada Lake metamorphic complex. *Geochim. Cosmochim. Acta* **40**, 163–175.

Hollister, L.S., Burruss, R.C., Henry, D.L. and Hendel, E.-M. (1979) Physical conditions during uplift of metamorphic terrains, as recorded by fluid inclusions. *Bull. Minér.* **102**, 555–561.

Hollister, L.S. and Crawford, M.L. (1981) *Short Course in Fluid Inclusions: Applications to Petrology.* Miner. Assoc. Canada, *Short Course Handbook* **6**.

Holloway, J.R. (1981) Compositions and volumes of supercritical fluids in the earth's crust. In *Short Course in Fluid Inclusions: Applications to Petrology*, eds. Hollister, L.S. and Crawford, M.L., Miner. Assoc. Canada, *Short Course Handbook* **6**, 13–38.

Holloway, J.R. (1984) Graphite–CH_4–H_2O–CO_2 equilibria at low-grade metamorphic conditions. *Geology* **12**, 455–458.

Holloway, J.R. and Reese, R.L. (1974) The generation of N_2–CO_2–H_2O fluids for use in hydrothermal experimentation. I. Experimental method and equilibrium calculations in the C–O–H–S system. *Amer. Miner.* **59**, 587–597.

Holtzapffel, T. and Ferrière, J. (1982) Minéraux argileux de roches anté-Crétacé supérieur d'Othrys (Grèce continentale): mise en évidence d'une diagenése. *Ann. Soc. Géol. Nord* (Lille) **52**, 25–32.

Hood, A., Gutjahr, C.C.M., and Heacock, R.L. (1975) Organic metamorphism and the generation of petroleum. *Bull. Amer. Assoc. Petrol. Geol.* **59**, 986–996.

Horváth, Z.A. (1983) Study on maturation process of huminitic organic matter by means of high-pressure experiments. *Acta Geol. Hung.* **26**, 137–148.

Hosterman, J.W. and Whitlow, S.I. (1983) Clay mineralogy in Devonian shales in the Appalachian basin. *US Geol. Surv. Prof. Paper* **1298**.

Hosterman, J.W., Wood, G.H. Jr, and Bergin, M.J. (1970) Mineralogy of underclays in the Pennsylvania anthracite region. *US Geol. Surv. Prof. Paper* **700-C**, C89–C97.

Houghton, B.F. (1982) Low-grade metamorphism of the Takitimu Group, western Southland, New Zealand. *N. Z. J. Geol. Geophys.* **25**, 1–19.

Howard, J.J. (1981) Lithium and potassium saturation of illite/smectite clays from interlaminated shales and sandstones. *Clays and Clay Minerals* **29**, 136–142.

Howard, J.J. and Roy, D.M. (1985) Development of layer charge and kinetics of experimental smectite alteration. *Clays and Clay Minerals* **33**(2), 81–88.

Hower, J.C., Eslinger, E.V., Hower, M.E. and Perry, E.A. (1976) Mechanism of burial metamorphism of argillaceous sediment: 1. Mineralogical and chemical evidence. *Bull. Geol. Soc. Amer.* **87**, 725–737.

Hower, J.C. (1978) Anisotropy of vitrinite reflectance in relation to coal metamorphism for selected United States coals. Ph. D. thesis, Pennsylvania State Univ.

Hower, J.C. and Davis, A. (1981) Vitrinite reflectance anisotropy as a tectonic fabric element. *Geology* **9**, 165–168.

Hoyer, P., Clausen, C.-D., Leuteritz, K., Teichmüller, R., and Thome, K.N. (1974) Ein Inkohlungsprofil zwischen dem Gelsenkirchener Sattel des Ruhrkohlenbeckens und dem Ostsauerländer Hauptsattel des Rheinischen Schiefergebirges. *Fortschr. Geol. Rheinld. Westf.* **24**, 161–172.

Huck, G. und Patteisky, K. (1964) Inkohlungsreaktionen unter Druck. *Fortschr. Geol. Rheinld. Westf.* **12**, 551–558.

Hufnagel, H. (1977) Das Fluoreszenzvermögen der Dinoflagellaten-Cysten—ein Inkohlungsparameter? *Geol. Jb.* **D23**, 59–65.

Huggett, J.M. (1984) An SEM study of phyllosilicates in a Westphalian Coal Measures sandstone using back-scattered electron imaging and wavelength dispersive spectral analysis. *Sediment. Geol.* **40**, 233–247.

Hughes, J.D. and Cameron, A.R. (1985) Lithology, depositional setting and coal rank-depth relationships in the Jurassic-Cretaceous Kootenay Group at Mount Allan, Cascade Coal Basin, Alberta. *Geol. Surv. Can. Paper* **81–11**.

Humphris, S.E. and Thompson, G. (1978a) Trace element mobility during hydrothermal alteration of oceanic basalts. *Geochim. Cosmochim. Acta*, **42**, 127–136.

Humphris S.E. and Thompson, G. (1978b) Hydrothermal alteration of oceanic basalts by seawater. *Geochim. Cosmochim. Acta* **42**, 107–125.

Hunt, J.M. (1979) *Petroleum Geochemistry and Geology*. Freeman, San Francisco.

Hunziker, J.C. (1974) Rb–Sr and K–Ar age determination and the Alpine tectonic history of the Western Alps. *Mem. Ist. Geol. Miner. Univ. Padova* **XXXI**.

Hunziker, J.C. (1979) Potassium argon dating. In *Lectures in Isotope Geology*, eds. Jäger, E. and Hunziker, J.C., Springer, Berlin etc., 52–76.

Hunziker, J.C. (1986) The evolution of illite to muscovite: an example of the behaviour of isotopes in low-grade metamorphic terrains. *Chem. Geol.* **57**, 31–40.

Hunziker, J.C., Frey, M., Clauer, N., Dallmeyer, R.D., Friedrichsen, H., Flehmig, W., Hochstrasser, K., Roggwiller, P. and Schwander, H. (1986) The evolution of illite to muscovite: Mineralogical and isotopic data from the Glarus Alps, Switzerland. *Contrib. Miner. Petrol.* **92**, 157–180.

Huon, S. (1985) Clivage ardoisier et réhomogénéisation isotopique K–Ar dans les schistes paléozoiques du Maroc. Thèse, Univ. Louis Pasteur, Strasbourg.

Hurford, A.J. (1986a) Cooling and uplift patterns in the Lepontine Alps, South Central Switzerland and an age of vertical movement on the Insubric Fault Line. *Contrib. Miner. Petrol.* **92**, 413–427.

Hurford, A.J. (1986b) Application of the fission track dating method to young sediments: principles, methodology and examples. In *Dating Young Sediments*, eds. Hurford, A.J., Jäger, E. and Ten Cate, J.A.M., Proc. Workshop, Beijing, People's Republic of China, CCOP-UNESCO, Bangkok.

Hutcheon, I., Oldershaw, A. and Ghent, E.D. (1980) Diagenesis of Cretaceous sandstones of the Kootenay Formation at Elk Valley (southeastern British Columbia) and Mt Allan (southwestern Alberta). *Geochim. Cosmochim. Acta* **44**, 1425–1435.

Hutcheon, J. (1983) Diagenesis 3. Aspects of the diagenesis of coarse-grained siliciclastic rocks. *Geosci. Canada* **10/1**, 4–14.

Hutton, A.C., Kantsler, A.J., Cook, A.C. and McKirdy, D.M. (1980) Organic matter in oil shales. *J. Austral. Petrol. Explor. Assoc.* **20**, 44–68.

Hutton, C.O. and Turner, F.J. (1936) Metamorphic zones in north-west Otago. *Trans. Proc. Roy. Soc. N. Zeal.* **65**, 405–406.

Ianovici, V., Neacsu, G. and Neacsu, V. (1981) Pyrophyllite occurrences and their genetic relations with the kaolin minerals in Romania. *Bull. Minéral.* **104**, 768–775.

Iijima, A. (1978) Geological occurrences of zeolites in marine environments. In *Natural Zeolites — Occurrence, properties, use*, eds. Sand, L.B. and Mumpton, F.A., Pergamon, Oxford, 175–198.

Iijima, A. and Matsumoto, R. (1982) Berthierine and chamosite in coal measures of Japan. *Clays and Clay Minerals* **30**, 264–274.

Iijima, A. and Utada, M. (1971) Present-day zeolitic diagenesis of the Neogene geosynclinal deposits in the Niigata oil field, Japan. In *Molecular Sieve Zeolites—I, Advances in Chemistry Series* **101**, 342–348.

International Committee for Coal Petrology (ed.) (1963, 1971, 1975) *International Handbook of Coal Petrography*, 2nd edn. (1963) 184 pp.; 1st suppl. to 2nd edn. (1971) 197 pp.; reprinted 1985; 2nd suppl. to 2nd. edn. (1975) 60 pp. Centre National de la Recherche Scientifique, Paris.

Ishiwatari, R., Ishiwatari, M., Rohrback, B.G., and Kaplan, I.R. (1977) Thermal alteration experiments on organic matter from recent marine sediments in relation to petroleum genesis. *Geochim. Cosmochim. Acta* **41**, 815–828.

Ishizuka, H. (1985) Prograde metamorphism of the Horokanai ophiolite in the Kamuikotan Zone, Hokkaido, Japan. *J. Petrol.* **26** (2), 391–417.

Islam, S. and Hesse, R. (1983) The P–T conditions of late-stage diagenesis and low-grade metamorphism in the Taconic belt of the Gaspé-Peninsula from fluid inclusions: Preliminary results. Curr. Res., Part B, *Geol. Surv. Can. Paper* **83–1B**, 145–150.

Islam, S., Hesse, R. and Chagnon, A. (1982) Zonation of diagenesis and low-grade metamorphism in Cambro-Ordovician flysch of the Gaspé Peninsula, Quebec Appalachians. *Can. Mineral.* **20**, 155–167.

Itaya, T. (1981) Carbonaceous material in pelitic schists of the Sanbagawa metamorphic belt in central Shikoku, Japan. *Lithos* **14**, 215–224.

Ivanova, N.V., Volkova, A.N., Rekshinskaya, L.G., and Konysheva, R.A. (1979) Pyroclastic material in coal measures of the Donets Basin and its diagnosis. *Litol. Polezn. Iskop.* **1979** (6), 71–80 (transl. in *Lithol. Miner. Resourc.* **1979** (6), 709–718).

Jacob, H. (1964) Neue Erkenntnisse auf dem Gebiet der Lumineszenzmikroskopie fossiler Brennstoffe. *Fortschr. Geol. Rheinld. Westf.* **12**, 569–588.

Jacob, H. (1985) Disperse solid bitumens as an indicator for migration and maturity in prospecting for oil and gas. *Erdöl und Kohle* **38**, 365.

Jacobs, G.K. and Kerrick, D.M. (1981) Methane: an equation of state with application to the ternary system H_2O-CO_2-CH_4. *Geochim. Cosmochim. Acta* **45**, 607–614.

Jansen, J.B.H. and Schuiling, R.D., (1976) Metamorphism on Naxos: Petrology and geothermal gradients. *Amer. J. Sci.* **276**, 1225–1253.

Jayko, A.S., Blake, M.C. Jr., and Brothers, R.N. (1986) Blueschist metamorphism of the eastern Franciscan belt, northern California. *Geol. Soc. Amer. Memoir* **164**.

Jehl, V., Poty, B., and Weisbrod, A. (1976) Hydrothermal metamorphism of the oceanic crust in the north Atlantic Ocean. *Amer. Geophys. Union Trans.* **8**, 597–598.

Jordan, H. and Koch, J. (1975) Inkohlungsuntersuchungen im Unterkarbon des Nordwestharzes. *Geol. Jb.* **A29**, 33–43.

Jordan, H. and Koch, J. (1979) Inkohlung der Unterkarbon- und Eifelschichten im Nordwestharz und ihre Ursache. *Geol. Jb.* **A51**, 39–55.

Juster, T.C. and Brown, P.E. (1984) Fluids in pelitic rocks during very low-grade metamorphism (abstract). *Geol. Soc. Amer. Abstr. Programs* **16**, 553.

Kalkreuth, W. (1976) Kohlenpetrologische und geochemische Untersuchungen an organischem Material Paläozoischer Sedimentgesteine aus der variskischen Geosynklinale. Dissertation, Rheinisch-Westfälische Technische Hochschule Aachen.

Kamineni, D.C. and Efthekhar-nezad, J. (1977) Mineralogy of the Permian laterite of north-western Iran. *Tschermaks Miner. Petrogr. Mitt.* **24**, 195–204.

Kantsler, A.J. and Cook, A.C. (1979) Maturation patterns in the Perth basin. *J. Austral. Petrol. Explor. Assoc.* **19**, 94–107.

Kantsler, A.J., Smith, G.C. and Cook, A.C. (1978) Lateral and vertical rank variation: implications for hydrocarbon exploration. *J. Austral. Petrol. Explor. Assoc.* **18**, 143–156.

Karpova, G.V. (1966) Paragonitic hydromicas in terrigenous rocks of the greater Doubas. *Doklady Acad. Sci. USSR, Earth Sci. Sect.* **171**, 185–187.

Karpova, G.V. (1967) Muscovitic hydromicas in coal-bearing polyfacies deposits (in Russian). *Litol. Polezn. Iskop.* **1967** (6), 15–27 (transl. in *Lithol. Miner. Resourc.* **1967** (6), 688–697).

Karpova, G.V. (1969) Clay mineral post-sedimentary ranks in terrigenous rocks. *Sedimentology* **13**, 5–20.

Karpova, G.V., Logvinenko, N.V., Orlova, L.V., and Belotserkovets (1981) Post-diagenetic changes in coal deposits of the Greater Don Basin. *Litol. Polezn. Iskop.* **1981** (6), 70–81 (transl. in *Lithol. Miner. Resourc.* **1981** (6), 594–604).

Karpova, G.V., Lukin, A.E., and Shevyakova, É.P. (1969) Catagenesis of Carboniferous deposits of the Dnepr-Donetsk basin (in Russian). *Litol. Polezn. Iskop.* **1969** (1), 16–31 (transl. in *Lithol. Miner. Resourc.* **1969** (1), 15–27).

Karweil, J. (1956) Die Metamorphose der Kohlen vom Standpunkt der physikalischen Chemie. *Z. dtsch. geol. Ges.* **107**, 132–139.

Kasig, W. and Spaeth, G. (1975) Neue Ergebnisse über die Geologie der Kern-und Mantelschich-ten des Hohen Venns auf Grund von Profilaufnahmen bei der Verlegung der Erdgasleitung Aachen-Rheinfelden. *Z. dtsch. geol. Ges.* **126**, 1–14.

Katagas, C. and Panagos, A.G. (1979) Pumpellyite–actinolite and greenschist facies metamorphism in Levos Island (Greece). *TMPM Tschermaks Mineral. Petrogr. Mitt.* **26**, 235–254.

Kawachi, Y. (1974) Geology and petrochemistry of weakly metamorphosed rocks in the Upper Wakatipu district, southern New Zealand. *N. Z. J. Geol. Geophys.* **17**, 169–208.

Kawachi, Y. (1975) Pumpellyite–actinolite and contiguous facies metamorphism in part of Upper Wakatipu district, South Island, New Zealand. *N. Z. J. Geol. Geophys.* **18**, 401–441.

Kazanskiy, Y.P., Ivanovskaya, A.V. and Sokolova, M.F. (1972) Clay minerals in the upper Precambrian of Siberia. *Doklady Acad. Sci. USSR, Earth Sci. Sect.* **200**, 240–241.

Kemp, A.E.S., Oliver, G.J.H., and Baldwin, J.R. (1985) Low-grade metamorphism and accretion tectonics: Southern Uplands terrain, Scotland. *Mineral. Mag.* **49**, 335–344.

Kerrick, D.M. and Ghent, E.D. (1979) $P-T-X_{CO_2}$ relations of equilibria in the system: CaO–Al_2O_3–SiO_2–CO_2–H_2O. In D.Z. Korzhinskii Memorial Vol. *Problems of Physicochemical Petrology*, **1**, 32–52.

Kerrick, D.M. and Jacobs, G.K. (1981) A modified Redlich-Kwong equation for H_2O and H_2O–CO_2 mixtures at elevated pressures and temperatures. *Amer. J. Sci.* **281**, 735–767.

Kinter, E.B. and Diamond, S. (1956) A new method for preparation and treatment of oriented aggregate specimens of soil clays for X-ray analysis. *Soil Sci.* **81**, 111–120.

Kirchner, E.C. (1980) Vulkanite aus dem Permoskyth der nördlichen Kalkalpen und ihre Metamorphose. *Mitt. Österr. geol. Ges.* **71**, 385–396.

Kisch, H.J. (1966) Zeolite facies and regional rank of bituminous coals. *Geol. Mag.* **103**, 414–422.

Kisch, H.J. (1968) Coal rank and lowest-grade regional metamorphism in the southern Bowen Basin, Queensland, Australia. *Geol. Mijnbouw* **47**, 28–36.

Kisch, H.J. (1969) Coal-rank and burial-metamorphic mineral facies. In *Advances in Organic Geochemistry 1968*, ed. Schenk, P.A. and Havenaar, I. Pergamon Press, Oxford, 407–424.

Kisch, H.J. (1974) Anthracite and meta-anthracite coal ranks associated with 'anchimetamorphism' and 'very-low-stage' metamorphism, I, II, III. *K. Ned. Akad. Wet. Amsterdam., Proc. Ser. B* **77** (2), 81–118.

Kisch, H.J. (1980a) Incipient metamorphism of Cambro-Silurian clastic rocks from the Jämtland Supergroup, central Scandinavian Caledonides, western Sweden: illite crystallinity and 'vitrinite' reflectance. *J. geol. Soc. London* **137**, 271–288.

Kisch, H.J. (1980b) Illite crystallinity and coal rank associated with lowest-grade metamorphism of the Taveyanne greywacke in the Helvetic zone of the Swiss Alps. *Eclogae geol. Helv.* **73**, 753–777.

Kisch, H.J. (1981a) Burial diagenesis in Tertiary 'flysch' of the external zones of the Hellenides in central Greece and the Olympos region, and its tectonic significance. *Eclogae geol. Helv.* **74**, 603–624.

Kisch, H.J. (1981b) Coal rank and illite crystallinity associated with the zeolite facies of Southland and the pumpellyite-bearing facies of Otago, southern New Zealand. *N.Z.J. Geol. Geophys.* **24**, 349–360.

Kisch, H.J. (1983) Mineralogy and petrology of burial diagenesis (burial metamorphism) and incipient metamorphism in clastic rocks. In *Diagenesis in Sediments and Sedimentary Rocks, 2*, eds. Larsen, G. and Chilingar, G.V., Elsevier, Amsterdam, 289–493 and 513–541 (Appendix B—literature published since 1976).

Klay, N. (1984) [39]Rückstoss- und Diffusionsartefakte bei der [40]Ar–[39]Ar-Datierung von Glaukonit. Diplomarbeit Max-Planck Kernphysik, Heidelberg.

Klementidis, R.E. and Mackinnar, I.D.R. (1986) High-resolution imaging of ordered mixed-layer clays. *Clays and Clay Minerals* **34**(2), 155–164.

Kligfield, R., Hunziker, J.C., Dallmeyer, R.D. and Schamel, S. (1986) Dating of deformation phases using K–Ar and [40]Ar/[39]Ar techniques: results from the northern Apennines. *J. struct. Geol.* **8**, 781–789.

Klug, H.P. and Alexander, L.E. (1974) *X-ray Diffraction Procedures.* 2nd edn., Wiley, New York.

Knipe, R.J. (1981) The interaction of deformation and metamorphism in slates. *Tectonophysics* **78**, 249–272.

Köppel, V. (1968) Age and history of the uranium mineralization of the Beaverlodge area, Saskatchewan. *Geol. Survey Paper Can.* **67–31**.

Korzhinskii, D.S. (1959) Physiochemical basis of the analysis of the paragenesis of minerals. New York: Consultants Bureau.

Kossovskaya, A.G. (1961) Specific nature of epigenetic alteration of terrigenous rocks in platform and geosynclinal regions. *Doklady Acad. Sci. USSR, Earth Sci. Sect.* **130**, 123–125.

Kossovskaya, A.G. and Shutov, V.D. (1955) Epigenesis zones in the terrigene complex of Mesozoic and Upper Paleozoic deposits of the western Upper Yana region (in Russian). *Dokl. Akad. Nauk SSSR* **103**, 1085–1088.

Kossovskaya, A.G. and Shutov, V.D. (1958) Zonality in the structure of terrigene deposits in platform and geosynclinal regions. *Eclogae geol. Helv.* **51**, 656–666.

Kossovskaya, A.G. and Shutov, V.D. (1961) The correlation of zones of regional epigenesis and metagenesis in terrigenous and volcanic rocks (in Russian). *Dokl. Akad. Nauk SSSR* **139**, 677–680 (transl. in *Dokl. Acad. Sci. USSR Earth Sci. Sec.* **139** (1963), 732–736).

Kossovskaya, A.G. and Shutov, V.D. (1963) Facies of regional epigenesis and metagenesis (in Russian). *Izv. Akad. Nauk SSSR, Ser. geol.* **1963**(7), 3–18 (transl. in *Int. Geol. Rev.* **7**(1965), 1157–1167).

Kossovskaya, A.G. and Shutov, V.D. (1970) Main aspects of the epigenesis problem. *Sedimentology* **15**, 11–40.

Kossovskaya, A.G., Shutov, V.D., and Alexandrova, V.A. (1964) Dependence on the mineral composition of the clays in the coal-bearing formations on the sedimentation conditions. *5me Congr. int. Stratigr. Géol. Carbonif., Paris 1963*, **II**, 519–529.

Kossovskaya, A.G., Logvinenko, N.V., and Shutov, V.D. (1957) Stages of formation and alteration in terrigenous rocks (in Russian). *Dokl. Akad. Nauk SSSR* **116** (2), 293–296.

Králík, J. (1984) Relationship between clay mineralogy and coalification-rank in sediments of the Ostrava-Karviná Coal Basin and the Nízký Jeseník Mts. *9th Conf. Clay Mineral. Petrol., Zvolen 1982*, 107–118.

Krader, T. (1985) Phasengleichgewichte und kritische Kurven des ternären systems H_2O-CH_4-NaCl bis 250 MPa und 800 K. Ph.D. Thesis, Univ. Karlsruhe.

Kramm, U. (1978) Die Metamorphose Mn-reicher Pelite der Wippraer Zone/Unterharz (abstract). *Fortschr. Miner.* **56**, Beih. 1, 67–68.

Kramm, U. (1980) Muskovit-Paragonit Phasenbeziehungen in niedriggradig metamorphen Schiefern des Venn-Stavelot Massivs, Ardennen. *Tschermaks Miner. Petrogr. Mitt.* **27**, 153–167.

Kramm, U., Spaeth, G., and Wolf, M. (1975) Variscan metamorphism in the NE Stavelot-Venn Massif, Ardennes: a new approach to the question of regional dynamothermal or contact metamorphism. *Neues Jb. Geol. Paläont. Abh.* **171**, 311–327.

Kreulen, R. (1977) CO_2-fluids during regional metamorphism on Naxos, a study on fluid inclusions and stable isotopes. Ph. D. Thesis, Univ. Utrecht.

Krevelen, D.W. van (1961) *Coal*. Elsevier, Amsterdam.

Krinsley, D.H., Pye, K. and Kearsley, A.T. (1983) Application of backscattered electron microscopy in shale petrology. *Geol. Mag.* **120**, 109–114.

Krishnaswami, S., Lal, D., Prabhu, N. and McDougall, D. (1974) Characteristics of fission tracks in zircon: applications to geochronology and cosmology. *Earth Planet. Sci. Lett.* **22**, 111–119.

Kristmandóttir, H. (1979) Alteration of basaltic rocks by hydrothermal activity 100–300 °C. In *International Clay Conference 1978*, ed. Mortland, M.M. and Farmer, V.C., Elsevier, Amsterdam, 359–367.

Kristmannsdottir, H. (1982) Alteration in the IRDP drill hole compared with other drill holes in Iceland. *J. Geophys. Res.* **87**, 6525–6531.

Kristmannsdottir, H. (1983) Chemical evidence from Icelandic geothermal systems as compared to submarine geothermal systems. In: *Hydrothermal Processes at Seafloor Spreading Centers*, ed. Rona, P.A. *et al.*, Plenum, New York, 291–320.

Kristmansdóttir, H. and Tómasson, J. (1978) Zeolite zones in geothermal areas in Iceland. In *Natural Zeolites, Occurrence, Properties, Use*, eds. Sand, L.B. and Mumpton, F.A., Pergamon, Oxford, 269–275.

Krumm, H. (1984) Anchimetamorphose im Anis und Ladin (Trias) der Nördlichen Kalkalpen zwischen Arlberg und Kaisergebirge—ihre Verbreitung und deren baugeschichtliche Bedeutung. *Geol. Rdsch.* **73**, 223–257.

Kübler, B. (1964) Les argiles, indicateurs de métamorphisme. *Rev. Inst. Franç. Pétrol.* **19**, 1093–1112.

Kübler, B. (1967a) La cristallinité de l'illite et les zones tout à fait supérieures du métamorphisme. In *Étages tectoniques, Colloque de Neuchâtel 1966*, À la Baconnière, Neuchâtel, Suisse, 105–121.

Kübler, B. (1967b) Anchimétamorphisme et schistosité. *Bull. Centre Rech. Pau—SNPA* **1**, 259–278.

Kübler, B. (1968) Evaluation quantitative du métamorphisme par la cristallinité de l'illite. *Bull. Centre Rech. Pau-SNPA* **2**, 385–397.

Kubler, B. (1970) Crystallinity of illite. Detection of metamorphism in some frontal parts of the Alps. *Fortschr. Mineral.* **47** (Beih. 1), 39–40.

Kübler, B. (1973) La corrensite, indicateur possible de milieux de sédimentation et du degré de transformation d'un sédiment. *Bull. Centre Rech. Pau—SNPA* **7**, 543–556.

Kübler, B. (1980). Les premiers stades de la diagenèse organique et de diagenèse minérale. Deuxième partie: Zonéographie par les transformations minéralogiques, comparaison avec la réflactance de la vitrinite, les extraits organiques et les gaz adsorbés. *Bull. Ver. schweiz. Petroleum-Geol. Ing.* **46** (110), 1–22.

Kübler, B. (1984) Les indicateurs des transformations physiques et chimiques dans la diagenèse, température et calorimétrie. In *Thérmométrie et barométrie géologiques*, ed. M. Lagache, Soc. Franç. Minér. Crist., Paris, 489–596.

Kübler, B., Martini, J., and Vuagnat, M. (1974) Very low grade metamorphism in the Western Alps. *Schweiz, Mineral. Petrogr. Mitt.* **54**, 461–469.

Kübler, B., Pittion, J.-L., Héroux, Y., Charollais, J., and Weidmann, M. (1979) Sur le pouvoir réflecteur de la vitrinite dans quelques roches du Jura, de la Molasse et des Nappes préalpines, helvétiques et penniques. *Eclogae geol. Helv.* **72**, 347–373.

Kulp, J.L. and Engels, J. (1963) Discordances in K–Ar and Rb–Sr isotopic ages. In *Radioactive Dating*, Int. Atomic Energy Agency, Vienna, 219–238.

Kuniyoshi, S. and Liou, J.G. (1976a) Contact metamorphism of the Karmutsen Volcanics, Vancouver Island, British Columbia. *J. Petrol.* **17**, 73–99.

Kuniyoshi, S. and Liou, J.G. (1976b) Burial metamorphism of the Karmutsen volcanic rocks, northeastern Vancouver Island, British Columbia. *Amer. J. Sci.* **276**, 1096–1119.

Kurylowicz, L.E., Ozimic, S., McKirdy, D.M., Kantsler, A.J. and Cook, A.C. (1976) Reservoir and source rock potential of the Larapinta Group, Amadeus Basin, Central Australia. *J. Austral. Petrol. Explor. Assoc.* **16**, 49–65.

Kuyl, O.S., Müller, J. and Waterbolk, H. Th. (1955) The application of palynology to oil geology with reference to western Venezuela. *Geol. Mijnb.* **17**, 49–76.

Kvenvolden, K.A. and Roedder, E. (1971) Fluid inclusions in quartz crystals from South-West Africa. *Geochim. Cosmochim. Acta* **35**, 1209–1229.

Kwiecinska, B. (1980) Mineralogy of natural graphites. *Prace Mineral.* **67**, 87 pp.

Kwiecinska, B., Murchison, D.G. and Scott, E. (1977) Optical properties of graphite. *J. Microsc.* **109**, 289–302.

Lafond, R. (1961) Etude minéralogique des argiles actuelles du bassin de la Vilaine. *C.R. Acad. Sci. Paris* **252**, 3614–3616.

Lafond, R. (1965) Précisions sur les minéraux argileux du Quaternaire de la Guyane française. *Bull. Soc. géol. France* **7**, 665–667.

Laird, J. and Albee, A.L. (1981) High-pressure metamorphism in mafic schist from northern Vermont. *Amer. J. Sci.* **281**, 97–126.

Lal, D., Murali, A.V., Rajan, R.S., Tamhane, A.S., Lorin, J.C. and Pellas, P. (1968) Techniques for proper revelation and viewing of etch tracks in meteoritic and terrestrial minerals. *Earth Planet. Sci. Lett.* **5**, 111–119.

Landais, P. and Dereppe, J.M. (1985) A chemical study of the carbonaceous material from the Carswell structure. In *The Carswell Structure Uranium Deposits, Saskatchewan*, eds. Laine, R., Alonso, D. and Svab, M., *Geol. Assoc. Canada, Spec. Paper* **29**, 165–174.

Landis, C.A. (1971) Graphitization of dispersed carbonaceous material in metamorphic rocks. *Contrib. Mineral. Petrol.* **30**, 34–45.

Landis, C.A. and Coombs, D.S. (1967) Metamorphic belts and orogenesis in southern New Zealand. *Tectonophysics* **4**, 501–518.

Lash, G.G. (1985) Accretion-related deformation of an ancient (early Paleozoic) trench-fill deposit, central Appalachian orogen. *Bull. Geol. Soc. Amer.* **96**, 1167–1178.

Lécolle, M. and Roger, G. (1976) Métamorphisme régional hercynien de 'faible degré' dans la province pyrito-cuprifère de Huelva (Espagne). Conséquence pétrologiques. *Bull. Soc. géol. France* **18**, 1687–1698.

Le Corre, C. (1975) Analyse comparée de la cristallinité dans le Briovérien et le Paléozoique centre-armoricains: zonéographie et structure d'un domaine épizonal. *Bull. Soc. géol. France* **17**, 547–553.

Lee, J.H. and Peacor, D.R. (1983) Intralayer transitions in phyllosilicates of Martinsburg shale. *Nature* **303**, 608–609.

Lee, J.H. and Peacor, D.R. (1985) Ordered 1:1 interstratification of illite and chlorite: a transmission and analytical electron microscopy study. *Clays and Clay Minerals* **33**, 463–467.

Lee, J.H., Ahn, J.H. and Peacor, D.R. (1985) Textures in layered silicates: Progressive changes through diagenesis and low-temperature metamorphism. *J. Sediment. Petrol.* **55**, 532–540.

Lee, J.H., Peacor, D.R., Lewis, D.D. and Wintsch, R.P. (1984) 'Chlorite-illite/muscovite interlayered and interstratified crystals: a TEM/STEM study. *Contrib. Miner. Petrol.* **88**, 372–385.

Lee, M. (1984) Diagenesis of the Permian Rotliegendes Sandstone, North Sea: K/Ar, O^{18}/O^{16}, and petrologic evidence. Unpublished thesis, Case Western Reserve University.

Lee, M., Aronson, J.L. and Savin, S.M. (1985) K–Ar dating of time of gas emplacement in Rotliegendes sandstone, Netherlands. *Bull. Amer. Assoc. Petrol. Geol.* **69**, 1381–1385.

Leitsch, E.C. (1975) Zonation of low grade regional metamorphic rocks, Nambucca Slate Belt, northeastern New South Wales. *J. geol. Soc. Austral.* **22**, 413–422.

Le Maitre, R.W. (1976) The chemical variability of some common igneous rocks. *J. Petrol.* **17**, 589–598.

Lemmlein, G.G. (1946) On the origin of flat quartzes with 'white band'. *Izdat. Akad. Nauk, SSSR, Moskwa*, 98–110. (in Russian; translated into German by Stalder, B. (1972) Ueber die Entstehung flacher Kristalle mit 'weissem Streifen'. *Schweizer Strahler* **2**, 430–437).

Lemmlein, G.G. (1956) Formation of fluid inclusions in minerals and their use in geological thermometry. *Geochemistry*, 630–642.

Lensch, G. (1963) Die Metamorphose der Kohle in der Bohrung Münsterland 1 auf Grund des optischen Reflexionsvermögen der Vitrinite. *Fortschr. Geol. Rheinld. Westf.* **11**, 197–204.

Leroy, J. (1979) Contribution a l'étalonnage de la pression interne des inclusions fluides lors de leur décrépitation. *Bull. Minér.* **102**, 584–593.

Levi, B., Aguirre L. and Nystrom J.O. (1982) Metamorphic gradients in burial metamorphosed vesicular lavas: Comparison of basalt and spilite in Cretaceous basic flows from central Chile. *Contrib. Miner. Petrol.* **80**, 49–58.

Levine, J.R. and Davis, A. (1984) Optical anisotropy of coals as an indicator of tectonic deformation, Broad Top Coal Field, Pennsylvania. *Bull. Geol. Soc. Amer.* **95**, 100–108.

Levinson, A.A. (1955) Studies in the mica group: polymorphism among illites and hydrous micas. *Amer. Miner.* **40**, 41–49.

Leythaeuser, D. and Welte, D.H. (1969) Relation between distribution of heavy n-paraffins and coalification in Carboniferous coals from the Saar District, Germany. In *Advances in Organic Geochemistry 1968*, eds. Schenck, P.A. and Havenaar, I., Pergamon, Oxford, 429–442.

Leythaeuser, D., Hagemann, H.W., Hollerbach, A. and Schaefer, R.G. (1980) Hydrocarbon generation in source beds as a function of type and maturation of their organic matter: a mass balance approach. *Proc. 10th World Petrol. Congr.* **2**, 31–41.

Liewig, M., Caron, J.M. and Clauer, N. (1981) Geochemical and K–Ar isotopic behaviour of Alpine sheet silicates during polyphased deformation. *Tectonophysics* **78**, 273–290.

Liewig, N., Clauer, N. and Sommer F. (1987) Rb–Sr and K–Ar dating of clay diagenesis in a Jurassic sandstone oil reservoir from the North Sea. *Bull. Amer. Assoc. Petrol. Geol.* (in press).

Lin, R., Davis, A., Bensley, D.F. and Derbyshire, F.J. (1986) Vitrinite secondary fluorescence: its chemistry and relation to the development of a mobile phase and thermoplasticity in coal. *Int. J. Coal Geol.* **6**, 215–228.

Liou, J.G. (1970) Synthesis and stability relations of wairakite, $CaAl_2Si_4O_{12}.2H_2O$. *Contrib. Miner. Petrol.* **27**, 259–282.

Liou, J.G. (1971a) P–T stabilities of laumontite, wairakite, lawsonite, and related minerals in the system $CaAl_2Si_2O_8–SiO_2–H_2O$. *J. Petrol.* **12**, 379–411.

Liou, J.G. (1971b) Synthesis and stability relations of prehnite, $Ca_3Al_2Si_3O_{10}(OH)_2$. *Amer. Miner.* **56**, 507–531.

Liou, J.G. (1971c) Stilbite–laumontite equilibrium. *Contrib. Miner. Petrol* **31**, 171–177.

Liou, J.G. (1971d) Analcime equilibria. *Lithos*, **4**, 389–402.

Liou, J.G. (1979) Zeolite facies metamorphism of basaltic rocks from the East Taiwan Ophiolite. *Amer. Miner.* **64**, 1–14.

Liou, J.G. (1981) Recent high CO_2 activity and Cenozoic progressive metamorphism in Taiwan. *Geol. Soc. China, Memoir* **4**, 451–501.

Liou, J.G., Kuniyoshi, S., and Ito, K. (1974) Experimental studies of the phase relations between greenschist and amphibolite in a basaltic system. *Amer. J. Sci.* **274**, 613–632.

Liou, J.G., Kim, H.S., and Maruyama, S. (1983) Prehnite-epidote equilibria and their petrologic applications. *J. Petrol.* **24**, 321–342.

Liou, J.G., Maruyama, S. and M. Cho (1985a) Phase equilibria and mineral parageneses of metabasites in low-grade metamorphism. *Miner. Mag.* **49**, 321–333.

Liou, J.G., Seki, Y. Guillemette, R. and H. Sakai (1985b) Compositions and parageneses of secondary minerals in the Onikobe geothermal system. Japan. *Chem. Geol.* **49**, 1–20.

Lippard, S. (1983) Cretaceous high pressure metamorphism in NE Oman and its relationship to subduction and ophiolite nappe emplacement. *J. Geol. Soc. London* **140**, 97–104.

Lippmann, F. (1977) Diagenese und beginnende Metamorphose bei Sedimenten. *Bull. Acad. serbe Sci. Arts; Classe Sci. nat. math.; Sciences nat.* **56**, No. 15, 49–67.

Lippmann, F. (1982) The thermodynamic status of clay minerals. In *Int. Clay Conference 1981*, ed. van Olphen, H. Developments in Sedimentology 35, Elsevier, Amsterdam, 475–485.

Lippmann, F. and Rothfuss, H. (1980) Tonminerale in Taveyannaz-Sandsteinen. *Schweiz. Mineral. Petrogr. Mitt.* **60**, 1–29.

Lippolt, H.J. (1971) Argon-Isotopen Anomalien in Gesteinen. *Arbeiten aus dem Lab. Geochronologie der Univ. Heidelberg*, 1–77.

Loeschke, J. and Weber, K. (1973) Geochemie und Metamorphose paläozoischer Tuffe und Tonschiefer aus den Karawanken (Oesterreich). *Neues Jb. Geol. Paläont. Abh.* **142**, 115–138.

Loewinson-Lessing, F.Y. and Struve, E.Z. (1937) *Petrographical Dictionary*, 2nd edn. (in Russian). Otni, Moscow–Leningrad.

Logvinenko, N.V. (1956) On the late diagenesis (epigenesis) of Carboniferous rocks of the Donbas (in Russian). *Dokl. Akad. Nauk SSSR* **106** (5), 889–892.

Logvinenko, N.B. (1957) Late diagenesis (epigenesis) of Donets Carboniferous rocks (in Russian). *Izv. Akad. Nauk SSSR. Ser. Geol.* **1957** (7), 64–86.

Logvinenko, N.V. (1959) The nature of alteration in Carboniferous rocks in the southeastern part of the greater Donets Basin (in Russian). *Dokl. Akad. Nauk SSSR* **126** (4), (transl. in *Dokl. Acad. Sci. USSR, Earth Sci. Sect.* **126**, 612–613).

Logvinenko, N.V. (1964) Mixed-layer phase in Silurian phyllitized shale of Nura-Tau. *Dokl. Acad. Sci. USSR, Earth Sci. Sect.* **157**, 99–101.

Lopatin, N.V. (1971) Temperature and geologic time as factors in coalification (in Russian). *Izv. Akad. Nauk SSSR. Ser. Geol.* **1971** (3), 95–106.

Lopatin, N.V. (1976a) The influence of temperature and geologic time on the catagenetic processes of coalification and petroleum and gas formation. In *Issledovaniya organicheskogo veshchestva sovremennykh i iskopayemykh osakdov*, Nauka Press, Moscow, 361–366 (Russian).

Lopatin, N.V. (1976b) Historical-genetic analysis of oil formation by use of a model of regular continuous subsidence of the source rock. *Akad. Nauk. SSSR, Ser. Geol. Izvestiya* **8**, 93–101.

Lopatin, N.V. and Bostick, N.H. (1974) The geologic factors in coal catagenesis. *Ill. State Geol. Surv., Repr. Ser.* **1974** Q, 1–16.

Lopez-Munoz, M. (1982) Aspects du métamorphisme dans la bordure méridionale de la Bretagne centrale: minéralogie, chimisme et conditions physiques. *Neues Jb. Miner. Mh.* **8**, 373–384.

Loughnan, F.C. and Ward, C.R. (1971) Pyrophyllite-bearing flint clay from the Cambewarra area, New South Wales. *Clay Miner.* **9**, 83–95.

Lucas, S.B. (1984) Low-grade metamorphism in the externides of Wopmay Orogen, N.W.T. Unpubl. B.A. Thesis, Queen's Univ. Kingston, Ontario.

Ludwig, V. (1972a) Die Paragenese Muscovit, Pyrophyllit, 7Å-Chlorit und Kaolinit im Silur des Frankenwaldes (NE-Bayern). *Neues Jb. Geol. Paläont. Mh.*, 303–305.

Ludwig, V. (1972b) Die Paragenese Chlorit, Muscovit, Paragonit und Margarit im 'Griffelschiefer' des Ordoviziums in NE-Bayern. (Mit einem Beitrag zum Problem der Illit-Kristallinität). *Neues Jb. Geol. Paläont. Mh.*, 546–560.

Ludwig, V. (1973) Zum Uebergang eines Tonschiefers in die Metamorphose: 'Griffelschiefer' im Ordovizium von NE-Bayern. *Neues Jb. Geol. Paläont. Abh.* **144**, 50–103.

Luzhnaya, N.P. and Vereshtchetina, I.P. (1946) Sodium, calcium, magnesium chlorides in aqueous solutions at − 57 to + 25 °C(polythermic solubility). *Zh. Prikl. Khimii* **19**, 723–733 (in Russian).

Lyons, P.C. and Chase, H.B. Jr (1981) Rank of coal beds of the Narragansett basin, Massachusetts and Rhode Island. *Int. J. Coal Geol.* **1**, 155–168.

McCartney, J.T. and Ergun, S. (1965) Electron microscopy of graphite crystallites in meta-anthracite. *Nature* **205**, 962–964.

McCartney, J.T. and Ergun, S. (1967) Optical properties of coal and graphite. *Bur. Mines Bull.* **641**.

McCulloh, T.H., Cashman, S.M., and Stewart, R.J. (1978) Diagenetic baselines for interpretative reconstructions of maximum burial depths and paleotemperatures in clastic sedimentary rocks. In *Low Temperature Metamorphism of Kerogen and Clay Minerals*, ed. Oltz, D.F., Pacific Section, Soc. Econ. Paleontol. Mineral., Los Angeles, CA, 18–46.

McDowell, S.D. and Elders, W.D. (1980) Authigenic layer silicate minerals in borehole Elmore 1, Salton Sea Geothermal Field, California, USA. *Contrib. Miner. Petrol.* **74**, 293–310.

McDowell, S.D. and Elders, W.A. (1983) Allogenic layer silicate minerals in borehole Elmore 1, Salton Sea geothermal field, California. *Amer. Miner.* **68**, 1146–1159.

McDowell, S.D. and Paces J.B. (1985) Carbonate alteration minerals in the Salton Sea geothermal system, California, USA. *Miner. Mag.* **49**, 469–479.

McMechan, M.E. and Price, R.A. (1982) Superimposed low-grade metamorphism in the Mount

Fisher area, southeastern British Columbia—implications for the East Kootenay orogeny. *Can. J. Earth Sci.* **19**, 476–489.

Malley, K., Juteau, T. and Blanco-Sanchez, J. (1983) Hydrothermal alteration of submarine basalts: from zeolitic to spilitic facies in the upper Triassic pillow-lavas of Antalya, Turkey. *Sci. Geol. Bull.* **36**, 139–163.

Manby, G.M. (1983) A reappraisal of chloritoid-bearing phyllites in the Forland Complex rocks of Prins Karls Forland, Spitsbergen. *Miner. Mag.* **47**, 311–318.

Maresch, W.V. (1977) Experimental studies on glaucophane: an analysis of present knowledge. *Tectonophysics* **43**, 109–125.

Marshak, S. and Engelder, T. (1985) Development of cleavage in limestones of a fold-thrust belt in eastern New York. *J. struct. Geol.* **7**(3/4), 345–359.

Martini, J. (1968) Etude pétrographique des Grès de Taveyanne entre Arve et Giffre (Haute-Savoie, France). *Bull. suisse Minéral. Pétrogr.* **48**, 539–654.

Martini, J. (1972) Le métamorphisme dans les chaînes alpines externes et ses implications dans l'orogenèse. *Bull. suisse Minéral. Pétrogr.* **52**, 257–275.

Martini J. and Vuagnàt, M. (1965) Présence du faciès à zéolites dans la formation des 'grès' de Taveyanne (Alpes franco-suisses). *Bull. suisse Minéral. Pétrogr.* **45**, 281–293.

Martini, J. and Vuagnat, M. (1970) Metamorphose niedrigst temperierten Grades in den Westalpen. *Fortschr. Miner.* **47**, 52–64.

Maruyama, S. and Liou, J.G. (1985) The stability of Ca–Na pyroxene in low-grade metabasites of high-pressure intermediate facies series. *Amer. Miner.* **70**, 16–29.

Maruyama, S. and Liou, J.G. (1987) Petrology of Franciscan metabasites along the jadeite-glaucophane type facies series, Cazadero, California, *J. Petrol.* **37**.

Maruyama, S., Cho, M. and Liou, J.G. (1986) Experimental investigations of blueschist-greenschist transition equilibria: Pressure dependence of Al_2O_3 contents in sodic amphiboles–a new geobarometer. *Geol. Soc. Amer. Memoir* **164**, 1–16.

Matter, A. and Ramseyer, K. (1985) Cathodoluminescence microscopy as a tool for provenance studies of sandstones. In *Provenance of Arenites*, ed. Zuffa, G.G., NATO ASI, Series C, 148, Reidel, 191–211.

Mattinson, J.M. (1981) U–Pb systematics and geochronology of blueschists: preliminary results. *Trans. Amer. Geophys. Union* **62**, 1059.

Maxwell, D.T. and Hower, J. (1967) High-grade diagenesis and low-grade metamorphism of illite in the Precambrian Belt Series. *Amer. Miner.* **52**, 843–857.

Maxwell, J.C. (1962) Origin of slaty and fracture cleavage in the Delaware Water Gap area, New Jersey and Pennsylvania. In *Petrologic Studies: a Volume in Honor of A.F. Buddington*, ed. Engel, A.E.J., James, H.L., and Leonard, B.F., Geol. Soc. Amer., 281–311.

Mehegan, J.M., Robinson, P.T., and Delaney, J.R. (1982) Secondary mineralization and hydrothermal alteration in the Reydarfjordur drill core, eastern Iceland. *J. Geophys. Res.* **87**, 6511–6524.

Melou, M. and Plusquellec, Y. (1967) Répartition de la pyrophyllite dans quelques niveaux du Briovérien et du Primaire armoricain. *C.R. Acad. Sci. Paris* **265**, 14–16.

Merriman, R.J. and Roberts, B. (1985) A survey of white mica crystallinity and polytypes in pelitic rocks of Snowdonia and Llŷn, North Wales. *Mineral. Mag.* **49**(3), 305–319.

Mevel, C. (1981) Occurrence of pumpellyite in hydrothermally altered basalts from the Vema Fracture zone. *Contrib. Miner. Petrol.* **76**, 386–393.

Miki, T. and Nakamuta, Y. (1985) Zeolitic diagenesis of the Paleogene formations in the Munakata coal field, Fukuoka Prefecture (in Japanese; Eng. abstr.) *Ganseki Kobutsu Kosho Gakkaishi* **80**(7), 283–291.

Millot, G. (1964) *Géologie des argiles*. Masson, Paris.

Millot, G. (1970) *Geology of Clays*. Springer Verlag, New York.

Mitra, G. and Yonkee, W.A. (1985) Relationship of spaced cleavage to folds and thrusts in the Idaho-Utah-Wyoming thrust belt. *J. struct. Geol.* **7**(3/4), 361–373.

Mitra, G, Yonkee, W.A., and Gentry, D.J. (1984) Solution cleavage and its relationship to major structures in the Idaho-Utah-Wyoming thrust belt. *Geology* **12**(6), 354–358.

Mitsui, K. (1975) Diagenetic alteration of some minerals in argillaceous sediments in western Hokkaido, Japan. *Sci. Rep. Tohoku Univ.* **13**, 13–65.

Miyashiro, A. (1961) Evolution of metamorphic belts. *J. Petrol.* **2**, 277–311.

Miyashiro, A. (1973) *Metamorphism and Metamorphic Belts.* Allen and Unwin, London.

Miyashiro, A. and Banno, S. (1958) Nature of glaucophane metamorphism. *Amer. J. Sci.* **256**, 97–110.

Miyashiro, A., Shido, F., and Ewing, M. (1971) Metamorphism in the mid-Atlantic ridge near 24° and 30°N. *Phil. Trans. Roy. Soc. London* **A268**, 589–603.

Monnier, F. (1982) Thermal diagenesis in the Swiss molasse basin: implications for oil generation. *Can. J. Earth Sci.* **19**, 328–342.

Moody, J.B., Meyer, D., Jenkins, J.Z. (1983) Quantitative characterization of the greenschist/amphibolite boundary in mafic systems. *Amer. J. Sci.* **283**, 48–92.

Moort, J.C. van (1971) A comparative study of the diagenetic alteration of clay minerals in Mesozoic shales from Papua, New Guinea, and in Tertiary shales from Louisiana, U.S.A. *Clays and Clay Minerals* **19**, 1–20.

Morikiyo, T. (1984) Carbon isotopic study on coexisting calcite and graphite in the Ryoke metamorphic rocks, northern Kiso district, central Japan. *Contrib. Miner. Petrol.* **87**(3), 251–256.

Morton, J.P. (1985) Rb–Sr evidence for punctuated illite/smectite diagenesis in the Oligocene Frio Formation, Texas Gulf Coast. *Bull. Geol. Soc. Amer.* **96**(1), 114–122.

Morton, J.P. and Long, L.E. (1980) Rb–Sr dating of Paleozoic glauconite from the Llano region, central Texas. *Geochim. Cosmochim. Acta* **44**, 663–672.

Mottl, M.J. (1983a) Hydrothermal processes at seafloor spreading centers: application of basalt-seawater experimental results. In *Hydrothermal processes at Seafloor Spreading Centers*, ed. Rona P.A. *et al.*, Plenum, New York, 199–224.

Mottl, M.J. (1983b) Metabasalts, axial hot springs and the structure of hydrothermal systems at mid-oceanic ridges. *Bull. Geol. Soc. Amer.* **94**, 161–180.

Muffler, L.J.P. and White, D.E. (1969) Active metamorphism of Upper Cenozoic sediments in the Salton Sea geothermal field and the Salton Trough, south-eastern California. *Bull. Geol. Soc. Amer.* **80**, 157–182.

Muir Wood, R. (1982) The Laytonville quarry (Mendocino County, California) exotic block: iron-rich blueschist-facies subduction-zone metamorphism. *Miner. Mag.* **45**, 87–99.

Mukhamet-Galeyev, A.P., Zotov, A.V., Pokrovskiy, V.A. and Kotova, Z.Y. (1986) Stability of the 1 M and 2 M$_1$ polytypic modifications of muscovite as determined from solubility at 300°C at saturation steam pressure. *Dokl. Acad. Sci. USSR, Earth Sci. Sect.* **278**, 140–143.

Mullis, J. (1975) Growth conditions of quartz crystals from Val d'Illiez (Valais, Switzerland). *Schweiz. Miner. Petrogr. Mitt.* **55**, 419–430.

Mullis, J. (1976a) Das Wachstumsmilieu der Quarzkristalle im Val d'Illiez (Wallis, Schweiz). *Schweiz. Miner. Petrogr. Mitt.* **56**, 219–268.

Mullis, J. (1976b) Die Quarzkristalle des Val d'Illiez—Zeugen spätalpiner Bewegungen. *Eclogae geol. Helv.* **69**, 343–357.

Mullis, J. (1979) The system methane-water as a geologic thermometer and barometer from the external part of the Central Alps. *Bull. Minér.* **102**, 526–536.

Mullis, J., Dubessy, J., Kosztolanyi, C. and Poty, B. (1983) Fluid evolution in Alpine fissures during prograde and retrograde metamorphism along the geotraverse: Lucerne—Bellinzona (Swiss Alps). *European Current Res. on Fluid Inclusions, Symposium*, 6–8 April 1983, Soc. Franç Minér. Crist., Orléans, 46.

Mullis, J., Poty, B. and Leroy, J. (1973) Nouvelles observations sur les inclusions à méthane des quartz du Val d'Illiez, Valais (Suisse). *C.R. Acad. Sci. Paris* **277**, Série D, 813–816.

Mullis, J., and Stalder, H.A. (1987) Salt-poor and salt-rich fluid inclusions in quartz from two boreholes in Northern Switzerland. *Chem. Geol.* (in press).

Mullis, J. (1983a) Evolution and migration of fluids in the Central Alps during the retrograde metamorphism. *European Current Res. on Fluid Inclusions. Symposium*, 6–8 April 1983, Soc. Franç. Minér. Crist., Orléans, 44.

Mullis, J. (1983b) Transported metamorphism of the external parts of the Central Alps, shown by fluid inclusions. *European Current Res. on Fluid Inclusions, Symposium*, 6–8 April 1983, Soc. Franç. Minér. Crist., Orléans, 45.

Murchison, D.G. (1978) Optical properties of carbonized vitrinites. In *Analytical Methods for Coal and Coal Products* 2, ed. Karr, C., Academic Press, New York, 415–464.

Murray, D.P., Hepburn, J.C., and Rehmer, J.A. (1979) Metamorphism of the Narragansett Basin. In *Evaluation of Coal Deposits in the Narragansett Basin, Massachusetts and Rhode Island*, rept. by Dept. Geol. Geophys., Boston Coll., Weston, Mass. for US Bur. Mines, 39–47.

Myers, G., Vrolijk, P. (1985) Fluid evolution in an underplated slate belt, Kodiak Islands, Alaska. *Trans. Amer. Geophys. Union* **66**, 1095.

Nadeau, P.H. and Reynolds, R.C. Jr (1981) Burial and contact metamorphism in the Mancos Shale. *Clays and Clay Minerals* **29**, 249–259.

Nadeau, P.H., Wilson, M.J., McHardy, W.J., and Tait, J.M. (1984) Interstratified clays as fundamental particles. *Science* **225**, 923–925.

Nadeau, P.H., Wilson, M.J., McHardy, W.J., and Tait, J.M. (1985) The conversion of smectite to illite during diagenesis: evidence from some illitic clays from bentonites and sandstones. *Mineral. Mag.* **49**(3), 393–400.

Naeser, C.W. (1979) Fission track dating and geologic annealing of fission tracks. In *Lectures in Isotope Geology*, eds. Jäger, E. and Hunziker, J.C., Springer, Heidelberg, 154–169.

Naeser, C.W. and Faul, H. (1969) Fission track annealing in apatite and sphene. *J. Geophys. Res.* **74**, 705–710.

Nakajima, T. (1983) Phase relations of pumpellyite-actinolite facies metabasites in the Sanbagawa metamorphic belt in central Shikoku, Japan. *Lithos* **15**, 267–280.

Nakajima, T., Banno, S., and Suzuki, T. (1977) Reactions leading to the disappearance of pumpellyite in low-grade metamorphic rocks of the Sanbagawa metamorphic belt in central Shikoku, Japan. *J. Petrol.* **18**, 263–284.

Nambu, M., Sato, T., Hayakawa, N. and Ohmori, Y. (1977) On the microanalysis of fluid inclusions with the ion microanalyzer (abstr.). *Mining Geol.* (Japan), **27**, 40 (in Japanese; English summary in *Fluid Inclusion Res., Proc. of COFFI*, **10**, 326–329).

Nelson, K.D. (1982) A suggestion for the origin of mesoscopic fabric in accretionary melange, based on features observed in the Chrystalls Beach Complex, South Island, New Zealand. *Bull. Geol. Soc. Amer.* **93**, 625–634.

Neruchev, S.G., Vassoievich, N.B. and Lopatin, N.V. (1976) The scale of catagenesis in connection with oil formation (in Russian). In *Fossil Fuels*, Nauka, Moscow, 47–62.

Neruchev, S.G., Zhukova, A.V., Fayzullina, Ye.M., and Kozlova, L.Ye. (1976) Diagenesis of the metamorphic catagenetic alteration of the non-bituminous part of the saproplanktonic dispersed organic matter (in Russian). *Akad. Nauk SSSR Sibirsk.* **330**, 30–78 (quoted by Price, 1983).

Newton, R.C. and Smith, J.V. (1967) Investigation concerning the breakdown of albite at depth in the earth. *J. Geol.* **75**, 268–286.

Nicot, E. (1981) Les phyllosilicates des terrains Précambriens du Nord-Ouest du Montana (USA) dans la transition anchizone-épizone. *Bull. Minéral.* **104**, 615–624.

Nicholls, J. and Crawford, M.L. (1985) Fortran programs for calculation of fluid properties from microthermometric data on fluid inclusions. *Computers geosci.* **11**, 619–645.

Niedermayr, G., Mullis, J., Niedermayr, E., and Schramm, J.-M. (1984) Zur Anchimetamorphose permo-skythischer Sedimentgesteine im westlichen Drauzug, Kärnten—Osttirol (Österreich). *Geol. Rundsch.* **73**(1), 207–221.

Niggli, E. and Niggli, C.R. (1965) Karten der Verbreitung einiger Mineralien der alpidischen Metamorpose in den Schweizer Alpen (Stilpnomelan, Alkali-Amphibol, Chloritoid, Staurolith, Disthen, Sillimanit). *Eclogae geol. Helv.* **58**, 335–368.

Nitsch, K.-H. (1971) Stabilitätsbeziehungen von Prehnit- und Pumpellyit-haltiger Paragenesen. *Contrib. Miner. Petrol.* **30**, 240–260.

Nitsch, K.-H. (1972) Das $P-T-X_{CO_2}$-Stabilitätsfeld von Lawsonit. *Contrib. Miner. Petrol.* **34**, 116–134.

Nyk, R. (1985) Illite crystallinity in Devonian slates of the Meggen mine (Rhenish Massif). *Neues Jb. Miner. Mh.* **1985**(6), 268–276.

Nyström, J.O. (1983) Pumpellyite-bearing rocks in central Sweden and extent of host-rock alteration as a control of pumpellyite composition. *Contrib. Miner. Petrol.* **83**, 159–168.

Oberlin, A., Boulmier, J.L. and Villey, M. (1980) Electron microscopy of kerogen microstructure. Selected criteria for determining the evolution path and evolution stage of kerogen. In *Kerogen*, ed. Durand, B., Technip, Paris, 191–241.

Oberlin, A., Goma, J. and Rouzaud, J.N. (1984) Techniques d'étude des structures et textures (microtextures) des matériaux carbonés. *J. Chim. Phys.* **81**, 701–710.

Odin, G.S., Hernandez, J. and Hunziker, J.C. (1986) Le volcanisme du 'Biarritziano' de Vénétie (Italie): Ages K–Ar sur basalte, plagioclase et céladonite. *Isotope Geosci.* **59**, 171–180.

Offler, R. and Prendergast, E. (1985) Significance of illite crystallinity and b_0 values of K-white mica in low-grade metamorphic rocks, North Hill End Synclinorium, New South Wales, Australia. *Miner. Mag.* **49**(3), 357–364.

Ogunyomi, O., Hesse, R., and Héroux, Y. (1980) Pre-orogenic and synorogenic diagenesis and anchimetamorphism in Lower Paleozoic continental margin sequences of the northern

Appalachian in and around Québec City, Canada. *Bull. Can. Petrol. Geol.* **28**, 559–577.

Ohmoto, H. and Kerrick. D.M. (1977) Devolatilization equilibria in graphitic systems. *Amer. J. Sci.* **277**, 1013–1044.

Ohta, Y., Hirajima, T., and Hiroi, Y. (1986) Caledonia high-pressure metamorphism in central western Spitzbergen. *Geol. Soc. Amer. Memoir* **164**, 205–216.

Oliver, G.J.H. (1978) Prehnite-pumpellyite facies metamorphism in County Cavan, Ireland. *Nature* **274**, 242–243.

Oliver, G.J.H. (1986) Arenig to Wenlock regional metamorphism in the paratectonic Caledonides of the British Isles: A review. In *The Caledonian–Appalachian Orogenesis*, ed. Harris, A.L., Spec. Publ. Geol. Soc. London (in press).

Oliver, G.J.H. and Leggett, J.K. (1980) Metamorphism in an accretionary prism: prehnite-pumpellyite facies metamorphism of the Southern Uplands of Scotland. *Trans. Roy. Soc. Edin., Earth Sci.* **71**, 235–246.

Oliver, G.J.H., Smellie, J.L., Thomas, L.J., Casey, D.M., Kemp., A.E.S., Evans, L.J., Baldwin, J.R. and Hepworth, B.C. (1984) Early Palaeozoic metamorphic history of the Midland Valley, Southern Uplands—Longford-Down Massif and the Lake District, British Isles. *Trans. Roy. Soc. Edin. (Earth Sci.)* **75**, 245–248.

Ottenjann, K. (1982) Verbesserungen bei der mikroskopischen Fluoreszenzmessung an Kohlen-mazeralen. *Zeiss Inform.* **26**, 40–46.

Ottenjann, K., Teichmüller, M. and Wolf, M. (1974) Spektrale Fluoreszenz-Messungen an Sporiniten mit Auflicht-Anregung, eine mikroskopische Methode zur Bestimmung des Inkohlungs-grades gering inkohlter Kohlen. *Fortschr. Geol. Rheinld. Westf.* **24**, 1–36.

Ottenjann, K., Teichmüller, M. and Wolf, M. (1975) Spectral fluorescence measurements of sporinites in reflected light and their applicability for coalification studies. In *Pétrographie organique et potentiel pétrolier*, ed. Alpern, B., Centre National de la Recherche Scientifique, Paris, 49–65.

Ottenjann, K., Wolf, M. und Wolff-Fischer, E. (1982) Das Fluoreszenzverhalten der Vitrinite zur Kennzeichnung der Kokungseigenschaften von Steinkohlen. *Glückauf Forschungsh.* **43**, 173–179.

Packham, G.H. and Crook, K.A.W. (1960) The principle of diagenetic facies and some of its implications. *J. Geol.* **68**, 392–407.

Padan, A., Kisch, H.J. and Shagam, R. (1982) Use of the lattice parameter b_0 of dioctahedral illite/muscovite for the characterization of P/T gradients of incipient metamorphism. *Contrib. Miner. Petrol.* **79**, 85–95.

Pagel, M. (1975) Détermination des conditions physico-chimiques de la silicification diagénétique des grès Athabasca (Canada) au moyen des inclusions fluides. *C.R. Acad. Sci. Paris* **280**, Série D, 2301–2304.

Pagel, M. and Poty, B. (1984) The evolution of composition, temperature and pressure of sedimentary fluids over time: a fluid inclusion reconstruction. In *Thermal Phenomena in Sedimentary Basins*, ed. Durand, B., Technip., Paris, 71–88.

Pagel, M., Walgenwitz, F. and Dubessy, J. (1986) Fluid inclusions in oil and gas-bearing sedimentary formations. In *Thermal Modeling in Sedimentary Basins*, ed. Burrus, J., Technip., Paris, 565–583.

Paradis, S., Velde, B., and Nicot, E. (1983) Chloritoid-pyrophyllite-rectorite rocks from Brittany, France. *Contrib. Miner. Petrol.* **83**, 342–347.

Parry, W.T. (1986) Estimation of X_{CO_2}, P, and fluid inclusion volume from inclusion temperature measurements in the system $NaCl–CO_2–H_2O$. *Econ. Geol.* **81**, 1009–1013.

Patrick, B.E., Evans, B.W., Dumoulin, J.A. and Harris, A.G. (1985) A comparison of carbonate mineral and conodont colour alteration index thermometry, Seward Peninsula, Alaska. *Geol. Soc. Amer. Abstr. Programs* **17**, 399.

Patteisky, K. and Teichmüller, M. (1960) Inkohlungs-Verlauf, Inkohlungs–Massstäbe und Klassifikation der Kohlen auf Grund von Vitrit-Analysen. *Brennstoff-Chemie* **41**, 79–84; 97–104; 133–137.

Pearce, J.A., and Cann, J.R. (1973) Tectonic setting of basic volcanic rocks determined using trace element analyses. *Earth Planet. Sci. Lett.* **19**, 290–300.

Peat, C.J., Muir, M.D., Plumb, K.A., McKirdy, D.M. and Norvick, M.S. (1978) Proterozoic microfossils from the Roper Group, Northern Territory, Australia. *J. Austral. Geol. Geophys.* **3**, 1–17.

Pécher, A. and Boullier, A.-M. (1984) Evolution à pression et température élevées d'inclusions fluides dans un quartz synthétique. *Bull. Minér.* **107**, 139–153.

Perchuk, L.L. and Aranovich, L. Ya (1981) Conditions in burial metamorphism. *Int. Geol. Rev.* **23**, 1210–1220.

Perkins, D., Westrum, E.F., and Essene, E.J. (1980) The thermodynamic properties and phase relations of some minerals in the system $CaO-Al_2O_3-SiO_2-H_2O$. *Geochim. Cosmochim. Acta.* **44**, 61–84.

Perry, E.A. and Hower, J. (1972) Late-stage dehydration in deeply buried pelitic sediments. *Bull. Amer. Assoc. Petrol. Geol.* **56**, 2013–2021.

Persoz, F. (1982) Inventaire minéralogique, diagenèse des argiles et minéralo-stratigraphie des séries jurassiques et crétacées inférieures du Plateau suisse et de la bordure sud-est du Jura entre les lacs d'Annecy et de Constance. *Beitr. geol. Karte Schweiz, Nouv. Sér.* **155**, 52pp.

Petrascheck, W.E. (1954) Zur optischen Regelung tektonisch beanspruchter Kohlen. *Tschermaks Miner. Petrogr. Mitt.* **4**, 232–239.

Pettijohn, F.J. (1957) *Sedimentary Rocks.* 2nd edn., Harper, New York.

Pevear, D.R. (1983) Illite/smectite diagenesis: relation to coal rank in Tertiary sediments of Pacific Northwest (abstr.). *Bull. Amer. Assoc. Petrol. Geol.* **67**(3), 533.

Pevear, D.R., Williams, V.E., and Mustoe, G.E. (1980) Kaolinite, smectite, and K-rectorite in bentonites: relation to coal rank at Tulameen, British Columbia. *Clays and Clay Mineral.* **28**, 241–254.

Phakey, P.P., Curtis, C.D. and Oertel, G. (1972) Transmission electron microscopy of fine-grained phyllosilicates in ultrathin rock sections. *Clays and Clay Minerals* **20**, 193–197.

Pichavant, M., Ramboz, C. and Weisbrod, A. (1982) Fluid immiscibility in natural processes: Use and misuse of fluid inclusion data. I. Phase equilibria analysis—a theoretical and geometrical approach. *Chem. Geol.* **37**, 1–27.

Piqué, A. (1975) Répartition des zones d'anchimétamorphisme dans les terrains dinantiens du Nord-Ouest du Plateau central (Meseta marocaine). *Bull. Soc. géol. Fr.* **17**(7), 416–420.

Piqué, A. (1982) Relations between stages of diagenesic and metamorphic evolution and the development of a primary cleavage in the northwestern Moroccan Meseta. *J. struct. Geol.* **4**, 491–500.

Piqué, A. (1984) La schistosité hercynienne et le métamorphisme associé dans la vallée de la Meuse entre Charleville et Namur (Ardennes franco-belges). *Bull. Soc. belge Géol.* **93**(1–2), 55–70.

Pollastro, R.M. (1985) Mineralogical and morphological evidence for the formation of illite at the expense of illite/smectite. *Clays and Clay Minerals* **33**(4), 265–274.

Pollastro, R.M. and Barker, C.E. (1986) Application of clay-mineral, vitrinite reflectance, and fluid inclusion studies to the thermal and burial history of the Pinedale anticline, Green River basin, Wyoming. In *Roles of Organic Matter in Sediment Diagenesis*, ed. Gautier, D.L., Soc. Econ. Paleontol. Mineral., Spec. Publ. **38**, 73–83.

Popp, R.K., and Gilbert, M.C. (1972) Stability of acmite–jadeite pyroxenes at low pressure. *Amer. Miner.* **57**, 1210–1231.

Potdevin, J.-L. and Caron, J.-M. (1986) Transfert de matière et déformation synmétamorphique dans un pli. I. Structures et bilans de matière. *Bull. Minéral.* **109**, 395–410.

Potter, II, R.W. and Brown, D.L. (1977) The volumetric properties of aqueous sodium chloride solutions from $0°$ to $500 °C$ and pressures up to 2000 bars based on a regression of available data in the literature. *US Geol. Surv. Bull.* **1421-C**.

Potter, II, R.W., Clynne, M.A. and Brown, D.L. (1987) Freezing point depression of aqueous sodium solutions. *Econ. Geol.* **73**, 284–285.

Poty, B. (1969) La croissance des cristaux de quartz dans les filons sur l'exemple du filon de la Gardette (Bourg d'Oisans) et des filons du massif du Mont Blanc. *Sci. de la Terre, Mém.* **17**.

Poty, B. and Stalder, H.A. (1970) Kryometrische Bestimmungen der Salz- und Gasgehalte eingeschlossener Lösungen in Quarzkristallen aus Zerrklüften der Schweizer Alpen. *Schweiz. Miner. Petrogr. Mitt.* **50**, 141–154.

Poty, B., Leroy, J. and Jachimowicz, L. (1976) Un nouvel appareil pour la mesure des températures sous le microscope: l'installation de microthermométrie Chaixmeca. *Bull. Minér.* **99**, 182–186.

Poty, B.P., Stalder, H.A. and Weisbrod, A.M. (1974) Fluid inclusions studies in quartz from fissures of Western and Central Alps. *Schweiz. Miner. Petrogr. Mitt.* **54**, 717–752.

Powell, R. (1978) *Equilibrium Thermodynamics in Petrology. An Introduction.* Harper and Row, London, etc., 284 pp.

Powell, R. and Holland, T.J.B. (1985) An internally consistent thermodynamic dataset with uncertainties and correlations: I. Methods and a worked examples. *J. metamorphic Geol.* **3**, 327–342.

Powell, R., Condliffe, D.M. and Condliffe, E. (1984) Calcite-dolomite geothermometry in the system $CaCO_3$–$MgCO_3$–$FeCO_3$: an experimental study. *J. metamorphic. Geol.* **2**, 33–41.

Powell, T.G., Foscolos, A.E., Gunther, P.R., and Snowdon, L.R. (1978) Diagenesis of organic matter and fine clay minerals: a comparative study. *Geochim. Cosmochim. Acta* **42**, 1181–1197.

Price, L.C. (1979) Aqueous solubility of methane at elevated pressures and temperatures. *Bull. Amer. Assoc. Petrol. Geol.* **63**, 1527–1533.

Price, L.C. (1981) Aqueous solubility of crude oil to 400 °C and 2,000 bars pressure in the presence of gas. *J. Petrol. Geol.* **4**, 195–223.

Price, L.C. (1982) Organic geochemistry of core samples from an ultra-deep hot well (300 °C, 7 km). *Chem. Geol.* **37**, 215–228.

Price, L.C. (1983) Geologic time as a parameter in organic metamorphism and vitrinite reflectance as an absolute paleogeothermometer. *J. Petrol. Geol.* **6**(1), 5–38.

Price, L.C. (1985) Geologic time as a parameter in organic metamorphism and vitrinite reflectance as an absolute paleogeothermometer. *J. Petrol. Geol.* **8**(2), 233–240.

Price, L.C., Clayton, J.L. and Rumen, L.L. (1981) Organic geochemistry of the 9.6 km Bertha Rogers No. 1 well, Oklahoma. *Org. Geochem.* **3**, 59–77.

Primmer, T.J. (1985) A transition from diagenesis to greenschist facies within a major Variscan fold/thrust complex in south-west England. *Miner. Mag.* **49**(3), 365–374.

Purdy, J.W. and Jäger, E. (1976) K–Ar ages on rock forming minerals from the Central Alps. *Mem. Ist. Geol. Miner. Univ. Padova*, **XXX**, 1–31.

Pye, K. and Krinsley, D.H. (1984) Petrographic examination of sedimentary rocks in the SEM using backscattered electron detectors. *J. sediment. Petrol.* **54**, 877–888.

Raben, J.D. and Gray, R.J. (1979a) The geology and petrology of anthracites and meta-anthracites in the Narragansett Basin, southeastern New England. In *Carboniferous Basins of Southeastern New England*, Field trip Guide Book 5, ed. Cameron, B., 9th Int. Congr. Carbonif. Strat. Geol. Amer. Geol. Inst. Falls Church, Va., 93–108.

Raben, J.D. and Gray, R.J. (1979b) The nature of highly deformed anthracites and meta-anthracites in southeastern New England. *9th Int. Congr. Carbonif. Strat. Geol. Abstr.*, Urbana, Illinois, 169.

Raben, J.D. and Gray, R.J. (1979c) Simplified classification and optical characterization of meta-coals in the Narragansett Basin, southeastern New England. *Geol. Soc. Amer. Ann. Meeting 1979, San Diego, Abstr. w. Progr.* **11**(7), 500.

Radoslovich, E.W. and Norrish, K. (1962) The cell dimensions and symmetry of layer-lattice silicates. I. Some structural considerations. *Amer. Miner.* **47**, 599–616.

Ragot, J.P. (1977) Contribution à l'étude de l'évolution des substances carbonées dans les formations géologiques. Thése Univ. P. Sabatier, Toulouse.

Ramboz, C., Pichavant, M. and Weisbrod, A (1982) Fluid immiscibility in natural processes: Use and misuse of fluid inclusion data. II. Interpretation of fluid inclusion data in terms of immiscibility. *Chem. Geol.* **37**, 29–48.

Ramboz, C., Schnapper, D. and Dubessy, J. (1985) The $P–\bar{V}–T–X–fO_2$ evolution of H_2O–CO_2–CH_4-bearing fluid in a wolframite vein: Reconstruction from fluid inclusion studies. *Geochim. Cosmochim. Acta* **49**, 205–219.

Ramsay, J.G. (1980) The crack-seal mechanism of rock deformation. *Nature* **284**, 135–139.

Reed, B.L. and Hemley, J.J. (1966) Occurrence of pyrophyllite in the Kekiktuk Conglomerate, Brooks Range, northeastern Alaska. *US Geol. Survey Prof. Paper* **550**–C, 162–165.

Rehmer, J., Hepburn, J.C., and Ostrowski, M. (1979) Illite crystallinity in subgreenschist argillaceous rocks and coal, Narragansett and Norfolk basins. USA (abstr.). *9th Int. Congr. Carbonif. Stratig. Geol., Urbana, Ill., 1979, Abstr. Paper*, 176.

Reuter, A. (1985) Korngrössenabhängigkeit von K–Ar Datierungen und Illitkristallinität anchizonaler Metapelite und assoziierter Metatuffe aus dem östlichen rheinischen Schieferge-birge. *Diss. Göttinger Arb. Geol. Paläont.* **27**, 1–91.

Reutter, K.-J., Teichmüller, M., Teichmüller, R. and Zanzucchi, G. (1983) The coalification pattern in the Northern Apennines and its palaeogeothermic and tectonic significance. *Geol. Rdsch.* **72**, 861–894.

Reynolds, R.C. (1963) Potassium-rubidium ratios and polymorphs in illites and microclines from the clay size fractions of Proterozoic carbonate rocks. *Geochim. Cosmochim. Acta* **27**, 1097–1112.

Riedel, D. (1966) Ein Beitrag zur Mineralogie und Chemie der Tone aus dem Tertiär der Niederrheinischen Bucht. Diss. Univ. Köln.

Rich, R.A. (1975) Fluid inclusions in metamorphosed Paleozoic rocks of eastern Vermont. Ph.D. thesis, Harvard University.

Richter, D.A. and Roy, D.C. (1976) Prehnite–pumpellyite facies metamorphism in central Aroostook County, Maine. *Geol. Soc. Amer. Memoir* **146**, 239–261.

Rickard, M.J. (1965) Taconic orogeny in the western Appalachians: Experimental application of microtextural studies to isotope dating. *Bull. Geol. Soc. Amer.* **76**, 523–536.

Roberson, H.E. and Lahann, R.W. (1981) Smectite to illite conversion rates: effects of solution chemistry. *Clays and Clay Mineral.* **29**, 129–135.

Robert, P. (1971) Étude pétrographique des matières organiques insolubles per la mesure de leur pouvoir réflecteur. Contribution à l'exploration pétrolière et à la connaissance des bassins sédimentaires. *Rev. Inst. fr. Pét. Ann. Combust. liq.* **26**, 105–135.

Robert, P. (1974) Analyse microscopique des charbons et des bitumes dispersés dans les roches et mesure de leur pouvoir réflecteur. Application à l'étude de la paléogéothermie des bassins sédimentaires et de la genèse des hydrocarbures. In *Advances in Organic Geochemistry 1973*, eds. Tissot, B. and Bienner, F., Technip, Paris, 549–569.

Robert, P. (1979) Classification des matières organiques en fluorescence. Application aux roches-mères pétrolières. *Bull. Centre Rech. Explor.-Prod. ELF-Aquitaine* **3**, 223–263.

Robert, P. (1985) Histoire géothermique et diagenèse organique. *Bull. Centre Rech. Explor. Prod. ELF-Aquitaine Mém.* **8**.

Roberts, B. (1981) Low grade and very low grade regional metabasic Ordovician rocks of Llŷn and Snowdonia, Gwynedd, North Wales. *Geol. Mag.* **118**, 189–200.

Roberts, B. and Merriman, R.J. (1985) The distinction between Caledonian burial and regional metamorphism in metapelites from North Wales: an analysis of isocryst patterns. *J. geol. Soc. London* **142**(4), 615–624.

Robinson, B.W. and Nickel, E.H. (1979) A useful new technique for mineralogy: the backscattered-electron/low vacuum mode of SEM operation. *Amer. Miner.* **64**, 1322–1328.

Robinson, D. and Bevins, R.E. (1986) Incipient metamorphism in the Lower Palaeozoic marginal basin of Wales. *J. metamorphic Geol.* **4**, 101–113.

Robinson, D., Nicholls, R.A., and Thomas, L.J. (1980) Clay mineral evidence for low-grade Caledonian and Variscan metamorphism in south-western Dyfed, south Wales. *Mineral. Mag.* **43**, 857–863.

Robinson, P.T., Hall, J.M., Christensen, N.I., Gibson, I.L., Fridleifsson, I.B., Schmincke, H.-U., and Schonharting, G. (1982) The Iceland Research Drilling Project: synthesis of results and implications for the nature of Icelandic and oceanic crust. *J. Geophys. Res.* **87** (B8), 6657–6667.

Rodionova, A.E. and Koval'skaya, M.S. (1974) Dickite distribution in coal-bearing formations of the Donets Basin. *Litologiya i Poleznye Iskopaemye* **6**, 132–137.

Roedder, E. (1962a) Ancient fluids in crystals. *Sci. Am.* **207**/4, 38–47.

Roedder, E. (1962b) Studies of fluid inclusions. I: Low temperature application of a dual-purpose freezing and heating stage. *Econ. Geol.* **57**, 1045–1061.

Roedder, E. (1970) Application of an improved crushing microscope stage to studies of the gases in fluid inclusions. *Schweiz. Miner. Petrogr. Mitt.* **50**, 41–58.

Roedder, E. (1971) Metastability in fluid inclusions. *Soc. Min. Geol. Japan, Spec. Issue* **3**, 327–334.

Roedder, E. (1981) Origin of fluid inclusions and changes that occur after trapping. In *Short course in fluid inclusions: Applications to Petrology*, eds. Hollister, L.S. and Crawford, M.L., Miner. Assoc. Canada, *Short Course Handbook* **6**, 101–137.

Roedder, E. (1984) Fluid inclusions. *Reviews in Mineralogy* **12**, Miner. Soc. Am., Washington.

Roedder, E. and Bodnar, R.J. (1980) Geologic pressure determinations from fluid inclusion studies. *Ann. Rev. Earth Planet. Sci.* **8**, 263–301.

Roedder, E., Ingram, B. and Hall, W.E. (1963) Studies of fluid inclusions III: Extraction and quantitative analysis of inclusions in the milligram range. *Econ. Geol.* **58**, 353–374.

Roedder, E. and Skinner, B.J. (1968) Experimental evidence that fluid inclusions do not leak. *Econ. Geol.* **63**, 715–730.

Roever, E.W.F. de (1977) Chloritoid-bearing metapelites associated with glaucophane rocks in W Crete. *Contrib. Miner. Petrol.* **60**, 317–319.

Roever, W.P., de, Roever, E.W., de, Beunk, F.F. and Lahaye, P.H.J. (1967) Preliminary note on ferrocarpholite from a glaucophane and lawsonite-bearing part of Calabria, Southern Italy. *Proc. Konf. Nederl. Akademie Wetenschappen*, Amsterdam, B70, No. 5, 534–537.

Rohde, A. (1980) Clay minerals and illite crystallinity of the Almesåkra Group. *Geol. Fören. Stockh. Förh.* **102**, 26.

Rosasco, G.J., Roedder, E. and Simmons, J.H. (1975) Laser-excited Raman spectroscopy for nondestructive partial analysis of individual phases in fluid inclusions in minerals. *Science* **190**, 557–560.

Rosenfeld, J.L. (1961) The contamination-reaction rules. *Amer. J. Sci.* **259**, 1–23.

Rossel, N.C. (1982) Clay mineral diagenesis in Rotliegend Aeolian sandstones of the southern North Sea. *Clay Minerals* **17**, 69–77.

Rouzaud, J.N. (1984) Relations entre la microtexture et les propriétés des matériaux carbonés—Application à la caractérisation des charbons. Thèse, Univ. Orléans.

Rouzaud, J.N. and Oberlin, A. (1983) Contribution of high resolution transmission electron microscopy (TEM) to organic materials, characterisation and interpretation of their reflectance. In *Thermal Phenomena in Sedimentary Basins*, Int. Colloq. Bordeaux, June 1983, Technip, Paris, 127–134.

Roy, A.B. (1978) Evolution of slaty cleavage in relation to diagenesis and metamorphism: A study from the Hunsrückschiefer. *Geol. Soc. Amer. Bull.* **89**, 1775–1785.

Rumeau, J.L. and Kulbicky, G. (1966) Evolution des minéraux argileux dans les dolomies et calcaires poreux du Crétacé supérieur de la plateforme d'Aquitaine. *Proc. Int. Clay Conf.*, Jerusalem, **2**, 103–117.

Rybach, L. and Bodmer, Ph. (1980) Die geothermischen Varhältnisse der Schweizer Geotraverse im Abschnitt Basel-Luzern. *Eclogae geol. Helv.* **73**, 501–512.

Sagon, J.-P. (1965) A propos du chloritoïde dans les schistes dévoniens du bassin de Châteaulin (région d'Uzel, Saint–Gilles–du–Vieux–Marché; Côtes–du– Nord). *C.R. Somm. Soc. Géol. France* **8**, 269–270.

Sagon, J.-P. (1970) Minéralogie des schistes paléozoiques du Bassin de Châteaulin (Massif armoricain): distribution de quelques minéraux phylliteux de métamorphisme. *C.R. Acad. Sci. Paris* **270**, Sér. D, 1853–1856.

Sajgó, C. (1979) Hydrocarbon generation in a super-thick Neogene sequence in SE Hungary. A study of the extractable organic matter. In *Advances in Organic Chemistry 1979*, eds. Douglas, A.G. and Maxwell, J.R., Pergamon Press, New York, 103–113.

Saliot, P., Guilhamou, N., and Barbillat, J. (1982) Les inclusions fluides dans les minéraux du métamorphisme à laumontite–prehnite–pumpellyite des Grès du Champsaur (Alpes du Dauphiné). Etude du méchansime de circulation des fluides. *Bull. minéral.* **105**, 648–657.

Sanner, W.S. (1967) Preparation characteristics of Pennsylvania anthracite from the Bottom Red Ash bed, Northern field. *US Bur. Mines Rept. Invest.* **6989**.

Sassi, F.P. (1972) The petrological and geological significance of the b_0 values of potassic white micas in low-grade metamorphic rocks. An application to the Eastern Alps. *Tschermaks Miner. Petrogr. Mitt.* **18**, 105–113.

Sassi, F.P. and Scolari, A. (1974) The b_0 value of the potassic white micas as a barometric indicator in low-grade metamorphism of pelitic schists. *Contrib. Miner. Petrol.* **45**, 143–152.

Sassi, F.P., Kräutner, H.G. and Zirpoli, G. (1976) Recognition of the pressure character in greenschist facies metamorphism. *Schweiz. Miner. Petrogr. Mitt.* **56**, 427–434.

Saupé, F., Dunoyer de Segonzac, G., and Teichmüller, M. (1977) Etude du métamorphisme régional dans la zone d'Almadén (Province de Cuidad Real, Espagne) par la cristallinité de l'illite et par le pouvoir réflecteur de la matière organique. *Sci. Terre* (Nancy) **21**, 251–269.

Schaer, J.-P. and Persoz, F. (1976) Aspects structuraux et pétrographiques du Haut Atlas calcaire de Midelt (Maroc). *Bull. Soc. géol. France* **18**, 1239–1250.

Schäfer, K. and Lax, E. (1962) *Landolt-Börnstein, Zahlenwerte und Funktionen aus Physik, Chemie, Astronomie, Geophysik und Technik.* 6. Aufl. II. Bd., 2. Teil, Bandteil b: Lösungsgleichgewichte I. Springer, Berlin.

Schamel, S. (1973) Eocene subduction in central Liguria, Italy. Unpubl. Ph.D. Thesis, Yale Univ.

Scherp, A., Stadler, G. and Schmidt, W. (1968) Die Pyrophyllit-führenden Tonschiefer des Ordoviziums im Ebbesattel und ihre Genese. *Neues Jb. Miner. Abh.* **108**, 142–165.

Scherrer, P. (1918) Bestimmung der Grösse und der inneren Struktur von Kolloidteilchen mittels Röntgenstrahlen. *Göttinger Nachr. Math. Phys.* **2**, 98–100.

Schiffman, P. and Liou, J.G. (1980) Synthesis and stability relations of Mg–Al pumpellyite, $Ca_4Al_5MgSi_6O_{21}(OH)_7$. *J. Petrol.* **21**, 441–474.

Schiffman, P. and Liou, J.G. (1983) Synthesis of Fe-pumpellyite and its stability relations with epidote. *J. metamorphic Geol.* **1**, 91–101.

Schiffman, P., Bird, D.K., and Elders, W.A. (1985) Hydrothermal mineralogy of calcareous sandstones from the Colorado River delta in the Cerro Prieto geothermal system, Baja California, Mexico. *Mineral. Mag.* **49** (3), 435–449.

Schiffman, P., Williams, A.E., and Evarts, R.C. (1984) Oxygen isotope evidence for submarine hydrothermal alteration of the Del Puerto ophiolite, California. *Earth Planet. Sci. Lett.* **70** (2), 207–220.

Schiffman, P., Elders, W.A., Williams, A.E., McDowell, S.D., and Bird, D.K. (1984) Active metasomatism in the Cerro Prieto geothermal system, Baja California, Mexico: A telescoped low-pressure, low-temperature metamorphic facies series. *Geology* **12**, 12–15.

Schmitz, H.-H. (1963) Untersuchungen am nordwestdeutschen Posidonienschiefer und seiner organischen Substanz. *Beih. geol. Jb.* **58**, 1–220.

Schoen, R. (1964) Clay minerals of the Silurian Clinton ironstones, New York State. *J. sediment. Petrol.* **34**, 855–863.

Schönlaub, H.P. and Zezula, G. (1975) Silur-Conodonten aus einer Phyllonitzone im Muralpen-Kristallin (Lungau/Salzburg). *Verh. Geol. Bundesanstalt Wien* **4**, 253–269.

Schramm, J.-M. (1974) Vorbericht über Untersuchungen zur Metamorphose im Raume Bischofshofen–Dienten–Saalfelden (Grauwackenzone/Nördliche Kalkalpen, Salzburg). *Anzeiger math. -naturw. Kl. Oesterr. Akad. Wiss.* **12**, 199–207.

Schramm, J.-M. (1977) Ueber die Verbreitung epi- und anchimetamorpher Sedimentgesteine in der Grauwackenzone und in den Nördlichen Kalkalpen (Oesterreich)—ein Zwischenbericht. *Geol. Paläont. Mitt. Innsbruck* **7**, 3–20.

Schramm, J.-M. (1978) Anchimetamorphes Permoskyth an der Basis des Kaisergebirges (Südrand der Nördlichen Kalkalpen zwischen Wörgl und St. Johann in Tirol, Oesterreich). *Geol. Paläont. Mitt. Innsbruck* **8**, 101–111.

Schramm, J.M. (1982*a*) Anchimetamorphose im klastischen Permoskyth der Schuppenzone von Göstling (Nördliche Kalkalpen, N.Ö.). *Verh. Geol. Bundesanst. Wien* **H.2**, 53–62.

Schramm, J.-M. (1982*b*) Zur Metamorphose des feinklastischen Permoskyth im Ostabschnitt der Nördlichen Kalkalpen (Ostösterreich). *Verh. Geol. Bundesanst. Wien* **H.2**, 63–72.

Schramm, J.-M. (1982*c*) Überlegungen zur Metamorphose des klastischen Permoskyth der Nördlichen Kalkalpen vom Alpenostrand bis zum Rätikon (Österreich). *Verh. Geol. Bundesanst. Wien* **1982**, 63–72.

Schramm, J.-M., von Gosen, W., Seeger, M. and Thiedig, F. (1982) Zur Metamorphose variszischer und post-variszischer Feinklastika in Mittel- und Ostkärnten (Oesterreich). *Mitt. Geol. Paläont. Inst. Univ. Hamburg* **53**, 169–179.

Schüller, A. (1961) Die Druck-Temperatur- und Energiefelder der Metamorphose. *Neues Jb. Miner. Abh.* **96**, 250–290.

Schultz, L.G. (1978) Mixed-layer clay in the Pierre Shale and equivalent rocks, northern Great Plains region. *US Geol. Surv. Prof. Paper* **1064-A**.

Scott, S.D. (1973) Experimental calibration of the sphalerite geobarometer. *Econ. Geol.* **68**, 466–4474.

Scott, S.D. (1976) Application of the sphalerite geobarometer to regionally metamorphosed terrains. *Amer. Miner.* **61**, 661–670.

Seidel, E. (1977) Lawsonite-bearing metasediments in the phyllite-quartzite series of SW-Crete (Greece). *Neues Jb. Miner. Abh.* **130**, 134–144.

Seidel, E. (1978) Zur Petrologie der Phyllit-Quarzit-Serie Kretas. Habil.–schrift Braunschweig.

Seki, Y. (1958) Glaucophanitic regional metamorphism in the Kanto Mountains, Central Japan. *Japan J. Geol. Geogr.* **29**, 233–258.

Seki, Y. (1961) Pumpellyite in low-grade regional metamorphism. *J. Petrol.* **2**, 407–423.

Seki, Y. (1969) Facies in low-grade metamorphism: *J. geol. Soc. Japan* **75**, 255–266.

Seki, Y. (1973) Metamorphic facies of pyrophilitic alteration. *J. geol. Soc. Japan* **79**, 771–780

Seki, Y. (1976) Comparison of CO_2 and O_2 in fluids attending the prehnite–pumpellyite facies metamorphism of the central Kii peninsula and the Tanzawa mountains, Japan. *Proc. 1st Int. Symp. on Water-Rock Interaction*, Prague, 230–235.

Seki, Y. and Liou, J.G. (1981) Recent study of low-grade metamorphism. *Geol. Soc. China, Memoir* **4**, 207–228.

Seki, Y., Oki, Y., Matsuda, T., Mikami, K. and Okumura, K. (1969) Metamorphism in the Tanazawa Mountains, central Japan. *J. Japan. Assn. Miner.* **61**, 1–75.

Seki, Y. Onuki, H., Oba, T., and Mori, R. (1971) Sanbagawa metamorphism in the central Kii Peninsula, *Japan. Geol. Geogr.* **16**, 65–78.

Seki, Y., Liou, J.G., Guillemette, R. and Sakai, H. (1983) Mineralogical and petrological investigation of drill hole core samples from the Onikobe geothermal system, Japan. *Hydrosci. and Geotechnol. Lab., Saitama Univ. Mem.* **3**.

Seki, Y., Liou, J.G., Guillemette, R., Sakai, H., Oki, Y., Hirano, T., and Onuki, H. (1983) Investigation of geothermal systems in Japan I. Onikobe geothermal area. *Hydrosci. and Geotechnol. Lab., Saitama Univ., Mem.* **3**.

Selverstone, J. and Spear, F. (1985) Metamorphic P–T paths from pelitic schists and greenstones from the south-west Tauern Window, eastern Alps. *J. metamorphic Geol.* **3**, 439–466.

Seyfried, W.E., Jr., Mottl, M.J., and Bischoff, J.L. (1979) Chemistry and mineralogy of spilites from the ocean floor: effect of seawater/basalt ratio. *Nature* **275**, 211–213.

Shelton, J.W. (1964) Authigenic kaolinite in sandstone. *J. sediment. Petrol.* **34**, 102–111.

Shepherd, T.J. (1981) Temperature-programmable heating-freezing stage for microthermometric analysis of fluid inclusions. *Econ. Geol.* **76**, 1244–1247.

Shepherd, T., Rankin, A.H. and Alderton, D.H.M. (1985) *A Practical Guide to Fluid Inclusion Studies*, Blackie, Glasgow and London.

Sheraton, J.W. (1985) Chemical changes associated with high-grade metamorphism of mafic rocks in the East Antarctic shield. *Chem. Geol.* **47**, 135–157.

Shirozu, H. (1978) Chlorite minerals. In *Clays and Clay Minerals of Japan*, eds. Sudo, T. and Shimoda, S., Elsevier, Amsterdam, 243–264.

Shikazono, N. (1985) Mineralogical and fluid inclusion features of rock alterations in the Seigoshi gold-silver mining district, western part of the Izu Peninsula, Japan. *Chem. Geol.* **49** (1/3), 213–230.

Shimoyama, T. and Iijima, A. (1976) Influence of temperature on coalification of Tertiary coal in Japan—summary. In *Circum-Pacific Energy and Mineral Resources. Amer. Assoc. Petrol. Geol. Mem.* **25**, 98–103.

Shutov, V.D., Aleksandrova, A.V. and Losievskaya, S.A. (1970) Genetic interpretation of the polymorphism of the kaolinite group in sedimentary rocks. *Sedimentology* **15**, 69–82.

Sicard, E., Potdevin, J.-L., and Caron, J.-M. (1984) Coexistences de lawsonite et de pseudomorphoses à pyrophyllite et kaolinite dans les Schistes lustrés corses: rôle des fluides. *C.R. Acad. Sci. Paris* **298**, 453–458.

Siddans, A.W.B. (1977) The development of slaty cleavage in a part of the French Alps. *Tectonophysics* **39**, 533–557.

Simanovich, I.M. (1972) Blastic transformation of quartzose rocks in various stages of postsedimentation alteration. *Dokl. Acad. Sci. USSR, Earth Sci. Sect.* **203**, 169–171.

Sivell, W.J. (1984) Low-grade metamorphism of the Brook Street volcanics, D'Urville Island, New Zealand. *N. Z. J. Geol. and Geophys.* **27**, 167–190.

Smart, G. and Clayton, T. (1985) The progressive illitization of interstratified illite-smectite from Carboniferous sediments of northern England and its relationship to organic maturity indicators. *Clay Miner.* **20**, 455–466.

Smith, G.C. and Cook, A.C. (1980) Coalification paths of exinite, vitrinite and inertite. *Fuel* **59**, 641–647.

Smith, J.W., Rigby, D., Gould, K.W. and Smyth, M. (1984) Exploration, characterization and assessment of fossil fuels. CSIRO Division of Fossil Fuels, Inst. Energy Earth Res., Rep. of Research 1982–1983, 5–8.

Smith, R.E. (1968) Redistribution of major elements in the alteration of some basic lavas during burial metamorphism. *J. Petrol.* **9**, 191–219.

Smith, R.E., and Smith, S.E. (1976) Comments on the use of Ti, Zr, Y, Sr, K, P and Nb in classification of basaltic magmas. *Earth planet. Sci. Lett.* **32**, 114–120.

Smith, R.E., Perdrix, J.L., and Parks, T.C. (1982) Burial metamorphism in the Hamersley Basin, Western Australia. *J. Petrol.* **23**, 75–102.

Smykatz-Kloss, W. and Althaus, E. (1974) Experimental investigation of the temperature dependence of the 'crystallinity' of illites and glauconites. *Bull. Groupe franç. Argiles* **26**, 319–325.

Sobolev, N.V., Dobretsov, N.L., Bakirov, A.B., and Shatsky, V.S. (1986) Eclogites from various types of metamorphic complexes in the USSR and the problems of their origin. *Geol. Soc. Amer. Memoir* **164**, 349–364.

Sommer, M.A., Yonover, R.N., Bourcier, W.L. and Gibson, E.K. (1985) Determination of H_2O and CO_2 concentrations in fluid inclusions in minerals using laser decrepitation and capacitance manometer analysis. *Anal. Chem.* **75**, 449–453.

Soom, M. (1986) Geologie und Petrographie von Ausserberg (VS). Kluftmineralisationen am Südwestrand des Aarmassivs. Unpubl. Lizentiatsarbeit Univ. Bern.

Spackman, W. and Moses, R.G. (1961) The nature and occurrence of ash-forming minerals in anthracite. *Miner. Ind. Exper. Stn., Penn. State Univ. Bull.* **75** (*Proc. Anthracite Conf.*), 1–15.

Spear., F.S., Ferry, J.M. and Rumble, D., III (1982) Analytical formulation of phase equilibria: The Gibbs method. In *Characterization of Metamorphism through Mineral Equilibria*, ed. Ferry, J.M., Reviews in Mineralogy **10**, Mineralogical Society of America, 105–152.

Spears, D.A. and Duff, P.McL.D. (1984) Kaolinite and mixed-layer illite-smectite in Lower Cretaceous bentonites from the Peace River coalfield, British Columbia. *Can. J. Earth Sci.* **21**(4), 465–476.

Środoń, J. (1979) Correlation between coal and clay diagenesis in the Carboniferous of the Upper Silesian Coal Basin. In *Int. Clay Conf. 1978*, ed. Mortland, M.M. and Farmer, V.C., Elsevier, Amsterdam, 251–260.

Stach, E., Mackowsky, M.-Th., Teichmüller, M., Taylor, G.H., Chandra, D., and Teichmüller, R. (1982) *Stach's Textbook of Coal Petrology*, 3rd edn., Gebrüder Borntraeger, Berlin.

Stadler, G. (1963) Die Petrographie und Diagenese der oberkarbonischen Tonsteine in der Bohrung Minsterland 1. *Fortschr. Geol. Rheinld. Westf.* **11**, 283–292.

Stadler, G. (1971) Die Kaolin-Kohlentonsteine aus dem Westfal C und B der Untertagebohrung 150 der Steinkohlenbergwerke Ibbenbüren und ihre Bedeutung für die Karbonstratigraphie Nordwest-deutschlands. *Fortschr. Geol. Rheinld. Westf.* **18**, 79–100.

Stadler, G. and Teichmüller, M. (1971) Die Umwandlung der Kohlen und die Diagenese der Tonund Sandsteine in der Untertagebohrung 150 der Steinkohlenbergwerke Ibbenbüren. *Fortschr. Geol. Rheinld. Westf.* **18**, 125–146.

Stadler, G. and Teichmüller, R. (1971) Zusammerfassender Überblick über die Entwicklung des Bramscher Massivs und des Niedersächsischen Tektogens. *Fortschr. Geol. Rheinld. Westf.* **18**, 547–564.

Stadler, G., Teichmüller, M. und Teichmüller, R. (1976) Zur geothermischen Geschichte des Karbons von Manno bei Lugano und des 'Karbons' von Falletti (Sesia Zone der Westalpen). *Neues Jb. Geol. Paläont. Abh.* **152**, 177–198.

Stalder, H.A. and Touray, J.C. (1970) Fensterquarze mit Methan-Einschlüssen aus dem westlichen Teil der schweizerischen Kalkalpen. *Schweiz. Miner. Petrogr. Mitt.* **50**, 109–130.

Stalder, P. (1979) Organic and inorganic metamorphism in the Taveyannaz Sandstone of the Swiss Alps and equivalent sandstones in France and Italy. *J. sediment. Petrol.* **49**, 463–482.

Staplin, F.L. (1969) Sedimentary organic matter, organic metamorphism, and oil and gas occurrences. *Bull. Can. Petrol. Geol.* **17**, 47–66.

Steiner, A. (1968) Clay minerals in hydrothermally altered rocks at Wairakei, New Zealand. *Clays and Clay Minerals* **16**, 193–213.

Stevaux, J. (1967) Etude géochimique et sédimentologique des Fermoy series de l'Eromanga sub-basin (Queensland–Australie). *Bull. Centre Rech. Pau-SNPA* **1**, 99–109.

Stone, I.J. and Cook, A.C. (1979) The influence of some tectonic structures upon vitrinite reflectance. *J. Geol.* **87**, 497–508.

Stopes, M.C. (1935) On the petrology of banded bituminous coals. *Fuel* **14**, 4–13.

Strakhov, N.M. (1957) Méthodes d'étude des roches sédimentaires. *Ann. Serv. Inform. Géol., Bur. Rech. Géol. Minière* 35,.

Strong, D.F., Dickson, W.L., and Pickerill, R.K. (1979) Chemistry and prehnite–pumpellyite facies metamorphism of calc-alkaline Carboniferous volcanic rocks of southeastern New Brunswick. *Can. J. Earth Sci.* **16**, 1071–1085.

Suggate, R.P. (1982) Low-rank sequences and scales of organic metamorphism. *J. Petrol. Geol.* **4**, 377–392.

Sultanov, R.C., Skripka, V.C. and Namiot, A. Yu. (1972) Solubility of methane in water at high temperatures and pressures. *Gazova Promyshlennost*, **17**, May, 6–7 (in Russian).

Surdam, R.C. (1969) Electron microprobe study of prehnite and pumpellyite from the Karmutsen Group. Vancouver Island, British Columbia. *Amer. Mineral.* **54**, 256–266.

Suzuki, S., Oda, Y., Karasawa, H., Hayakawa, N., and Nambu, M. (1979) Relation between vitrinite reflectivity and illite-montmorillonite mixed-layer minerals of the Miocene sediments in the west of Sannohe district, Japan (in Japanese; Eng. abstr.). *Mining Geol.* (Japan) **29**, 363–372.

Suzuki, S., Oda, Y., Karasawa, H., and Nambu, M. (1980) Thermal alteration of vitrinite in the Miocene sediments of the Ani area, northeast Honshu, Japan (in Japanese; Eng. abstr.). *Mining Geol.* **30**, 299–307.

Swanenberg, H.E.C. (1979) Phase equilibria in carbonic systems, and their application to freezing studies of fluid inclusions. *Contrib. Miner. Petrol.* **68**, 303–306.

Swanenberg, H.E.C. (1980) Fluid inclusions in high-grade metamorphic rocks from S.W. Norway. *Geol. Ultraiectina Univ.Utrecht* **25**.

Takenouchi, S. and Kennedy, G.C. (1964) The binary system H_2O–CO_2 at high temperatures and pressures. *Amer. J. Sci.* **262**, 1055–1074.

Taylor, G.H. (1971) Carbonaceous matter: a guide to the genesis and history of ores. *Soc. Min. Geol. Japan, spec. Issue* **3**, 283–288.

Taylor, H.P. and Coleman, R.G. (1968) O^{18}/O^{16} ratios of coexisting minerals in glaucophane-bearing metamorphic rocks. *Bull Geol. Soc. Amer.* **79**, 1727–1756.

Teichmüller, M. (1954) Combination of the current methods of coal petrography by examination of polished thin sections—a means for better international cooperation in coal petrography. *Proc. Int. Comm. Coal Petrol.* **1**, 25–29.

Teichmüller, M. (1973) Zur Petrographie und Genese der Naturkokse im Flöz Präsident/Helene der Zeche Friedrich Heinrich bei Kamp-Lintfort (Linker Niederrhein). *Geol. Mitt.* **12**, 219–254.

Teichmüller, M. (1974a) Entstehung und Veränderung bituminöser Substanzen in Kohlen in Beziehung zur Entstehung und Umwandlung des Erdöls. *Fortschr. Geol. Rheinld. Westf.* **24**, 65–112.

Teichmüller, M. (1974b) Generation of petroleum-like substances in coal seams as seen under the microscope. In *Advances in Organic Geochemistry 1973*, eds. Tissot, B. and Bienner, F., Technip, Paris, 321–348.

Teichmüller, M. (1978) Nachweis von Graptolithen-Periderm in geschieferten Gesteinen mit Hilfe kohlenpetrologischer Methoden. *Neues Jb. Paläont. Mh.*, 430–447.

Teichmüller, M. (1982/84) Fluoreszenzmikroskopische Änderungen von Liptiniten und Vitriniten mit zunehmendem Inkohlungsgrad und ihre Beziehungen zu Bitumenbildung und Verkokungsverhalten. Geol. Landesamt Nordrhein-Westfalen, Krefeld. English translation by N. Bostick appeared as Spec. Publ. Soc. Organic Petrol. 1, Houston, 1984.

Teichmüller, M. (1986) Organic petrology of source rocks, history and state of the art. In *Advances in Organic Geochemistry 1985*, eds. Leythaueser, D. and Rullkötter, J., Pergamon, Oxford (in press).

Teichmüller, M. and Ottenjann, K. (1977) Art und Diagenese von Liptiniten und lipoiden Stoffen in einem Erdölmuttergestein auf Grund fluoreszenzmikroskopischer Untersuchungen. *Erdöl und Kohle* **30**, 387–398.

Teichmüller, M. and Teichmüller, R. (1949) Inkohlungsfragen im Ruhrkarbon. *Z. dtsch. geol. Ges.* **99**, 40–77.

Teichmüller, M. and Teichmüller, R. (1968) Geological aspects of coal metamorphism. In *Coal and Coal-Bearing Strata*, eds. Murchison, D.G. and Westoll, T.S., Oliver and Boyd, Edinburgh, 233–267.

Teichmüller, M. and Teichmüller, R. (1979a) Diagenesis of coal (coalification). In *Diagenesis in sediments and sedimentary rocks*, Vol. 1, eds. Larsen, G. and Chilingar, G.V., Developments in Sedimentology 25A, Elsevier, Amsterdam, 207–246.

Teichmüller, M. and Teichmüller, R. (1979b) Ein Inkohlungsprofil entlang der linksrheinischen Geotraverse von Schleiden nach Aachen und die Inkohlung in der Nord–Süd–Zone der Eifel. *Fortschr. Geol. Rheinld. Westf.* **27**, 323–355.

Teichmüller, M. and Teichmüller, R. (1981) The significance of coalification studies to geology—a review. *Bull. Centre Rech. Expl.—Prod. ELF-Aquitaine* **5**, 491–534.

Teichmüller, M. und Teichmüller, R. (1982) Das Inkohlungsbild des Lippstädter Gewölbes. *Fortschr. Geol. Rheinld. Westf.* **30**, 223–239.

Teichmüller, M. und Teichmüller, R. (1984) Verbreitung und Eigenschaften tiefliegender Steinkohlen in der Bundesrepublik Deutschland. *Glückauf Forschungsh.* **45**, 140–153.

Teichmüller, M. and Teichmüller, R. (1986) Relations between coalification and palaeogeothermics in Variscan and Alpidic foredeeps of western Europe. In *Paleogeothermics*, ed. Buntebarth, G. and Stegena, L., Springer Verlag, Berlin, 53–78.

Teichmüller, M., Teichmüller, R., and Weber, K. (1979) Inkohlung und Illit-Kristallinität—Vergleichende Untersuchungen im Mesozoikum und Paläozoikum von Westfalen. *Fortschr. Geol. Rheinld. Westf.* **27**, 201–276.

Teichmüller, M., Teichmüller, R. and Lorenz, V. (1983) Inkohlung und Inkohlungsgradienten im Permokarbon der Saar–Nahe–Senke. *Z. dtsch. geol. Ges.* **134**, 153–210.

Teodorovich, G.I. (1961) *Authigenic Minerals in Sedimentary Rocks.* Consultants Bureau, New York.

Teodorovich, G.I. and Konyukhov, A.I. (1970) Mixed-layer minerals in sedimentary rocks as indicators of the depth of their catagenetic alteration (in Russian). *Dokl. Akad. Nauk SSSR* **191**, 1123–1126 (transl. in *Dokl. Acad. Sci. USSR, Earth Sci. Sec.* **191**, 174–176).

Thompson, A.B. (1971) P_{CO_2} in low-grade metamorphism: zeolite, carbonate, clay mineral, prehnite relations in the system $CaO-Al_2O_3-SiO_2-CO_2-H_2O$. *Contrib. Miner. Petrol.* **33**, 145–161.

Thompson, A.B. (1976) Mineral reactions in pelitic rocks: I. Prediction of P–T–X(Fe–Mg) phase relations. *Amer. J. Sci* **276**, 401–424.

Thompson, G. (1983) Basalt–seawater interaction. In *Hydrothermal Processes at Seafloor Spreading Centers*, ed. Rona, P.A. *et al.*, Plenum, New York, 225–278.

Thompson, J.B., Jr. and Norton, S.A. (1968) Palaeozoic regional metamorphism in New England and adjacent areas. In *Studies of Appalachian Geology: Northern and Maritime*, ed. Zen, E-an, Interscience, New York, 319–327.

Thöni, M. (1981) Degree and evolution of the Alpine metamorphism in the Austroalpine unit W of the Hohe Tauern in the light of K/Ar and Rb/Sr age determinations on micas. *Jb. geol. Bundesanst. Wien* **124**, 111–174.

Thöni, M. (1986) The Rb–Sr thin slab isochron method—an unreliable geochronologic method for dating geologic events in polymetamorphic terrains? *Mem. Sci. Geol. Padova* **XXXVIII**, 283–352.

Thum, I. and Nabholz, W. (1972) Zur Sedimentologie und Metamorphose der penninischen Flysch- und Schieferabfolgen im Gebiet Prättigau-Lenzerheide-Oberhalbstein. *Beitr. Geol. Karte Schweiz*, NF 144.

Ting, F.C. (1978) New techniques for measuring maximum reflectance of vitrinite and dispersed vitrinite in sediments. *Fuel* **57**, 717–721.

Tissot, B.P. and Espitalié, J. (1975) L' évolution thermique de la matière organique des sédiments: application d'une simulation mathématique. *Rev. Inst. franç. Pétrole* **30**, 743–777.

Tissot, B.P. and Welte, D.H. (1978) *Petroleum Formation and Occurrence.* Springer, Berlin etc.

Tobschall, H.J. (1969) Eine Subfaziesfolge der Grünschieferfazies in den Mittleren Cévennen (Dép. Ardèche) mit Pyrophyllit aufweisenden Mineralparagenesen. *Contrib. Miner. Petrol.* **24**, 76–91.

Toselli, A.J. (1982) Criterios de definición del metamorfismo de muy bajo grado. Con especial énfasis en el perfil de Falda Ciénaga, Punta de Catamarca. *Asoc. Geol. Argent., Rev.* **37**, 205–213.

Toselli, A.J. and Toselli, J.N. Rossi de (1982) Metamorfismo de la Formación Puncoviscana en las províncias de Salta y Tucumán, Argentina. *5° Congr. Latinoam. Geol., Argentina, 1982, Actas* **II**, 37–52.

Toselli, A.J. and Weber, K. (1982) Anquimetamorfirsmo en rocas del Paleozoico inferior en el noroeste de Argentina–Valor de la cristalinidad de la illita como índice. *Acta Geol. Lilloana (Argentina)* **14**, 187–200.

Touray, J.-C. (1970) Analyse microcryoscopique des inclusions gazeuses des 'quartz à fenêtres'. Exemples d'homogénéisation au voisinage de la température critique du méthane ($-82,5\,°C$). *C.R. Acad. Sci., Paris Sér. D.*, **270**, 2613–2615.

Touray, J.-C. (1976) Activation analysis for liquid inclusion studies: a brief review. *Bull. Minér.* **99**, 162–164.

Touray, J.-C. and Barlier, J. (1975) Liquid and gaseous hydrocarbon inclusions in quartz monocrystals from 'Terres Noires' and 'Flysch à helminthoides' (French Alps). *Fortschr. Mineral.* **52** (Spec. issue), 419–426.

Touray, J.-C. and Lantelme, F. (1966) Analyse des gaz inclus dans des minéraux: méthode du chauffage progressif. *Bull. Soc. franç. Minér. Crist.* **89**, 394–398.

Touray, J.-C. and Sagon, J.P. (1967) Inclusions à méthane dans les quartz des marnes de la région de Mauléon (Basses-Pyrénées). *C.R. Acad. Sci., Paris Sér. D*, **265**, 1269–1272.

Touray, J.-C., Beny, C., Dubessy, J. and Guillaumou, N. (1985) Micro characterization of fluid inclusions in minerals by Raman microprobe. *Scanning Electron Microscopy* **I**, 103–118.

Touray, J.-C., Vogler, M. and Stalder, H.A. (1970) Inclusions à hydrocarbures liquéfiés dans les quartz de Zingel/Seewen (Suisse). *Schweiz. Miner. Petrogr. Mitt.* **50**, 131–139.

Touret, J. (1977) The significance of fluid inclusions in metamorphic rocks. In *Thermodynamics in Geology*, ed. Fraser, D.G., Reidel, Dordrecht, 203–227.

Triplehorn, D.M. (1970) Clay mineral diagenesis in Atoka (Pennsylvanian) sandstones, Crawford County, Arkansas. *J. sediment. Petrol.* **40**, 838–847.

Tröhler, B. (1966) Geologie der Glockhaus-Gruppe. *Beitr. Geol. Schweiz, Geotech. Ser.* 13/10.

Tsui, T.-F. and Holland, H.D. (1979) The analysis of fluid inclusions by laser microprobe. *Econ. Geol.* **74**, 1647–1653.

Turner, F.J. (1938) Progressive regional metamorphism in southern New Zealand. *Geol. Mag.* **75**, 160–174.

Turner, F.J. (1948) Mineralogical and structural evolution of the metamorphic rocks. *Geol. Soc. Amer. Memoir* **30**.

Turner, F.J. (1958) Mineral assemblages of individual metamorphic facies. In *Metamorphic Reactions and Metamorphic Facies*, eds. Fyfe, W.S., Turner, F.J. and Verhoogen, J., *Geol. Soc. Amer. Memoir* **73**, 199–239.

Turner, F.J. (1968) *Metamorphic Petrology*. 1st edn., McGraw-Hill, New York..

Turner, F.J. (1981) *Metamorphic Petrology–Mineralogical, Field, and Tectonic Aspects*, 2nd edn., McGraw-Hill, New York.

Turner, F.J. and Verhoogen, J. (1960) *Igneous and Metamorphic Petrology*. McGraw-Hill, New York.

Turner, G. (1970) Thermal histories of meteorites by the $^{40}Ar/^{39}Ar$ method. In *Meteorite Research*, ed. Millman, P., Reidel, Dordrecht, 407–417.

Tzeng, Shih-Ying and Lidiak, E.D. (1976) Low-grade metamorphism in east-central Puerto Rico. *Geol. Soc. Amer. Ann. Meeting, 1976, Denver. Abstr. w. Progr.* **8**(6) 1150.

Umegaki, Y., and Ogawa, T. (1965) A note on occurrence of zeolites in the Miocene formation in Shimane Prefecture, Japan. *J. Sci. Hiroshima Univ. Sect. C, Geol. Miner.* **4**, 479–497.

Utada, M. (1965) Zonal distribution of authigenic zeolites in the Tertiary pyroclastic rocks in Mogami district, Yamagata Prefecture. *Tokyo Uni. Coll. Gen. Educ. Sci. Pa.* **15**, 173–216.

Utada, M. and Vine, J.D. (1984) Zonal distribution of zeolites and authigenic plagioclase, Spanish Peaks region, southern Colorado. *Proc. 6th Int. Zeolite Conf.* (Reno, 1983) 604–615.

Uyeda, S and Miyashiro, A. (1974) Plate tectonics and the Japanese Islands: A synthesis. *Bull. Geol. Soc. Amer.* **85**, 1159–1170.

Vassoievich, N.B. (1962) On the definition of stages in the process of lithogenesis (in Russian). *Tr. Vses. Nauchn. Issled. Geologorazved. Inst. SSSR* **190**, 220–243.

Vassoyevich, N.B., Korchagina, Yu.I., Lopatin, N.V., and Chernyshev, V.V. (1969) Principal phase of oil formation (in Russian). *Vestnik Mosk. Univ.* **1969**, 3–27 (transl. in *Z. angew. Geol.* **15**, 611–621 and *Int. Geol. Rev.* **12**, 1276–1296)

Velde, B. (1965) Experimental determination of muscovite polymorph stabilities. *Amer. Miner.* **50**, 436–449.

Velde, B. (1977) *Clays and Clay Minerals in Natural and Synthetic Systems*. Developments in Sedimentology 21, Elsevier, Amsterdam.

Velde, B. (1983) Diagenetic reactions in clays. In *Sedimentary Diagenesis*, eds. Parker, A. and Sellwood, B.W., Reidel, Dordrecht, 215–268.

Velde, B. (1985a) *Clay Minerals. A Physico-Chemical Explanation of their Occurrence*. Elsevier, Amsterdam.

Velde, B. (1985b) Possible chemical controls of illite/smectite composition during diagenesis. *Mineral. Mag.* **49**(3), 387–391.

Velde, B. and Hower, J. (1963) Petrological significance of illite polymorphism in Paleozoic sedimentary rocks. *Amer. Miner.* **48**, 1239–1254.

Velde, B. and Nicot, E. (1985) Diagenetic clay mineral composition as a function of pressure, temperature, and chemical activity. *J. sediment. Petrol.* **55**(4), 541–547.

Velde, B. and Odin, G.S. (1975) Further information related to the origin of glauconite. *Clays and Clay Minerals* **23**, 376–381.

Velde, B., Raoult, J.-F. and Leikine, M. (1974) Metamorphosed berthierine pellets in mid-Cretaceous rocks from north-eastern Algeria. *J. sediment. Petrol.* **44**, 1275–1280.

Venturelli, G. and Frey, M. (1977) Anchizone metamorphism in sedimentary sequences of the Northern Apennines. *Rend. Soc. Ital. Miner. Petrol.* **33**, 109–123.

Veselovskaya, M.M. (1967) Characteristics of argillaceous strata in the stage of catagenesis and initial metagenesis as illustrated by a study of the Wendian and Riphean of the Russian platform. *Dokl. Acad. Sci. USSR, Earth Sci. Sect.* **176**, 189–190.

Viereck, L.G., Griff, B.J., Schmincke, H., and Pritchard, R.G. (1982) Volcaniclastic rocks of the Reydarfjordur Drill Hole, eastern Iceland: 2, alteration. *J. Geophys. Res.* **87**, 6459–6474.

Viswanathan, K. and Seidel, E. (1979) Crystal chemistry of Fe–Mg-carpholites. *Contrib. Miner. Petrol.* **70**, 41–47.

Voll, G. (1976) Recrystallization of quartz, biotite and feldspars from Erstfeld to the Leventina nappe, Swiss Alps, and its geological significance. *Schweiz. Miner. Petrogr. Mitt.* **56**, 641–647.

Vrolijk, P. (1985a) Rapid and cyclic pressure drops in a continuously deforming melange: Fluid inclusion data (abstr.). *Trans. Amer. Geophys. Union* **66**, 1095.

Vrolijk, P. (1985b) Fluid and paleohydrogeologic evolution during progressive melange formation. *Geol. Soc. Amer. Abst. Programs* **17**, 415.

Vrolijk, P. (1987) Rapid tectonically-driven fluid flow in the Kodiak accretionary complex, Alaska. (*Geology*, in press).

Vrolijk, P., Moore, J.C. and O'Neil, J. (1985) Fluid flow in accretionary prisms. *Geol. Soc. Amer. Abst. Programs* **17**, 741.

Wagner, G.A. (1968) Fission track dating of apatites. *Earth Planet. Sci. Lett.* **4**, 411–415.

Wagner, G.A., Reimer, G.M. and Jäger, E. (1977) Cooling ages derived by apatite fission track, mica Rb/Sr and K/Ar dating: uplift and cooling history of the Central Alps. *Mem. Ist. Geol. Mineral. Padova* **XXX**.

Walker, G.P.L. (1960) Zeolite zones and dyke distribution in relation to the structure of the basalts of eastern Iceland. *J. Geol.* **68**, 515–528.

Wall, V.J. and Kesson, S. (1969) Pyrophyllite bearing rocks in a regionally altered volcanic sequence (abstract). *Ann. Meeting Geol. Soc. Amer. Abst. Programs*, 232.

Waples, D.W. (1980) Time and temperature in petroleum formation: Application of Lopatin's method to petroleum exploration. *Bull. Amer. Assoc. Petrol. Geol.* **64**, 916–929.

Wardlaw, B.R. and Harris, A.G. (1984) Conodont-based thermal maturation of Paleozoic rocks in Arizona. *Bull. Amer. Assoc. Petrol. Geol.* **68**, 1101–1106.

Wardlaw, B.R., Harris, A.G. and Schindler, K.S. (1984) Thermal maturation values (conodont colour alteration indices) for Paleozoic rocks in Arizona. *U.S. geol. Surv., Open-File Rep.* 83–819.

Watanabe, T. and Kobayashi, H. (1984) Occurrence of lawsonite in pelitic schists from the Sanbagawa metamorphic belt, central Shikoku, Japan. *J. metamorphic Geol.* **2**, 365–369.

Watson, S.W. (1976) The sedimentary geochemistry of the Moffat shales: a carbonaceous sequence in the Southern Uplands of Scotland. Unpubl. Ph.D. Thesis, Univ. St Andrews.

Weaver, C.E. (1960) Possible uses of clay minerals in search for oil. *Bull. Amer. Assoc. Petrol. Geol.* **44**, 1505–1518.

Weaver, C.E. (1961) Clay minerals of the Ouachita structural belt and the adjacent foreland. In *The Ouachita Belt*, eds. Flawn, P.T., Goldstein, A. Jr. King, P.B., and Weaver, C.E., Univ. Texas Publ. **6120**, 147–160.

Weaver, C.E. (1984) *Shale–Slate Metamorphism in Southern Appalachians*. Developments in Petrology 10, Elsevier, Amsterdam.

Weaver, C.E. and Beck, K.C. (1971) Clay water diagenesis during burial: How mud becomes gneiss. *Geol. Soc. Amer. Spec. Pap.* 134.

Weaver, C.E. and Broekstra, B.R. (1984) Illite-Mica. In *Shale–Slate Metamorphism in Southern Appalachians*, eds. Weaver, C.E. *et al.*, Developments in Petrology 10, Elsevier, Amsterdam, 67–97.

Weber, K. (1972a) Notes on the determination of illite crystallinity. *Neues Jb. Mineral. Mh.* **1972**, 267–276.

Weber, K. (1972b) Kristallinität des Illits in Tonschiefern und andere Kriterien schwacher Metamorphose im nordöstlichen Rheinischen Schiefergebirge. *Neues Jb. Geol. Paläontol. Abh.* **141**, 333–363.

Weber, K. (1976) Gefügeuntersuchungen an transversalgeschieferten Gesteinen aus dem östlichen Rheinischen Schiefergebirge (Ein Beitrag zur Genese der transversalen Schieferung). *Geol. Jahrb., Reihe D* (Hannover), **15**.

Weber, F., Dunoyer de Segonzac, G. and Economou, C. (1976) Une nouvelle expression de la 'cristallinité' de l'illite et des micas. Notion d''épaisseur apparente' des cristallites. *C.R. somm. Soc. Géol. Fr.* **5**, 225–227.

Welsch, H. (1973) Die Systeme Xenon-Wasser und Methan-Wasser bei hohen Drücken und Temperaturen. Ph.D. thesis, Univ. Karlsruhe.

Werre, Jr., R.W., Bodnar, R.J., Bethke, P.M. and Barton, Jr., B.P. (1979) A novel gas-flow fluid inclusion heating/freezing stage (abstr.) *Geol. Soc. Amer. Abst. Programs* **11**, 539.

White, S.H., Huggett, J.M. and Shaw, H.F. (1985) Electron-optical studies of phyllosilicate intergrowths in sedimentary and metamorphic rocks. *Miner. Mag.* **49**, 413–423.

White, S.H., Shaw, H.F. and Huggett, J.M. (1984) The use of back-scattered electron imaging for the petrographic study of sandstones and shales. *J. sediment. Petrol.* **54**, 487–494.

Wiebe, R. and Gaddy, V.L. (1940) The solubility of carbon dioxide in water at various temperatures from 12 to 40° and at pressures to 500 atmospheres. Critical phenomena. *J. Amer. Chem. Soc.* **62**, 815–817.

Wieland, B. (1979) Zur Diagenese und schwachen Metamorphose eozaener siderolithischer Gesteine des Helvetikums. *Schweiz. Miner. Petrogr. Mitt.* **59**, 41–66.

Wilkins, R.W.T. and Barkas, J.P. (1978) Fluid inclusions, deformation and recrystallization in granite tectonites. *Contrib. Miner. Petrol.* **65**, 293–299.

Wilson, M.J. and Bain, D.C. (1970) The clay mineralogy of the Scottish Dalradian meta-limestone. *Contrib. Miner. Petrol.* **26**, 285–295.

Winkler, H.G.F. (1967) *Die Genese der metamorphen Gesteine.* Springer Verlag, Berlin.

Winkler, H.G.F. (1976) *Petrogenesis of Metamorphic Rocks.* 4th edn., Springer Verlag, New York.

Winkler, H.G.F. (1979) *Petrogenesis of Metamorphic Rocks.* 5th edn., Springer Verlag, New York.

Wise, D.U., Dunn, D.E., Engelder, J.T., Geiser, P.A., Hatcher, R.D., Kish, S.A., Odom, A.L. and Schamel, S. (1984) Fault-related rocks: Suggestions for terminology. *Geology* **12**, 391–394.

Wolf, M. (1972) Beziehungen zwischen Inkohlung und Geotektonik im nördlichen Rheinischen Schiefergebirge. *Neues Jb. Geol. Paläontol. Abh.* **141**, 222–257.

Wolf, M. (1975) Ueber die Beziehungen zwischen Illit-Kristallinität und Inkohlung. *Neues Jb. Geol. Paläont. Mh.* 437–447.

Wood, D.A., Joron, J.L. and Treuil, M. (1979) A re-appraisal of the use of trace elements to classify and discriminate between magma series erupted in different tectonic settings. *Earth planet. Sci. Lett.* **45**, 326–336.

Woodland, B.G. (1985) Relationship of concretions and chlorite-muscovite porphyroblasts to the development of domainal cleavage in low-grade metamorphic deformed rocks from north-central Wales, Great Britain. *J. struct. Geol.* **7**(2), 205–215.

Wopenka, B. and Pasteris, J.D. (1986) Limitations to quantitative analysis of fluid inclusions in geological samples by laser Raman microprobe spectroscopy. *Appl. Spectrosc.* **40**, 144–151.

Wybrecht, E., Duplay, J., Piqué, A., and Weber, F. (1985) Mineralogical and chemical evolution of white micas and chlorites, from diagenesis to low-grade metamorphism; data from various size fractions of greywackes (Middle Cambrian, Morocco). *Mineral. Mag.* **49**(3), 401–411.

Yanatieva, O.K. (1946) Solubility polytherms in the systems $CaCl_2$–$MgCl_2$–H_2O and $CaCl_2$–$NaCl$–H_2O. *Zh. Prikl. Khimii* **19**, 707–722 (in Russian).

Yermakov, N.P. (1949) About the primary-secondary inclusions in minerals. *Miner. sb. L'vov. geolog. obshch.* **3**, 21–27 (in Russian).

Yermakov, N.P. (1965) *Research on the Nature of Mineral-Forming Solutions with Special Reference to Data from Fluid Inclusions.* Pergamon, Oxford, etc.

Yoder, H.S. and Eugster, H.P. (1955) Synthetic and natural muscovites. *Geochim. Cosmochim. Acta* **8**, 225–280.

Yokoyama, K., Brothers, R.N., and Black, P.M., (1986) Regional eclogite facies in the high-pressure metamorphic belt of New Caledonia. *Geol. Soc. Amer. Mem.* **164**, 407–423.

Zagoruchenko, V.A. and Zhuravlev, A.M. (1970) Thermophysical properties of gaseous and liquid methane. *Israel Progr. Sci. Transl.*, Jerusalem.

Zaporozhtseva, A.S., Vishnevskaya, T.N., and Glushinskii, P.I. (1963) Zeolites from Cretaceous formations in northern Yakutia (in Russian). *Litol. Polezn. Iskop.* **1963**(2), 161–177.

Zen, E-an (1960) Metamorphism of lower Paleozoic rocks in the vicinity of the Taconic range in West-central Vermont. *Amer. Miner.* **45**, 129–175.

Zen, E-an (1961a) Mineralogy and petrology of the system Al_2O_3–SiO_2–H_2O in some pyrophyllite deposits of North Carolina. *Amer. Miner.* **46**, 52–66.

Zen, E-an (1961b) The zeolite facies: An interpretation. *Amer. J. Sci.* **259**, 401–409.

Zen, E-an (1966) Construction of pressure-temperature diagrams for multi-component systems after the method of Schreinemakers—a geometrical approach. *US Geol. Surv. Bull.* **1225**.

Zen, E-an (1972) Gibbs free energy, enthalpy, and entropy of ten rock-forming minerals: Calculations, discrepancies, implications. *Amer. Mineral.* **12**, 445–455.

Zen, E-an (1974) Prehnite and pumpellyite-bearing mineral assemblages, west side of the Appalachian metamorphic belt, Pennsylvania to Newfoundland. *J. Petrol.* **15**, 197–242.

Zen, E-an and Thompson, A.B. (1974) Low grade regional metamorphism: Phase equilibrium relations. *Ann. Rev. Earth Planet. Sci.* **2**, 179–212.

Zeng, Y. and Liou, J.G. (1982) Experimental investigation of yugawaralite-wairakite equilibrium. *Amer. Miner.* **67**, 937–943.

Zhang, Z.M., Liou, J.G. and Coleman, R.G. (1984) An outline of plate tectonics of China. *Bull. Geol. Soc. Amer.* **95**, 295–312.

Zheliniskii, V.M. (1980) Catagenesis of terrigenous rocks and metamorphism of coals in south Yakutia (in Russian). *Litol. Polezn. Iskop.* **1980**(2), 99–114 (transl. in *Lithol. Miner. Resourc.* **1980**, 187–199).

Ziegenbein, D. and Johannes, W. (1980) Graphite in C–O–H fluids: an unsuitable compound to buffer fluid composition at temperatures up to 700 °C. *Neues Jb. Miner. Mh.* 289–305.

Zimmermann, J.-L. (1966) Etude par spectrometrie de masse des fluides occlus dans quelques échantillons de quartz. *C.R. Acad. Sci., Paris* **263**, *Série D*, 461–464.

Zingg, A., Hunziker, J.C., Frey, M., and Ahrendt, H. (1976) Age and degree of metamorphism of the Canavese Zone and of the sedimentary cover of the Sesia Zone. *Schweiz. Mineral. Petrogr. Mitt.* **56**, 361–375.

Index

References to tables are printed in italic;
references to illustrations are in bold face
type.